MODAL CONTROL

Theory and applications

MODAL CONTROL

Theory and applications

BRIAN PORTER
and
ROGER CROSSLEY

University of Salford

TAYLOR & FRANCIS LTD
LONDON

BARNES & NOBLE BOOKS
NEW YORK

1972

First published 1972 by Taylor & Francis Ltd London and Barnes & Noble books (a division of Harper & Row Publishers, Inc.)

Taylor & Francis ISBN 0 85066 057 2
Barnes & Noble Books ISBN 06 4956563

Printed and bound in Great Britain by
Taylor & Francis Ltd
10–14 Macklin Street, London WC2B 5NF

" but there is one system of coordinates which is especially suitable . . . . "

Lord Rayleigh, 'Theory of Sound'.

CONTENTS

CHAPTER 3

Eigenvalue and eigenvector sensitivities in linear systems theory

CHAPTER 4

Controllability and observability characteristics of linear dynamical systems

CHAPTER 5

Single-input modal control systems

CHAPTER 9

Sensitivity characteristics of modal control systems

PART 2: APPLICATIONS

CHAPTER 10

Synthesis of aircraft lateral autostabilisation systems

CHAPTER 11

Synthesis of aircraft longitudinal autostabilisation systems

PREFACE

The modal analysis of dynamical systems is, of course, one of the best known and most extensively cultivated fields of classical physics. It is therefore not surprising to find that, in the relatively recent explosive development of control theory, the modal approach to the analysis and synthesis of multivariable control systems has led to many interesting theoretical developments and important practical applications. It is the principal purpose of this book to present a unified treatment of modal control theory as it relates to continuous-time and discrete-time lumped-parameter linear systems, and also to indicate the practical relevance and scope of the theory by exhibiting *in extenso* its applications to a selection of systems of various types: the theory (leavened by many illustrative numerical examples) is expounded in the nine chapters which constitute Part I of the book, and the selected applications are discussed in the seven independent chapters which constitute Part II. This structure has been adopted in the hope that the book will not only be useful as a work of reference for practising control engineers and systems theorists, but also as a textbook for undergraduate and graduate students of control engineering or linear systems theory: the sets of problems which are given at the end of the book are primarily included for those who use the book in the latter mode.

In writing this book we were greatly encouraged by our colleagues at the University of Salford and would particularly like to thank Professor H. M. Power, Dr. A. Bradshaw, Dr. T. Learney, Dr. M. L. Tatnall, and Dr. M. A. Woodhead for their interest: we would also like to express our gratitude to the staff of the Library of the University of Salford for their unfailing helpfulness. The various drafts of this book were all typed with singular expertise and great forbearance by Mrs. Beryl Hickling whom we would like both to thank and to congratulate for the excellence of her contribution.

Finally, it has been a most pleasurable experience to work with the Directors and Staff of Taylor and Francis Ltd., during the production of this book, and we would particularly like to thank Mr. Charles W. Wheeler, Mr. Stanley A. Lewis, and Dr. J. Thomson in this connection.

Only we, of course, are responsible for the nature of the technical contents of this book.

B. Porter

T. R. Crossley.

PART 1 THEORY

CHAPTER 1

Introduction

1.1. Mathematical models

THE presentation of modal control theory contained in this book is concerned with the analysis and synthesis of deterministic linear time-invariant systems of the following two classes :

(i) lumped-parameter *continuous-time* systems,

(ii) lumped-parameter *discrete-time* systems.

The outlines of a theory of modal control as applied to distributed-parameter systems have been developed [1], [2], but these outlines are felt to be not yet definitive enough to warrant inclusion in the present book.

The continuous-time systems discussed in this book are assumed to be such that their dynamical behaviour can be adequately modelled by a vector-matrix differential *state equation* of the form

$$\dot{\mathbf{x}}(t) = \mathbf{A}\mathbf{x}(t) + \mathbf{B}\mathbf{z}(t) \tag{1.1}$$

and an associated algebraic *output equation* of the form

$$\mathbf{y}(t) = \mathbf{C}'\mathbf{x}(t), \tag{1.2}$$

where the dot in eqn. (1.1) denotes differentiation with respect to time, t, and the prime in eqn. (1.2) denotes matrix transposition. In eqns. (1.1) and (1.2), the $n \times 1$ vector $\mathbf{x}(t)$ defines the state of the system at any instant, t, and the $q \times 1$ and $r \times 1$ vectors $\mathbf{y}(t)$ and $\mathbf{z}(t)$ define respectively the output and input of the system at the same instant : $\mathbf{A}$, $\mathbf{B}$, and $\mathbf{C}'$ are respectively the constant $n \times n$, $n \times r$, and $q \times n$ plant matrix, input matrix, and output matrix of the continuous-time system.

The discrete-time systems discussed in this book are, in general, assumed to arise from continuous-time systems† as a result of the way in which systems of the latter class are controlled and observed, and therefore to be such that their

† The economic system discussed in § 15.3 and the manufacturing system discussed in Chapter 16 are, however, modelled *ab initio* as discrete-time systems.

dynamical behaviour can be adequately modelled by a vector-matrix difference *state equation* of the form

$$\mathbf{x}\{(k+1)T\}=\boldsymbol{\Psi}(T)\mathbf{x}(kT)+\boldsymbol{\Delta}(T)\mathbf{z}(kT) \tag{1.3}$$

and an associated algebraic *output equation* of the form

$$\mathbf{y}(kT)=\boldsymbol{\Gamma}'(T)\mathbf{x}(kT). \tag{1.4}$$

In eqns. (1.3) and (1.4), the $n\times 1$ vector $\mathbf{x}(kT)$ defines the state of the system at the discrete instants $t=kT$ $(k=0, 1, 2, \ldots)$, and the $q\times 1$ and $r\times 1$ piecewise-constant vectors $\mathbf{y}(kT)$ and $\mathbf{z}(kT)$ define respectively the output and input of the system on the intervals $kT\leqslant t<(k+1)T$ $(k=0, 1, 2, \ldots)$, where T is the so-called *sampling period* : $\boldsymbol{\Psi}(T)$, $\boldsymbol{\Delta}(T)$, and $\boldsymbol{\Gamma}'(T)$ are respectively the $n\times n$, $n\times r$, and $q\times n$ plant matrix, input matrix, and output matrix of the discrete-time system which, as indicated, are constants for any fixed value of the sampling period T. In fact, it is shown in § 2.3.1. that the matrices in eqns. (1.3) and (1.4) are related to the corresponding matrices in eqns. (1.1) and (1.2) by the equations

$$\boldsymbol{\Psi}(T)=\exp(\mathbf{A}T), \tag{1.5}$$

$$\boldsymbol{\Delta}(T)=\left\{\int_0^T \exp(\mathbf{A}t)\,dt\right\}\mathbf{B}, \tag{1.6}$$

and

$$\boldsymbol{\Gamma}'(T)=\mathbf{C}'. \tag{1.7}$$

In view of these relationships, it is not surprising to find that the modal theories of continuous-time and discrete-time lumped-parameter systems are essentially homomorphic : it is accordingly the usual practice in this book first to derive a result in the continuous-time domain and then (if desired) to pass by analogy to the discrete-time domain by using the appropriate homomorphism.

1.2. Modal control

The central concept of modal control is very simple : it is merely that of generating the input vector of a system by linear feedback of the state vector in such a way that prescribed eigenvalues are associated with the dynamical modes of the resulting closed-loop system.

Thus, consider a simple continuous-time system governed by the first-order scalar state equation

$$\dot{x}(t)=ax(t)+bz(t), \tag{1.8}$$

where a and b are real constants. It is obvious that in the absence of control (i.e. when $z\equiv 0$), the state of the system at any time t is given by the formula

$$x(t)=x(0)\exp(at), \tag{1.9}$$

where $\exp(at)$ defines the single dynamical mode of the system (1.8). Now, if a is positive, it is obvious from eqn. (1.9) that the uncontrolled system will be unstable : also, although the system will be asymptotically stable if a is negative, the decay $x(t)\to 0$ may not be sufficiently rapid. However, if linear feedback of state according to the control law

$$z(t)=gx(t) \tag{1.10}$$

Fig. 1-1

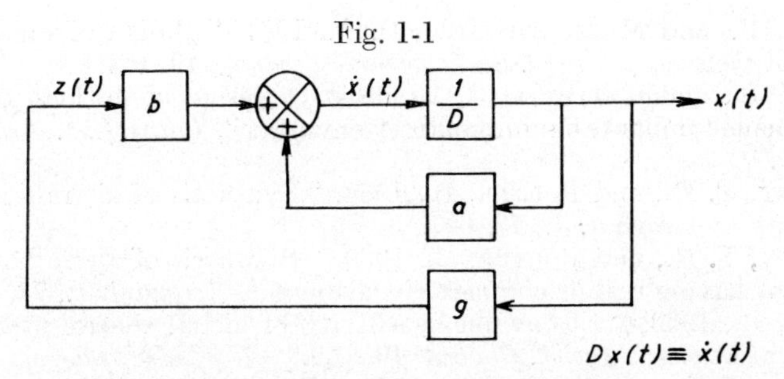

Simple system with state feedback.

is introduced (see fig. 1-1), eqn. (1.8) clearly assumes the closed-loop form

$$\dot{x}(t) = (a + bg)x(t), \tag{1.11}$$

where g is an arbitrary real constant. Since the solution of eqn. (1.11) may be written in the form

$$x(t) = x(0) \exp(\rho t), \tag{1.12}$$

where

$$\rho = a + bg, \tag{1.13}$$

it is obvious that ρ may be assigned any arbitrary real negative value simply by choosing g in the feedback law (1.10) according to the formula

$$g = (\rho - a)/b. \tag{1.14}$$

The principal results presented in this book are merely a generalization of this method of modal eigenvalue assignment so as to embrace systems of arbitrary order. Sometimes, of course, only the output vector, $\mathbf{y}(t)$, of a system is available for control purposes and it is then necessary to extend modal control theory in such a way that a closed-loop system having prescribed modal characteristics can still be realized: this topic is discussed in Chapter 7.

1.3. Historical background

Most of the results presented in this book were obtained by Crossley and Porter [3]–[25] after the original adumbration of modal control theory given by Rosenbrock [26]. Inevitably, however, certain of these results have been obtained independently by other systems theorists: in particular, some interesting results relating to single-input systems were obtained by Ellis and White [27] and the work of Simon and Mitter [28] constitutes a major contribution to modal control theory. Also, although this book is concerned only with lumped-parameter systems, it is essential to mention the pioneering work of Murray-Lasso [1], [2] and Gould [2] in connection with the modal control of distributed-parameter systems.

References

[1] Murray-Lasso, M. A., 1965, " The modal analysis and synthesis of linear distributed control systems ", Sc.D. dissertation, M.I.T., Dept. of Elec. Engineering.

[2] Gould, L. A., and Murray-Lasso, M. A., 1966, " On the modal control of distributed systems with distributed feedback ", *I.E.E.E. Trans autom. Control*, **11,** 79.

[3] Porter, B., and Micklethwaite, D. A., 1967, " Design of multi-loop modal control systems ", *Trans. Soc. Instrum. Technol.*, **19,** 143.
[4] Porter, B., and Carter, J. D., 1968, " Design of multi-loop modal control systems for plants having complex eigenvalues ", *Trans. Inst. Measmnt Control*, **1,** T61.
[5] Crossley, T. R., and Porter, B., 1968, " Synthesis of aircraft modal control systems ", *Aeronaut. J.*, **72,** 697.
[6] Crossley, T. R., and Porter, B., 1969, " Synthesis of aircraft modal control systems having real or complex eigenvalues ", *Aeronaut. J.*, **73,** 138.
[7] Porter, B., 1969, " Eigenvalue sensitivity of modal control systems to loop-gain variations ", *Int. J. Control*, **10,** 159.
[8] Crossley, T. R., and Porter, B., 1969, " Eigenvalue and eigenvector sensitivities in linear systems theory ", *Int. J. Control*, **10,** 163.
[9] Crossley, T. R., and Porter, B., 1969, " High-order eigenproblem sensitivity methods : theory and application to the design of linear dynamical systems ", *Int. J. Control*, **10,** 315.
[10] Porter, B., and Crossley, T. R., 1969, " Modal control of linear time-invariant discrete-time systems ", *Trans. Inst. Measmnt Control*, **2,** T133.
[11] Porter, B., and Crossley, T. R., 1970, " Modal control of cascaded-vehicle systems ", *Int. J. Systems Sci.*, **1,** 111.
[12] Crossley, T. R., and Porter, B., 1970, " Properties of the mode-controllability matrix ", *Int. J. Control*, **12,** 289.
[13] Porter, B., 1970, " Synthesis of control policies for economic models : a continuous-time multiplier model ", *Int. J. Systems Sci.*, **1,** 1.
[14] Porter, B., and Crossley, T. R., 1970, " Synthesis of control policies for economic models : a discrete-time multiplier model ", *Int. J. Systems Sci.*, **1,** 9.
[15] Crossley, T. R., and Porter, B., 1970, " Mode-controllability structure of a multivariable linear system ", *Electron. Lett.*, **6,** 342.
[16] Crossley, T. R., and Porter, B., 1970, " Modal control of multivariable time-invariant linear systems by output feedback ", *Electron. Lett.*, **6,** 630.
[17] Crossley, T. R., and Porter, B., 1970, " Direct digital modal control of cascaded-vehicle systems ", *Int. J. Systems Sci.*, **1,** 323.
[18] Crossley, T. R., and Porter, B., 1970, " Synthesis of helicopter stabilization systems using modal control theory ", AIAA Guidance, Control & Flight Mechanics Conference, Santa Barbara, California, August.
[19] Porter, B., and Crossley, T. R., 1970, " Mode-observability structure of a multivariable linear system ", *Electron. Lett.*, **6,** 664.
[20] Porter, B., and Crossley, T. R., 1971, " Dyadic modal control of a class of multi-input time-invariant linear systems ", *Electron. Lett.*, **7,** 26.
[21] Porter, B., and Power, H. M., 1971, " Dyadic modal control of multi-input time-invariant linear systems incorporating integral feedback ", *Electron. Lett.*, **7,** 18.
[22] Porter, B., and Power, H. M., 1970, " Mode-controllability matrices of multi-variable linear systems incorporating integral feedback ", *Electron. Lett.*, **6,** 809.
[23] Crossley, T. R., 1971, " Synthesis of modal control systems having confluent eigenvalues ", *Int. J. Control*, **13,** 821.
[24] Porter, B., and Crossley, T. R., 1971, " Synthesis of control policies for economic models : a continuous-time multiplier-accelerator model ", *Int. J. Systems Sci.*, **1,** 191.
[25] Porter, B., and Crossley, T. R., 1971, " Synthesis of control policies for manufacturing systems using eigenvalue-assignment techniques ", 3rd CIRP Int. Seminar on Optimization of Manufacturing Systems, Pisa, June.
[26] Rosenbrock, H. H., 1962, " Distinctive problems of process control ", *Chem. Engng Prog.*, **58,** 43.
[27] Ellis, J. K., and White, G. W. T., 1965, " An introduction to modal analysis and control ", *Control*, **9,** 193, 262, 137.
[28] Simon, J. D., and Mitter, S. K., 1968, " A theory of modal control ", *Inf. Control*, **13,** 316.

Modal analysis of linear dynamical systems

CHAPTER 2

2.1. Introduction

In this chapter the essential dynamical characteristics of continuous-time and discrete-time systems of the type described in Chapter 1 are expressed in terms of the eigenproperties of the plant matrices, $\mathbf{A}$ and $\mathbf{\Psi}(T)$, which occur in eqns. (1.1) and (1.3), respectively. In particular, the concept of a dynamical mode of a system is discussed in detail : this concept is, of course, fundamental to the whole theme of this book.

2.2. Dynamical characteristics of continuous-time systems

2.2.1 *Free-response characteristics*

In the absence of any input, the state equation of the continuous-time system governed by eqn. (1.1) assumes the form

$$\dot{\mathbf{x}}(t) = \mathbf{A}\mathbf{x}(t), \tag{2.1}$$

where $\mathbf{A}$ is a real $n \times n$ plant matrix and $\mathbf{x}(t)$ is an $n \times 1$ state vector whose variation with time defines the free motion of the system. The precise nature of the free motion of the continuous-time system following any disturbance can be described very simply in terms of the eigenvalues and eigenvectors of the plant matrix $\mathbf{A}$.

Thus, it is well known [1] that if $\mathbf{A}$ has n distinct eigenvalues λ_i ($i = 1, 2, \ldots, n$) then it will also have n corresponding linearly independent $n \times 1$ eigenvectors $\mathbf{u}_i$ ($i = 1, 2, \ldots, n$) which are related by the equations

$$\mathbf{A}\mathbf{u}_i = \lambda_i \mathbf{u}_i \quad (i = 1, 2, \ldots, n). \tag{2.2}$$

The set of vectors $\{\mathbf{u}_1, \mathbf{u}_2, \ldots, \mathbf{u}_n\}$ accordingly constitutes a *basis*[1] for an n-dimensional vector space, S_n, which can therefore be used as a *state space* [1] for a system modelled by eqn. (2.1). This means that the motion of such a system can be characterized by the trajectory generated in the space S_n by the state vector $\mathbf{x}(t)$ as the components of this vector vary with time.

Now eqns. (2.2) can clearly be written in the alternative form

$$[\mathbf{A}-\lambda_i\mathbf{I}_n]\mathbf{u}_i=\mathbf{0} \quad (i=1, 2, ..., n), \tag{2.3}$$

where $\mathbf{I}_n$ is the $n \times n$ unit matrix. These equations will have non-null solutions $\mathbf{u}_i$ $(i=1, 2, . . ., n)$ provided that

$$|\mathbf{A}-\lambda_i\mathbf{I}_n|=0 \quad (i=1, 2, ..., n). \tag{2.4}$$

It is therefore evident that the n eigenvalues of $\mathbf{A}$ are given by the n roots of the so-called *characteristic equation*

$$|\mathbf{A}-\lambda\mathbf{I}_n|=0. \tag{2.5}$$

The eigenvectors of $\mathbf{A}$ are obtained by solving eqns. (2.2) or (2.3) for each of the $\mathbf{u}_i$, after substitution of the corresponding eigenvalue λ_i into the appropriate equation : it will be noted that the $\mathbf{u}_i$ are thus determined only to within a scalar multiplier.

In addition to the eigenproperties of $\mathbf{A}$, the corresponding properties of the transposed plant matrix, $\mathbf{A}'$, play an important role in the modal analysis of continuous-time systems. Thus, if $\mathbf{A}'$ has n distinct eigenvalues μ_j $(j=1, 2, . . ., n)$ and n corresponding $n \times 1$ eigenvectors $\mathbf{v}_j$ $(j=1, 2, . . ., n)$, these quantities are related by the equations

$$\mathbf{A}'\mathbf{v}_j=\mu_j\mathbf{v}_j \quad (j=1, 2, ..., n) \tag{2.6}$$

which are clearly analogous to eqns. (2.2). It may therefore be inferred that the n eigenvalues of $\mathbf{A}'$ are given by the n roots of the equation

$$|\mathbf{A}'-\mu\mathbf{I}_n|=0. \tag{2.7}$$

Since $\mathbf{I}_n$ is symmetric and since

$$|\mathbf{X}'|=|\mathbf{X}|$$

for any square matrix $\mathbf{X}$, eqn. (2.7) implies that

$$|\mathbf{A}-\mu\mathbf{I}_n|=0. \tag{2.8}$$

It is therefore clear by comparing eqns. (2.5) and (2.8) that $\mathbf{A}'$ and $\mathbf{A}$ have the same eigenvalues

$$\mu_j=\lambda_j \quad (j=1, 2, ..., n). \tag{2.9}$$

However, the fact that $\mathbf{A}$ and $\mathbf{A}'$ each has the same set of eigenvalues does not, of course, imply that the corresponding sets of eigenvectors will in general be equal. Nevertheless, since $\mu_j \neq \mu_k$ $(j \neq k\,;\, j, k=1, 2, . . ., n)$, the n corresponding $n \times 1$ eigenvectors $\mathbf{v}_j$ $(j=1, 2, . . ., n)$ given by eqns. (2.6) will certainly be linearly independent in this case.

Now, in view of eqn. (2.9), it is evident that eqn. (2.6) may be written in the form

$$\mathbf{A}'\mathbf{v}_j=\lambda_j\mathbf{v}_j \quad (j=1, 2, ..., n). \tag{2.10}$$

If each member of each of eqns. (2.10) is transposed and then post-multiplied by an eigenvector $\mathbf{u}_i$ $(i \neq j)$ of $\mathbf{A}$, it follows that

$$\mathbf{v}_j'\mathbf{A}\mathbf{u}_i=\lambda_j\mathbf{v}_j'\mathbf{u}_i \quad (i \neq j\,;\; j=1, 2, ..., n). \tag{2.11}$$

Similarly, if each member of each of eqn. (2.2) is pre-multiplied by a transposed eigenvector $\mathbf{v}_j'$ $(j \neq i)$ of $\mathbf{A}'$, it follows that

$$\mathbf{v}_j'\mathbf{A}\mathbf{u}_i=\lambda_i\mathbf{v}_j'\mathbf{u}_i \quad (i \neq j\,;\; i, j=1, 2, ..., n). \tag{2.12}$$

Since the left-hand members of eqn. (2.11) and (2.12) are identical, subtraction of the former of these equations from the latter yields the equation

$$(\lambda_i - \lambda_j)\mathbf{v}_j'\mathbf{u}_i = 0 \quad (i \neq j\,;\ i, j = 1, 2, \ldots, n) \tag{2.13}$$

which in turn implies that

$$\mathbf{v}_j'\mathbf{u}_i = 0 \quad (i \neq j\,;\ i, j = 1, 2, \ldots, n). \tag{2.14}$$

This means that eigenvectors of $\mathbf{A}$ and $\mathbf{A}'$ corresponding to different eigenvalues are orthogonal. However, in the case of eigenvectors corresponding to the same eigenvalue, it is evident that

$$\mathbf{v}_i'\mathbf{u}_i = c_i \quad (i = 1, 2, \ldots, n), \tag{2.15}$$

where the c_i are non-zero constants. Since the eigenvectors are determined only to within a scalar multiplier it is convenient to normalize these vectors so that eqns. (2.15) assume the form

$$\mathbf{v}_i'\mathbf{u}_i = 1 \quad (i = 1, 2, \ldots, n). \tag{2.16}$$

Equations (2.14) and (2.16) may then be combined to form the equations

$$\mathbf{v}_j'\mathbf{u}_i = \mathbf{u}_i'\mathbf{v}_j = \delta_{ij} \quad (i, j = 1, 2, \ldots, n), \tag{2.17}$$

where δ_{ij} is the Kronecker delta. In view of eqns. (2.17) the set of vectors $\{\mathbf{v}_1, \mathbf{v}_2, \ldots, \mathbf{v}_n\}$ is said to constitute a *reciprocal basis* [1] for the vector space S_n.

It is convenient at this stage to introduce a number of matrices in terms of which the eigenproperties of $\mathbf{A}$ and $\mathbf{A}'$ can be expressed very succinctly. These matrices are the $n \times n$ *modal matrix* of $\mathbf{A}$

$$\mathbf{U} = [\mathbf{u}_1, \mathbf{u}_2, \ldots, \mathbf{u}_n], \tag{2.18}$$

the $n \times n$ *modal matrix* of $\mathbf{A}'$

$$\mathbf{V} = [\mathbf{v}_1, \mathbf{v}_2, \ldots, \mathbf{v}_n], \tag{2.19}$$

and the $n \times n$ *eigenvalue matrix* of $\mathbf{A}$ and $\mathbf{A}'$

$$\mathbf{\Lambda} = \begin{bmatrix} \lambda_1, & 0, & 0, & \ldots, & 0 \\ 0, & \lambda_2, & 0, & \ldots, & 0 \\ 0, & 0, & \lambda_3, & \ldots, & 0 \\ \cdots & \cdots & \cdots & \cdots & \cdots \\ 0, & 0, & 0, & \ldots, & \lambda_n \end{bmatrix} = \operatorname{diag}\{\lambda_1, \lambda_2, \lambda_3, \ldots, \lambda_n\}. \tag{2.20}$$

In terms of these matrices, eqns. (2.2), (2.6), and (2.17) may be written in the respective forms

$$\mathbf{A}\mathbf{U} = \mathbf{U}\mathbf{\Lambda}, \tag{2.21}$$

$$\mathbf{A}'\mathbf{V} = \mathbf{V}\mathbf{\Lambda}, \tag{2.22}$$

and

$$\mathbf{V}'\mathbf{U} = \mathbf{I}_n. \tag{2.23}$$

It follows readily from these equations that

$$\mathbf{V}' = \mathbf{U}^{-1}, \tag{2.24}$$

$$\mathbf{V} = (\mathbf{U}^{-1})', \tag{2.25}$$

$$\mathbf{U}^{-1}\mathbf{A}\mathbf{U} = \mathbf{\Lambda}, \tag{2.26}$$

and

$$\mathbf{U}\mathbf{\Lambda}\mathbf{U}^{-1} = \mathbf{A}. \tag{2.27}$$

The equations describing the free motion of the system (2.1) can now be readily derived using these results. Thus, if a new state vector, $\boldsymbol{\xi}(t)$, is introduced into eqn. (2.1) by the transformation

$$\mathbf{x}(t)=\mathbf{U}\boldsymbol{\xi}(t), \tag{2.28}$$

where $\mathbf{U}$ is the modal matrix of $\mathbf{A}$, the new state equation has the form

$$\mathbf{U}\dot{\boldsymbol{\xi}}(t)=\mathbf{A}\,\mathbf{U}\boldsymbol{\xi}(t). \tag{2.29}$$

It follows from eqn. (2.29) that

$$\dot{\boldsymbol{\xi}}(t)=\mathbf{U}^{-1}\mathbf{A}\,\mathbf{U}\boldsymbol{\xi}(t)$$

and therefore that

$$\dot{\boldsymbol{\xi}}(t)=\boldsymbol{\Lambda}\boldsymbol{\xi}(t) \tag{2.30}$$

in view of eqn. (2.26). The importance of eqn. (2.30) as compared with eqn. (2.1) is that $\boldsymbol{\Lambda}$ is a diagonal matrix whereas $\mathbf{A}$ is, in general, non-diagonal.

Equation (2.30) clearly implies that

$$\dot{\xi}_i(t)=\lambda_i\xi_i(t) \quad (i=1, 2, \ldots, n), \tag{2.31}$$

where the $\xi_i(t)$ $(i=1, 2, \ldots, n)$ are the n components of the vector $\boldsymbol{\xi}(t)$. Equation (2.31) are uncoupled because of the diagonal nature of $\boldsymbol{\Lambda}$, and the solutions of these equations are obviously given by the formulae

$$\xi_i(t)=\xi_i(0)\exp(\lambda_i t) \quad (i=1, 2, \ldots, n), \tag{2.32}$$

where the $\xi_i(0)$ $(i=1, 2, \ldots, n)$ are the initial values of the components of $\boldsymbol{\xi}(t)$. Now (2.28) indicates that the original state vector, $\mathbf{x}(t)$, is given by the equation

$$\mathbf{x}(t)=\mathbf{U}\boldsymbol{\xi}(t)=[\mathbf{u}_1, \mathbf{u}_2, \ldots, \mathbf{u}_n]\begin{bmatrix}\xi_1(t)\\ \xi_2(t)\\ \vdots\\ \xi_n(t)\end{bmatrix}$$

which implies that

$$\mathbf{x}(t)=\mathbf{u}_1\xi_1(0)\exp(\lambda_1 t)+\mathbf{u}_2\xi_2(0)\exp(\lambda_2 t)+\ldots+\mathbf{u}_n\xi_n(0)\exp(\lambda_n t). \tag{2.33}$$

It follows by putting $t=0$ in eqn. (2.33) that

$$\mathbf{x}(0)=\mathbf{u}_1\xi_1(0)+\mathbf{u}_2\xi_2(0)+\ldots+\mathbf{u}_n\xi_n(0) \tag{2.34}$$

which, together with eqns. (2.17), yields the formulae

$$\xi_i(0)=\mathbf{v}_i'\mathbf{x}(0) \quad (i=1, 2, \ldots, n). \tag{2.35}$$

Equation (2.33) may therefore be expressed in the form

$$\mathbf{x}(t)=\mathbf{u}_1\mathbf{v}_1'\mathbf{x}(0)\exp(\lambda_1 t)+\mathbf{u}_2\mathbf{v}_2'\mathbf{x}(0)\exp(\lambda_2 t)+\ldots$$
$$\ldots+\mathbf{u}_n\mathbf{v}_n'\mathbf{x}(0)\exp(\lambda_n t) \tag{2.36}$$

or, more compactly, in the form

$$\mathbf{x}(t)=\sum_{i=1}^{n}[\exp(\lambda_i t)]\mathbf{u}_i\mathbf{v}_i'\mathbf{x}(0). \tag{2.37}$$

Equation (2.37) shows clearly that the free motion of the continuous-time system governed by (2.1) is a linear combination of n functions of the form $[\exp(\lambda_i t)]\mathbf{u}_i$ $(i=1, 2, \ldots, n)$ which are said to describe the n *dynamical modes* of the system. Thus, the ‘ shape ’ of a mode is described by its associated *eigenvector*, $\mathbf{u}_i$, and its time-domain characteristics by its associated *eigenvalue*, λ_i : in particular, it is evident from eqn. (2.37) that the equilibrium state $\mathbf{x}=\mathbf{0}$ of the system (2.1) will be asymptotically stable in the sense that $\mathbf{x}(t) \rightarrow \mathbf{0}$ as $t \rightarrow \infty$ if and only if

$$\text{Re } \lambda_i < 0 \quad (i = 1, 2, \ldots, n). \tag{2.38}$$

In geometric terms, the condition (2.38) is equivalent to the requirement that all the eigenvalues of the matrix $\mathbf{A}$ lie in the open left half of the complex plane.

2.2.2. *Transition matrices*

In the present context, eqn. (2.37) is of paramount importance because this equation clearly exhibits the modal characteristics of the free response of the continuous-time system (2.1). However, eqn. (2.37) can be written in an important alternative form which involves the so-called *transition matrix*, $\mathbf{\Phi}(t)$, of the system (2.1). In order to obtain this alternative expression, it is necessary to introduce the *matrix exponential function* which is defined by the infinite series

$$\exp(\mathbf{A}t) = \sum_{k=0}^{\infty} \mathbf{A}^k \frac{t^k}{k!}. \tag{2.39}$$

Since it can be shown that the series (2.39) is convergent for all square matrices, $\mathbf{A}$, and all scalars, t, differentiation of this series is permissible and indicates that

$$\frac{d}{dt}[\exp(\mathbf{A}t)] = \mathbf{A}[\exp(\mathbf{A}t)] = [\exp(\mathbf{A}t)]\mathbf{A}. \tag{2.40}$$

It is clear from (2.40) that the function $[\exp(\mathbf{A}t)]\mathbf{x}(0)$ is such that

$$\frac{d}{dt}\{[\exp(\mathbf{A}t)]\mathbf{x}(0)\} = \mathbf{A}\{[\exp(\mathbf{A}t)]\mathbf{x}(0)\}. \tag{2.41}$$

Since the solution of eqn. (2.1) is unique for a given initial state, it may be concluded from a comparison of eqns. (2.1) and (2.41) that this solution is given by the formula

$$\mathbf{x}(t) = [\exp(\mathbf{A}t)]\mathbf{x}(0). \tag{2.42}$$

This equation has the form

$$\mathbf{x}(t) = \mathbf{\Phi}(t)\mathbf{x}(0) \tag{2.43}$$

where, by definition, the transition matrix of the system (2.1) is given by the equation

$$\mathbf{\Phi}(t) = \exp(\mathbf{A}t). \tag{2.44}$$

Now it is a well-known result in matrix algebra [2] that if

$$\mathbf{A} = \mathbf{U}\Lambda\mathbf{U}^{-1},$$

then

$$\mathbf{f}(\mathbf{A}) = \mathbf{U}\mathbf{f}(\Lambda)\mathbf{U}^{-1}, \tag{2.45}$$

where $\mathbf{f}$ is a matrix-valued function. Therefore, in the particular case when $\mathbf{f}(\mathbf{A}) = \exp(\mathbf{A}t)$, eqn. (2.45) (together with eqn. (2.44)) yields the alternative formula

$$\mathbf{\Phi}(t) = \mathbf{U}[\exp(\Lambda t)]\mathbf{U}^{-1} \tag{2.46}$$

for the transition matrix $\mathbf{\Phi}(t)$. It is of interest to note that eqn. (2.37) has the form

$$\mathbf{x}(t) = \mathbf{U}[\text{diag}\{\exp(\lambda_1 t), \exp(\lambda_2 t), \ldots, \exp(\lambda_n t)\}]\mathbf{U}^{-1}\mathbf{x}(0)$$

which, together with eqns. (2.43) and (2.46), implies that

$$\exp(\boldsymbol{\Lambda}t) = \operatorname{diag}\{\exp(\lambda_1 t), \exp(\lambda_2 t), \ldots, \exp(\lambda_n t)\}. \tag{2.47}$$

The alternative formulae for the transition matrix given in eqns. (2.44) and (2.46) clearly express $\boldsymbol{\Phi}(t)$ as a function of $\mathbf{A}$ and as an explicit function of the eigenproperties of $\mathbf{A}$, respectively.

It is appropriate to note at this point that it follows immediately from eqns. (2.39) and (2.44) that $\boldsymbol{\Phi}(t)$ has the following important properties :

$$\text{(i)} \quad \boldsymbol{\Phi}(0) = \mathbf{I}_n, \tag{2.48 a}$$

$$\text{(ii)} \quad \boldsymbol{\Phi}(t_1)\boldsymbol{\Phi}(t_2) = \boldsymbol{\Phi}(t_1 + t_2), \tag{2.48 b}$$

$$\text{(iii)} \quad \boldsymbol{\Phi}^{-1}(t) = \boldsymbol{\Phi}(-t). \tag{2.48 c}$$

2.2.3. *Forced-response characteristics*

In the presence of an input vector, $\mathbf{z}(t)$, the state equation of the continuous-time system described in § 1.1 assumes the complete form

$$\dot{\mathbf{x}}(t) = \mathbf{A}\mathbf{x}(t) + \mathbf{B}\mathbf{z}(t), \tag{2.49}$$

where the dimensions of the various vectors and matrices in this equation are as defined in § 1.1. The forced motion of the system resulting from the presence of the input $\mathbf{z}(t)$ can be readily obtained by determining a particular integral of eqn. (2.49).

Thus, let a particular integral of eqn. (2.49) be expressed in the form

$$\mathbf{x}_{\mathrm{p}}(t) = \boldsymbol{\Phi}(t)\mathbf{q}(t), \tag{2.50}$$

where $\boldsymbol{\Phi}(t)$ is the transition matrix of the system and $\mathbf{q}(t)$ is a vector-valued function of time. It follows immediately from eqns. (2.49) and (2.50) that

$$\dot{\boldsymbol{\Phi}}(t)\mathbf{q}(t) + \boldsymbol{\Phi}(t)\dot{\mathbf{q}}(t) = \mathbf{A}\boldsymbol{\Phi}(t)\mathbf{q}(t) + \mathbf{B}\mathbf{z}(t), \tag{2.51}$$

which may clearly be reduced to the simpler equation

$$\boldsymbol{\Phi}(t)\dot{\mathbf{q}}(t) = \mathbf{B}\mathbf{z}(t) \tag{2.52}$$

since $\boldsymbol{\Phi}(t)$ satisfies eqn. (2.1). Now since

$$\boldsymbol{\Phi}^{-1}(t) = \boldsymbol{\Phi}(-t),$$

it may be deduced from eqn. (2.52) that

$$\mathbf{q}(t) = \int_0^t \boldsymbol{\Phi}(-\tau)\mathbf{B}\mathbf{z}(\tau)\, d\tau. \tag{2.53}$$

Since also

$$\boldsymbol{\Phi}(t)\boldsymbol{\Phi}(-\tau) = \boldsymbol{\Phi}(t-\tau),$$

the particular integral of eqn. (2.49) defined by eqn. (2.50) may be expressed in the form

$$\mathbf{x}_{\mathrm{p}}(t) = \int_0^t \boldsymbol{\Phi}(t-\tau)\mathbf{B}\mathbf{z}(\tau)\, d\tau. \tag{2.54}$$

The right-hand member of eqn. (2.54) thus gives the forced response of the system (2.49) to the input $\mathbf{z}(t)$.

In view of the fact that the complementary function of eqn. (2.49) is given by the right-hand member of eqn. (2.43), it may therefore be concluded that the complete solution of eqn. (2.49) is

$$\mathbf{x}(t) = \mathbf{\Phi}(t)\mathbf{x}(0) + \int_0^t \mathbf{\Phi}(t-\tau)\mathbf{B}\mathbf{z}(\tau)\, d\tau. \tag{2.55}$$

Since $\mathbf{\Phi}(t) = \exp(\mathbf{A}t)$, this solution may obviously be written in the alternative form

$$\mathbf{x}(t) = [\exp(\mathbf{A}t)]\mathbf{x}(0) + \int_0^t [\exp\{\mathbf{A}(t-\tau)\}]\mathbf{B}\mathbf{z}(\tau)\, d\tau. \tag{2.56}$$

2.3. Dynamical characteristics of discrete-time systems

2.3.1. *Governing equations of discrete-time systems*

In the case of continuous-time systems governed by equations of the form (1.1) and (1.2), consider that the input vector, $\mathbf{z}(t)$, and the output vector, $\mathbf{y}(t)$, are piecewise-constant functions defined by the equations

$$\mathbf{z}(t) = \mathbf{z}(kT) \tag{2.57 a}$$

and

$$\mathbf{y}(t) = \mathbf{y}(kT) \tag{2.57 b}$$

$$(kT \leqslant t < (k+1)T\ ;\ k = 0, 1, 2, \ldots),$$

and that only the values of the state vector, $\mathbf{x}(t)$, at the discrete instants corresponding to $t = kT$ $(k = 0, 1, 2, \ldots)$ are of interest. It may be deduced from eqn. (2.56) that, over an interval $kT \leqslant t \leqslant (k+1)T$, the states $\mathbf{x}\{(k+1)T\}$ and $\mathbf{x}(kT)$ are related by the difference equation

$$\mathbf{x}\{(k+1)T\} = [\exp(\mathbf{A}T)]\mathbf{x}(kT) + \left\{\int_0^T [\exp(\mathbf{A}T)]\, dt\right\} \mathbf{B}\mathbf{z}(kT). \tag{2.58}$$

This equation may clearly be written in the form

$$\mathbf{x}\{(k+1)T\} = \mathbf{\Psi}(T)\mathbf{x}(kT) + \mathbf{\Delta}(T)\mathbf{z}(kT), \tag{2.59}$$

where

$$\mathbf{\Psi}(T) = \exp(\mathbf{A}T) \tag{2.60}$$

and

$$\mathbf{\Delta}(T) = \left\{\int_0^T [\exp(\mathbf{A}t)]\, dt\right\} \mathbf{B}. \tag{2.61}$$

The state eqn. (2.59) is identical to eqn. (1.3) since the matrices $\mathbf{\Psi}(T)$ and $\mathbf{\Delta}(T)$ defined in eqns. (2.60) and (2.61) are identical to those defined in eqns. (1.5) and (1.6).

The discrete-time form of the continuous-time output eqn. (1.2) is clearly

$$\mathbf{y}\{(k+1)T\} = \mathbf{\Gamma}'(T)\mathbf{x}\{(k+1)T\}, \tag{2.62}$$

where obviously

$$\mathbf{\Gamma}'(T) = \mathbf{C}'. \tag{2.63}$$

The output eqn. (2.62) is identical to eqn. (1.4) since the matrix $\mathbf{\Gamma}'(T)$ defined in eqn. (2.63) is identical to that defined in eqn. (1.7).

2.3.2. *Free-response characteristics*

In the absence of any input, the state equation of the discrete-time system governed by eqn. (1.3) assumes the form

$$\mathbf{x}\{(k+1)T\}=\mathbf{\Psi}(T)\mathbf{x}(kT), \tag{2.64}$$

where $\mathbf{\Psi}(T)$ is a real $n \times n$ plant matrix and $\mathbf{x}(kT)$ is an $n \times 1$ state vector whose variation with k ($k=0, 1, 2, \ldots$) defines the free motion of the system. The precise nature of the free motion of the discrete-time system following any disturbance can be described very simply in terms of the eigenproperties of the plant matrix $\mathbf{\Psi}(T)$, as was the case with the continuous-time system (2.1).

Thus, since $\mathbf{\Psi}(T)=\exp(\mathbf{A}T)$, it follows from eqns. (2.44) and (2.46) that

$$\mathbf{\Psi}(T)=\mathbf{U}[\exp(\mathbf{\Lambda}T)]\mathbf{U}^{-1}, \tag{2.65}$$

where $\mathbf{U}$ and $\mathbf{\Lambda}$ are respectively the modal matrix and the eigenvalue matrix of the continuous-time plant matrix $\mathbf{A}$. It may therefore be inferred from eqn. (2.65) that the $n \times n$ modal matrix of $\mathbf{\Psi}(T)$ is

$$\mathbf{U}=[\mathbf{u}_1, \mathbf{u}_2, \ldots, \mathbf{u}_n], \tag{2.66}$$

the $n \times n$ modal matrix of $\mathbf{\Psi}'(T)$ is

$$\mathbf{V}=[\mathbf{v}_1, \mathbf{v}_2, \ldots, \mathbf{v}_n], \tag{2.67}$$

and the $n \times n$ eigenvalue matrix of $\mathbf{\Psi}(T)$ and $\mathbf{\Psi}'(T)$ is

$$\exp(\mathbf{\Lambda}T)=\begin{bmatrix} \exp(\lambda_1 T), & 0, & 0, & \ldots, & 0 \\ 0, & \exp(\lambda_2 T), & 0, & \ldots, & 0 \\ 0, & 0, & \exp(\lambda_3 T), & \ldots, & 0 \\ \cdots & \cdots & \cdots & \cdots & \cdots \\ 0, & 0, & 0, & \ldots, & \exp(\lambda_n T) \end{bmatrix}$$

$$=\operatorname{diag}\{\exp(\lambda_1 T), \exp(\lambda_2 T), \exp(\lambda_3 T), \ldots, \exp(\lambda_n T)\} \tag{2.68}$$

where, in eqns. (2.66) and (2.67), the $\mathbf{u}_i$ and the $\mathbf{v}_j$ are the eigenvectors of $\mathbf{A}$ and $\mathbf{A}'$, and in eqn. (2.68), the λ_i are the eigenvalues of $\mathbf{A}$ and $\mathbf{A}'$. It is therefore evident that the eigenproperties of the discrete-time plant matrix, $\mathbf{\Psi}(T)$, are the same as those of the continuous-time plant matrix, $\mathbf{A}$, except that the respective eigenvalue matrices are $\exp(\mathbf{\Lambda}T)$ and $\mathbf{\Lambda}$: however, the eigenvalues of $\mathbf{\Psi}(T)$ will be distinct if and only if the eigenvalues of $\mathbf{A}$ are distinct and

$$\exp(\lambda_j T) \neq \exp(\lambda_k T) \quad (j, k=1, 2, \ldots, n).$$

In view of this intimate relationship between the two sets of eigenproperties, the discrete-time analogues of eqns. (2.21) to (2.27) are clearly

$$\mathbf{\Psi}(T)\mathbf{U}=\mathbf{U}[\exp(\mathbf{\Lambda}T)], \tag{2.69}$$

$$\mathbf{\Psi}'(T)\mathbf{V}=\mathbf{V}[\exp(\mathbf{\Lambda}T)], \tag{2.70}$$

$$\mathbf{V}'\mathbf{U}=\mathbf{I}_n, \tag{2.71}$$

$$\mathbf{V}'=\mathbf{U}^{-1}, \tag{2.72}$$

$$\mathbf{V} = (\mathbf{U}^{-1})', \tag{2.73}$$

$$\mathbf{U}^{-1}\mathbf{\Psi}(T)\mathbf{U} = \exp(\mathbf{\Lambda}T), \tag{2.74}$$

and

$$\mathbf{U}[\exp(\mathbf{\Lambda}T)]\mathbf{U}^{-1} = \mathbf{\Psi}(T). \tag{2.75}$$

The equations describing the free motion of the discrete-time system (2.64) can now be derived using these results. Thus, if a new state vector $\boldsymbol{\xi}(kT)$ $(k=0, 1, 2, \ldots)$ is introduced into eqn. (2.64) according to the transformation

$$\mathbf{x}(kT) = \mathbf{U}\boldsymbol{\xi}(kT), \tag{2.76}$$

where $\mathbf{U}$ is the modal matrix of $\mathbf{\Psi}(T)$, the resulting state equation has the form

$$\mathbf{U}\boldsymbol{\xi}\{(k+1)T\} = \mathbf{\Psi}(T)\mathbf{U}\boldsymbol{\xi}(kT). \tag{2.77}$$

It follows from eqn. (2.77) that

$$\boldsymbol{\xi}\{(k+1)T\} = \mathbf{U}^{-1}\mathbf{\Psi}(T)\mathbf{U}\boldsymbol{\xi}(kT)$$

and therefore that

$$\boldsymbol{\xi}\{(k+1)T\} = [\exp(\mathbf{\Lambda}T)]\boldsymbol{\xi}(kT) \tag{2.78}$$

in view of eqn. (2.74). Equation (2.78) is the discrete-time analogue of the continuous-time eqn. (2.30) and is important because $\exp(\mathbf{\Lambda}T)$ (like $\mathbf{\Lambda}$) is a diagonal matrix.

Equation (2.78) obviously implies that

$$\xi_i\{(k+1)T\} = [\exp(\lambda_i T)]\xi_i(kT) \quad (i=1, 2, \ldots, n), \tag{2.79}$$

where the $\xi_i(kT)$ $(i=1, 2, \ldots, n)$ are the n components of the vector $\boldsymbol{\xi}(kT)$. The solutions of the uncoupled difference eqns. (2.79) are clearly given by the formulae

$$\xi_i(kT) = \xi_i(0)[\exp(k\lambda_i T)] \quad (i=1, 2, \ldots, n\,;\; k=0, 1, 2, \ldots), \tag{2.80}$$

where the $\xi_i(0)$ $(i=1, 2, \ldots, n)$ are the initial values of the components of $\boldsymbol{\xi}(kT)$. In view of the fact that eqns. (2.80) and (2.32) are precisely analogous, the reasoning which leads from eqn. (2.32) to eqn. (2.37) in the continuous-time case may be repeated in relation to the discrete-time eqn. (2.80). The resulting equation can be written in the form

$$\mathbf{x}(kT) = \sum_{i=1}^{n} [\exp(k\lambda_i T)]\mathbf{u}_i\mathbf{v}_i{}'\mathbf{x}(0) \quad (k=0, 1, 2, \ldots). \tag{2.81}$$

Equation (2.81) shows clearly that the free motion of the discrete-time system governed by eqn. (2.64) is a linear combination of n functions of the form $[\exp(k\lambda_i T)]\mathbf{u}_i$ $(i=1, 2, \ldots, n)$ which (as in the case of the continuous-time system) are said to describe the n *dynamical modes* of the system. Thus, the 'shape' of a mode is again described by its associated *eigenvector*, $\mathbf{u}_i$, and its characteristics in the discrete time-domain by its associated *eigenvalue*, $\exp(\lambda_i T)$: in particular, it is clear from eqn. (2.81) that the equilibrium state

$\mathbf{x}=\mathbf{0}$ of the system (2.64) will be asymptotically stable in the sense that $\mathbf{x}(kT)\to\mathbf{0}$ as $k\to\infty$ if and only if

$$|\exp(\lambda_i T)|<1 \quad (i=1, 2, \ldots, n). \tag{2.82}$$

In geometric terms, the inequality (2.82) is equivalent to the requirement that all the eigenvalues of the matrix $\mathbf{\Psi}(T)$ lie inside the unit circle centred at the origin of the complex plane.

2.3.3. *Transition matrices*

In the present context, eqn. (2.81) is of central importance because this equation clearly exhibits the *modal* characteristics of the free response of the discrete-time system (2.64). However, as is the case with the continuous-time system (see § 2.2.2), it is possible to write the equation which determines the free response in terms of a *transition matrix.*

Now, since the plant matrix of the discrete-time system is given by $\mathbf{\Psi}(T)=\exp(\mathbf{A}T)$ (see eqn. (2.60)), it may be inferred from the state eqn. (2.64) that

$$\mathbf{x}(kT)=\mathbf{\Psi}(kT)\mathbf{x}(0) \tag{2.83}$$

and, equivalently, that

$$\mathbf{x}(kT)=[\exp(k\mathbf{A}T)]\mathbf{x}(0). \tag{2.84}$$

It is also evident that eqn. (2.84) can be written in the form

$$\mathbf{x}(kT)=\mathbf{\Phi}(kT)\mathbf{x}(0) \tag{2.85}$$

in view of eqn. (2.44). Equation (2.85) indicates that the transition matrix of the discrete-time system is $\mathbf{\Phi}(kT)$, where $\mathbf{\Phi}(t)$ is the transition matrix of the underlying continuous-time system.

It is of interest to derive one further expression for the transition matrix of the discrete-time system which clearly expresses this matrix as an explicit function of the eigenproperties of the system. This expression can be derived by noting that eqn. (2.81) has the form

$$\mathbf{x}(kT)=\mathbf{U}\,\mathrm{diag}\,\{\exp(k\lambda_1 T), \exp(k\lambda_2 T), \ldots, \exp(k\lambda_n T)\}\mathbf{U}^{-1}\mathbf{x}(0)$$

which implies that

$$\mathbf{x}(kT)=\mathbf{U}[\exp(k\mathbf{\Lambda}T)]\mathbf{U}^{-1}\mathbf{x}(0). \tag{2.86}$$

Thus, it is evident from eqns. (2.83) to (2.86) that the four equivalent expressions for the transition matrix of the discrete-time system (2.64) derived in this section are

$$\mathbf{\Phi}(kT)=[\exp(k\mathbf{A}T)]=\mathbf{\Psi}(kT)=\mathbf{U}[\exp(k\mathbf{\Lambda}T)]\mathbf{U}^{-1}. \tag{2.87}$$

2.3.4. *Forced-response characteristics*

In the presence of a piecewise-continuous input vector, $\mathbf{z}(kT)$, the state equation of the discrete-time system described in § 1.1 assumes the complete form

$$\mathbf{x}\{(k+1)T\}=\mathbf{\Psi}(T)\mathbf{x}(kT)+\mathbf{\Delta}(T)\mathbf{z}(kT), \tag{2.88}$$

where the dimensions of the various vectors and matrices in this equation are as defined in § 1.1.

The forced motion of the system resulting from the presence of the input $\mathbf{z}(kT)$ defined by eqn. (2.57 *a*) can be readily obtained by determining a particular integral of eqn. (2.88). In fact, by direct substitution into eqn. (2.88), it may be readily verified that such a particular integral is given by the formula

$$\mathbf{x}_{\mathrm{p}}(kT)=\sum_{l=0}^{k-1}\mathbf{\Phi}[(k-1-l)T]\mathbf{\Delta}(T)\mathbf{z}(lT). \tag{2.89}$$

Since the complementary function of eqn. (2.88) is given by the right-hand member of eqn. (2.85), it may accordingly be concluded that the complete solution of eqn. (2.88) is

$$\mathbf{x}(kT)=\mathbf{\Phi}(kT)\mathbf{x}(0)+\sum_{l=0}^{k-1}\mathbf{\Phi}[(k-1-l)T]\mathbf{\Delta}(T)\mathbf{z}(lT)\ (k=0,\ 1,\ 2,\ \ldots). \tag{2.90}$$

This solution may obviously be written in a number of alternative forms by using the alternative expressions for the transition matrix, $\mathbf{\Phi}(kT)$, given in eqn. (2.87).

2.4. Plant matrices with confluent eigenvalues

It is important to recall that in the case of both the continuous-time and discrete-time systems discussed in this chapter, it has hitherto been assumed that their respective $n\times n$ plant matrices, $\mathbf{A}$ and $\mathbf{\Psi}(T)$, each has a spectrum of *distinct* eigenvalues associated with corresponding sets of *linearly independent* eigenvectors, $\mathbf{u}_j$ $(j=1,\ 2,\ \ldots,\ n)$. It has accordingly been possible, for both classes of system, to use a set of such vectors $\{\mathbf{u}_1,\ \mathbf{u}_2,\ \ldots,\ \mathbf{u}_n\}$ as a basis for an n-dimensional state space and also as the columns of a non-singular $n\times n$ modal matrix of the form

$$\mathbf{U}=[\mathbf{u}_1,\ \mathbf{u}_2,\ \ldots,\ \mathbf{u}_n]. \tag{2.91}$$

It has also transpired that these modal matrices have the very important property that they may be used to diagonalize the plant matrices of both the continuous-time and discrete-time system : thus it has been shown that

$$\mathbf{U}^{-1}\mathbf{A}\mathbf{U}=\mathbf{\Lambda} \tag{2.92 a}$$

and

$$\mathbf{U}^{-1}\mathbf{\Psi}(T)\mathbf{U}=\exp(\mathbf{\Lambda}T), \tag{2.92 b}$$

where $\mathbf{\Lambda}$ and $\exp(\mathbf{\Lambda}T)$ are the diagonal eigenvalue matrices which constitute the right-hand members of eqns. (2.36) and (2.74), respectively. Now eqns. (2.92) are special cases of the general equation

$$\mathbf{T}^{-1}\mathbf{\Pi}\mathbf{T}=\mathbf{\Sigma}, \tag{2.93}$$

where $\mathbf{\Pi}$ is any $n\times n$ plant matrix and $\mathbf{T}$ is any $n\times n$ non-singular matrix. Two matrices $\mathbf{\Pi}$ and $\mathbf{\Sigma}$ related by an equation of the form (2.93) are said to be *similar* and to be connected by a *similarity transformation* with *transforming matrix* $\mathbf{T}$: it is a fundamental theorem in matrix algebra that similar matrices have the same eigenvalues [2]. Indeed, if $\mathbf{\Pi}$ has n distinct eigenvalues associated with n linearly independent eigenvectors, eqns. (2.92) indicate that a transforming matrix exists (namely, the modal matrix of $\mathbf{\Pi}$) such that, in eqn. (2.93), $\mathbf{\Sigma}$ is a diagonal matrix with the eigenvalues of $\mathbf{\Pi}$ on its leading diagonal.

The situation is more complicated, however, if some of the eigenvalues of $\mathbf{\Pi}$ are *confluent* (i.e. non-distinct) since such a matrix may have fewer than n linearly independent eigenvectors. This fact implies that a non-singular matrix of the form (2.91) may no longer exist and that the plant matrix, $\mathbf{\Pi}$, concerned will therefore no longer be diagonalizable. Nevertheless, it is always possible to transform a matrix, $\mathbf{\Pi}$, with confluent eigenvalues to a similar matrix, $\mathbf{J}$, known as the Jordan canonical form of $\mathbf{\Pi}$ by means of a *generalized modal matrix*.

In order to define $\mathbf{J}$, it is necessary to introduce the $k \times k$ *Jordan block matrix* associated with an eigenvalue, λ, of $\mathbf{\Pi}$. This block matrix is defined by the equation

$$\mathbf{J}_k(\lambda) = \begin{bmatrix} \lambda, & 1, & 0, \ldots, 0, & 0 \\ 0, & \lambda, & 1, \ldots, 0, & 0 \\ \cdots & \cdots & \cdots & \cdots \\ \cdots & \cdots & \cdots & \cdots \\ 0, & 0, & 0, \ldots, \lambda, & 1 \\ 0, & 0, & 0, \ldots, 0, & \lambda \end{bmatrix} \tag{2.94}$$

and it is obvious that $\mathbf{J}_k(\lambda)$ has k confluent eigenvalues λ. The *Jordan canonical form* of $\mathbf{\Pi}$ can now be defined as the block-diagonal matrix, $\mathbf{J}$, given by the *direct sum* [3]

$$\mathbf{J} = \mathbf{J}_{\alpha_1}(\lambda_1) \oplus \mathbf{J}_{\alpha_2}(\lambda_1) \oplus \ldots \oplus \mathbf{J}_{\alpha_f}(\lambda_1) \oplus \mathbf{J}_{\beta_1}(\lambda_2) \oplus \mathbf{J}_{\beta_2}(\lambda_2) \oplus \ldots$$
$$\ldots \oplus \mathbf{J}_{\beta_g}(\lambda_2) \oplus \ldots \oplus \mathbf{J}_{\omega_p}(\lambda_s), \tag{2.95}$$

where the sum of the orders of the Jordan blocks associated with λ_l is equal to its multiplicity k_l: in other words

$$\left.\begin{aligned} \alpha_1 + \alpha_2 + \ldots + \alpha_f &= k_1, \\ \beta_1 + \beta_2 + \ldots + \beta_g &= k_2, \\ \cdots\cdots&\cdots\cdots \\ \omega_1 + \omega_2 + \ldots + \omega_p &= k_s. \end{aligned}\right\} \tag{2.96}$$

The Jordan canonical form, $\mathbf{J}$, of a matrix, $\mathbf{\Pi}$, is *unique* apart from the ordering of the blocks, although there are infinitely many transforming matrices, $\mathbf{T}$.

In order to obtain the Jordan canonical form, $\mathbf{J}$, of an $n \times n$ plant matrix, $\mathbf{\Pi}$, it is necessary to determine the elementary divisors of $\mathbf{\Pi}$ over the complex field, C. This determination in turn involves the introduction of the concepts of the determinantal divisors and invariant polynomials of an $m \times n$ *polynomial matrix* [2] (or *lambda matrix*) $\mathbf{P}(\lambda)$ whose elements are polynomials in λ with coefficients over C [4]. Thus, if the rank of $\mathbf{P}(\lambda)$ is r, let $\mathrm{d}_j(\lambda)$ be the greatest common divisor of all jth-order minors of $\mathbf{P}(\lambda)$ $(j = 1, 2, \ldots, r)$: the polynomials, $d_j(\lambda)$, are called the *determinantal divisors* of $\mathbf{P}(\lambda)$ and it can clearly be assumed without loss of generality that each $d_j(\lambda)$ is a monic polynomial. Furthermore, it can be shown [2] that, with $d_0(\lambda)$ taken as unity, then $d_j(\lambda)$ is divisible by $d_{j-1}(\lambda)$ $(j = 1, 2, \ldots, r)$. The quotients

$$i_j(\lambda) = d_j(\lambda)/d_{j-1}(\lambda) \quad (j = 1, 2, \ldots, r) \tag{2.97}$$

will clearly also be monic polynomials and are known as the *invariant polynomials* of $\mathbf{P}(\lambda)$. Each of the latter polynomials may be written as a product of the form

$$i_j(\lambda) = (\lambda - c_1)^{n_{j1}}(\lambda - c_2)^{n_{j2}} \dots (\lambda - c_m)^{n_{jm}} \quad (j = 1, 2, \dots, r). \tag{2.98}$$

In eqn. (2.98), each n_{jl} is a non-negative integer, and $(\lambda - c_1)$, $(\lambda - c_2)$, . . ., $(\lambda - c_m)$ are known as the distinct *linear factors* of $i_j(\lambda)$ if $n_{jl} > 0$ $(l = 1, 2, \dots, m)$: those factors $(\lambda - c_k)^{n_{jl}}$ which are not unity are called the *elementary divisors* of $\mathbf{P}(\lambda)$.

Now it can be readily verified that two square matrices $\mathbf{\Pi}$ and $\mathbf{\Sigma}$ are similar if and only if their *characteristic matrices* $(\mathbf{\Pi} - \lambda\mathbf{I})$ and $(\mathbf{\Sigma} - \lambda\mathbf{I})$ have the same rank and the same elementary divisors. It therefore follows that the elementary divisors of $(\mathbf{\Pi} - \lambda\mathbf{I})$ are the same as the elementary divisors of $(\mathbf{J} - \lambda\mathbf{I})$, where $\mathbf{J}$ is the Jordan canonical form of $\mathbf{\Pi}$. Furthermore, if an $n \times n$ matrix, $\mathbf{X}$, can be expressed as the direct sum

$$\mathbf{X} = \mathbf{X}_1 \oplus \mathbf{X}_2 \oplus \dots \oplus \mathbf{X}_k, \tag{2.99}$$

where $\mathbf{X}_i$ $(i = 1, 2, \dots, k)$ is an $n_i \times n_i$ matrix over C, then it may be inferred that the set of elementary divisors of $(\mathbf{X} - \lambda\mathbf{I}_n)$ comprises all the elementary divisors of all the matrices $(\mathbf{X}_i - \lambda\mathbf{I}_{ni})$, where

$$\sum_{i=1}^{n} n_i = n. \tag{2.100}$$

Therefore, because the elementary divisor of $[\mathbf{J}_k(\lambda_i) - \lambda\mathbf{I}_k]$ is simply $(\lambda - \lambda_i)^k$, the elementary divisors of $(\mathbf{J} - \lambda\mathbf{I})$ (and therefore, of course, those of $(\mathbf{\Pi} - \lambda\mathbf{I})$) are the elementary divisors of the constituent Jordan blocks of $\mathbf{J}$. It is now evident that $\mathbf{J}$ can be simply written down if the elementary divisors of $\mathbf{\Pi}$ are known.

Thus, for example, if the elementary divisors of $\mathbf{\Pi}$ are $(\lambda - \lambda_1)^3$, $(\lambda - \lambda_1)$, and $(\lambda - \lambda_2)^2$, then the corresponding Jordan canonical form is (in the notation of (2.94) and (2.95))

$$\mathbf{J} = \mathbf{J}_3(\lambda_1) \oplus \mathbf{J}_1(\lambda_1) \oplus \mathbf{J}_2(\lambda_2). \tag{2.101}$$

Similarly, if the elementary divisors of $\mathbf{\Pi}$ are

$$(\lambda - \lambda_1)^2, \ (\lambda - \lambda_1)^2, \ (\lambda - \lambda_2),$$

then

$$\mathbf{J} = \mathbf{J}_2(\lambda_1) \oplus \mathbf{J}_2(\lambda_1) \oplus \mathbf{J}_1(\lambda_2), \tag{2.102}$$

and if the elementary divisors are $(\lambda - \lambda_1)^3$, $(\lambda - \lambda_1)$, and $(\lambda - \lambda_2)$, then

$$\mathbf{J} = \mathbf{J}_3(\lambda_1) \oplus \mathbf{J}_1(\lambda_1) \oplus \mathbf{J}_1(\lambda_2). \tag{2.103}$$

The matrices with the Jordan canonical forms displayed in eqns. (2.102) and (2.103) are interesting in that, although each matrix has an eigenvalue λ_1 of multiplicity four and an eigenvalue λ_2 of multiplicity unity, the two matrices are nevertheless *not* similar.

The actual computation of the Jordan canonical form of a matrix can be conveniently illustrated by considering the matrix

$$\mathbf{\Pi} = \begin{bmatrix} 6, & 2, & 2 \\ -2, & 2, & 0 \\ 0, & 0, & 2 \end{bmatrix}. \tag{2.104}$$

The characteristic matrix of $\mathbf{\Pi}$ is clearly

$$\mathbf{\Pi}-\lambda\mathbf{I}=\begin{bmatrix}6-\lambda, & 2, & 2\\ -2, & 2-\lambda, & 0\\ 0, & 0, & 2-\lambda\end{bmatrix} \tag{2.105}$$

and its characteristic equation is

$$|\mathbf{\Pi}-\lambda\mathbf{I}|=\begin{vmatrix}6-\lambda, & 2, & 2\\ -2, & 2-\lambda, & 0\\ 0, & 0, & 2-\lambda\end{vmatrix}=(4-\lambda)^2(2-\lambda)=0. \tag{2.106}$$

It is evident from eqn. (2.106) that $\mathbf{\Pi}$ has eigenvalues $\lambda_1=2$ and $\lambda_2=4$ of multiplicities one and two, respectively. The determinantal divisors of the polynomial matrix $(\mathbf{\Pi}-\lambda\mathbf{I})$ defined in eqn. (2.105) can be readily computed and are found to be

$$d_0=1\ ;\ d_1=1\ ;\ d_2=1\ ;\ d_3=(\lambda-2)(\lambda-4)^2. \tag{2.107}$$

It then follows from the definition (2.97) that the invariant polynomials of $(\mathbf{\Pi}-\lambda\mathbf{I})$ are

$$i_1=d_1/d_0=1\ ;\ i_2=d_2/d_1=1\ ;\ i_3=d_3/d_2=(\lambda-2)(\lambda-4)^2. \tag{2.108}$$

Finally, the definition associated with eqns. (2.98) indicates that the elementary divisors of $(\mathbf{\Pi}-\lambda\mathbf{I})$ in this example are

$$(\lambda-2)\ ;\ (\lambda-4)^2. \tag{2.109}$$

The Jordan canonical form of the matrix $\mathbf{\Pi}$ given in eqn. (2.104) therefore has the structure

$$\mathbf{J}_1(\lambda_1)\oplus\mathbf{J}_2(\lambda_2)=\begin{bmatrix}2, & 0, & 0\\ 0, & 4, & 1\\ 0, & 0, & 4\end{bmatrix}. \tag{2.110}$$

It will be recalled that there are infinitely many transforming matrices, $\mathbf{T}$, which will transform the matrix $\mathbf{\Pi}$ defined in eqn. (2.104) into its Jordan canonical form given in eqn. (2.110). There are also many ways of determining a suitable matrix, $\mathbf{T}$ [5]. However, in the case of low-order matrices it is feasible to compute a value for $\mathbf{T}$ simply by inserting the appropriate values of $\mathbf{\Pi}$ and $\mathbf{J}$ into the equation

$$\mathbf{\Pi}\mathbf{T}=\mathbf{T}\mathbf{J} \tag{2.111}$$

and then solving the resulting simultaneous linear equations for the elements of $\mathbf{T}$. Thus, if the matrices $\mathbf{\Pi}$ and $\mathbf{J}$ given in eqns. (2.104) and (2.110) are substituted into eqn. (2.111), it may be readily verified that the matrix

$$\mathbf{T}=\begin{bmatrix}0, & 4, & 1\\ -1, & -4, & 1\\ 1, & 0, & 0\end{bmatrix} \tag{2.112}$$

is one of the infinitely many matrices which will effect the transformation of $\mathbf{\Pi}$ into its Jordan canonical form. The matrix given in eqn. (2.112) may therefore be regarded as a generalized modal matrix of a system whose plant matrix is given in eqn. (2.104). The columns of this matrix may be used as a basis for the state space concerned : a set of vectors which constitute a reciprocal basis for this space may readily be computed by determining any set of vectors which satisfy equations of the form (2.17).

The nature of the transition matrix associated with a continuous-time system whose plant matrix has a non-diagonal Jordan canonical form can be inferred by first considering the 2×2 Jordan block

$$\mathbf{J}_2(\lambda) = \begin{bmatrix} \lambda, & 1 \\ 0, & \lambda \end{bmatrix}. \tag{2.113}$$

In this case,

$$\exp(\mathbf{J}_2 t) = \exp \begin{bmatrix} \lambda t, & t \\ 0 & \lambda t \end{bmatrix} = [\exp(\lambda \mathbf{I} t)][\exp \mathbf{C}], \tag{2.114}$$

where

$$\mathbf{C} = \begin{bmatrix} 0, & t \\ 0, & 0 \end{bmatrix}.$$

Since $\mathbf{C}^2$, $\mathbf{C}^3$, . . . are all null, it follows from the definition (2.39) of the exponential matrix that

$$\exp(\mathbf{C}) = \mathbf{I} + \mathbf{C} = \begin{bmatrix} 1, & t \\ 0, & 1 \end{bmatrix}. \tag{2.115}$$

which, together with eqn. (2.114), implies that

$$\exp(\mathbf{J}_2 t) = \begin{bmatrix} e^{\lambda t}, & 0 \\ 0, & e^{\lambda t} \end{bmatrix} \begin{bmatrix} 1, & t \\ 0, & 1 \end{bmatrix} = \begin{bmatrix} e^{\lambda t}, & t\,e^{\lambda t} \\ 0, & e^{\lambda t} \end{bmatrix}. \tag{2.116}$$

In the same way, in the case of a 3×3 Jordan block

$$\mathbf{J}_3(\lambda) = \begin{bmatrix} \lambda, & 1, & 0 \\ 0, & \lambda, & 1 \\ 0, & 0, & \lambda \end{bmatrix}, \tag{2.117}$$

it is found that

$$\exp(\mathbf{J}_3 t) = \begin{bmatrix} e^{\lambda t}, & t\,e^{\lambda t}, & t^2\,e^{\lambda t}/2! \\ 0, & e^{\lambda t} & t\,e^{\lambda t} \\ 0, & 0, & e^{\lambda t} \end{bmatrix}. \tag{2.118}$$

Equations (2.116) and (2.118) give rise to the conjecture that, in the case of the $k \times k$ Jordan block defined in eqn. (2.94),

$$\exp(\mathbf{J}_k t) = \begin{bmatrix} e^{\lambda t}, & t\,e^{\lambda t}, & t^2\,e^{\lambda t}/2!, & t^3\,e^{\lambda t}/3!, & \ldots, & t^{k-1}\,e^{\lambda t}/(k-1)! \\ 0, & e^{\lambda t}, & t\,e^{\lambda t}, & t^2\,e^{\lambda t}/2!, & \ldots, & t^{k-2}\,e^{\lambda t}/(k-2)! \\ 0, & 0, & e^{\lambda t} & t\,e^{\lambda t}, & \ldots, & t^{k-3}\,e^{\lambda t}/(k-3)! \\ \cdots & \cdots & \cdots & \cdots & \cdots & \cdots \\ \cdots & \cdots & \cdots & \cdots, & e^{\lambda t}, & t\,e^{\lambda t} \\ 0, & 0, & 0, & 0, & \ldots, & 0, \quad e^{\lambda t} \end{bmatrix} \tag{2.119}$$

It can be shown by mathematical induction that this conjecture is correct.

These results indicate that, when the appropriate plant matrix has confluent eigenvalues, the free response of a continuous-time system of the class (2.1) will include terms of the form

$$\frac{t^{m-1}}{(m-1)!} \exp(\lambda_i t) \tag{2.120}$$

and that the free response of a discrete-time system of the class (2.64) will include terms of the appropriate analogous form. It is evident from eqn. (2.120) that a continuous-time system will still be asymptotically stable if and only if

$$\operatorname{Re} \lambda_i < 0 \quad (i = 1, 2, \ldots, n)$$

in spite of the presence of the terms involving powers of t. It is also evident that the condition (2.82) for the asymptotic stability of a discrete-time system will similarly be unchanged.

References

[1] Zadeh, L. A., and Desoer, C. A., 1963, *Linear System Theory* (McGraw-Hill).
[2] Gantmacher, F. R., 1959, *The Theory of Matrices* (Chelsea Publishing Co.).
[3] Barnett, S., and Storey, C., 1970, *Matrix Methods in Stability Theory* (Nelson).
[4] Lancaster, P., 1966, *Lambda-matrices and Vibrating Systems* (Pergamon Press).
[5] Gupta, S. C., 1966, *Transform and State Variable Methods in Linear Systems* (Wiley).

Eigenvalue and Eigenvector Sensitivities in Linear Systems Theory

CHAPTER 3

3.1. Introduction

In this chapter simple and explicit derivations are given of expressions for the first- and second-order eigenvalue and eigenvector sensitivity coefficients for the fundamental eigenproblem associated with the behaviour of linear systems governed by equations of the form (2.1) and (2.64). Such expressions relating changes in the eigenvalues and eigenvectors of the appropriate plant matrix to changes in its elements have been given by numerous authors since the early work of Jacobi [1]. These expressions, which are of great importance in modal control theory, have been given in various forms and from different points of view, and have also been derived by a number of different methods. Thus, the sensitivity eigenproblem has been viewed as a problem in numerical analysis (Faddeev and Faddeeva [2], Wilkinson [3]), as a problem in perturbation theory (Bellman [4], Wilkinson [3]), and as a problem in linear systems theory (Laughton [5], Mann and Marshall [6], Mann, Marshall and Nicholson [7], Van Ness, Boyle and Imad [8], Rosenbrock [9], Morgan [10], Nicholson [11]).

The sensitivity coefficients derived in this chapter can be used in the design of modal control systems to solve a number of related problems of which the following are typical :

(i) the selection of system parameters so that the eigenvalues and eigenvectors associated with the plant matrix are as insensitive as possible to changes in such parameters ;

(ii) the determination of good approximations to the eigenvalues and eigenvectors associated with various plant matrices in the vicinity of a datum matrix having known eigenvalues and eigenvectors without having to compute the eigenvalues and eigenvectors *ab initio* for each new plant matrix, thus facilitating the choice of appropriate system parameters.

3.2. Continuous-time systems

3.2.1. *Introduction*

As was indicated in Chapter 2, one of the central results in linear systems theory is that the free motions of a continuous-time dynamical system governed by a state equation of the form

$$\dot{\mathbf{x}}(t) = \mathbf{A}\mathbf{x}(t) \tag{3.1}$$

are given by the expression

$$\mathbf{x}(t) = \sum_{i=1}^{n} [\exp(\lambda_i t)]\mathbf{u}_i \mathbf{v}_i' \mathbf{x}(0). \tag{3.2}$$

In eqn. (3.1), $\mathbf{x}(t)$ is the $n \times 1$ state vector of the system, $\mathbf{A}$ is an $n \times n$ matrix with distinct eigenvalues λ_i $(i = 1, 2, \ldots, n)$: in (3.2), $\mathbf{u}_i$ $(i = 1, 2, \ldots, n)$ are the linearly independent eigenvectors of $\mathbf{A}$ which satisfy

$$\mathbf{A}\mathbf{u}_i = \lambda_i \mathbf{u}_i \tag{3.3}$$

and $\mathbf{v}_j$ $(j = 1, 2, \ldots, n)$ are the corresponding eigenvectors of $\mathbf{A}'$ which satisfy

$$\mathbf{A}'\mathbf{v}_j = \lambda_j \mathbf{v}_j, \tag{3.4}$$

where the prime denotes transposition. The $\mathbf{u}_i$ and $\mathbf{v}_j$ satisfy the equations

$$\mathbf{u}_i'\mathbf{v}_j = \mathbf{v}_j'\mathbf{u}_i = \delta_{ij} \quad (i, j = 1, 2, \ldots, n), \tag{3.5}$$

where δ_{ij} is the Kronecker delta.

The corresponding fundamental problem in sensitivity theory is to determine the sensitivities of λ_i, $\mathbf{u}_i$, $\mathbf{v}_i$ $(i = 1, 2, \ldots, n)$ to changes in the elements of the matrix $\mathbf{A}$. The remainder of this chapter is devoted to the determination of these sensitivities to a first- and second-order of approximation : in addition, the application of the results to the corresponding discrete-time sensitivity problem is discussed.

3.2.2. *First-order eigenvalue sensitivities*

If

$$\mathbf{A} = [a_{kl}]$$

and the generic element a_{kl} is perturbed due to changes in system parameters, then the eigenvalues and eigenvectors of $\mathbf{A}$ will change. Indeed, partial differentiation of eqn. (3.3) with respect to a_{kl} indicates that

$$\frac{\partial \mathbf{A}}{\partial a_{kl}} \mathbf{u}_i + \mathbf{A} \frac{\partial \mathbf{u}_i}{\partial a_{kl}} = \frac{\partial \lambda_i}{\partial a_{kl}} \mathbf{u}_i + \lambda_i \frac{\partial \mathbf{u}_i}{\partial a_{kl}}. \tag{3.6}$$

Pre-multiplication of eqn. (3.6) by $\mathbf{v}_i'$ then gives

$$\mathbf{v}_i' \frac{\partial \mathbf{A}}{\partial a_{kl}} \mathbf{u}_i + \lambda_i \mathbf{v}_i' \frac{\partial \mathbf{u}_i}{\partial a_{kl}} = \mathbf{v}_i' \frac{\partial \lambda_i}{\partial a_{kl}} \mathbf{u}_i + \lambda_i \mathbf{v}_i' \frac{\partial \mathbf{u}_i}{\partial a_{kl}}$$

which reduces to the set of scalar equations

$$\frac{\partial \lambda_i}{\partial a_{kl}} = v_i^k u_i^l \quad (i, k, l = 1, 2, \ldots, n) \tag{3.7}$$

in view of eqns. (3.5) and the fact that

$$\frac{\partial \mathbf{A}}{\partial a_{kl}} = [\alpha_{ij}] = \delta_{ik}\delta_{jl}. \tag{3.8}$$

In eqn. (3.7), v_i^k and u_i^l are respectively the kth element of $\mathbf{v}_i$ and the lth element of $\mathbf{u}_i$. The $\partial\lambda_i/\partial a_{kl}$ are, of course, the desired eigenvalue sensitivity coefficients which relate changes in the λ_i to changes in the a_{kl}. These coefficients may be viewed as the elements of a set of n eigenvalue sensitivity matrices

$$\mathbf{S}_i = \left[\frac{\partial \lambda_i}{\partial a_{kl}}\right] = \mathbf{v}_i \mathbf{u}_i' \quad (i, k, l = 1, 2, \ldots, n). \tag{3.9}$$

These matrices are clearly all of unit rank and therefore singular.

The sensitivity matrices have a number of interesting properties which are useful both in theoretical work and in checking numerical calculations. Firstly, it follows immediately from eqns. (3.5) and (3.9) that

$$\mathbf{S}_i\mathbf{S}_j = \mathbf{v}_i\mathbf{u}_i'\mathbf{v}_j\mathbf{u}_j' = \delta_{ij}\mathbf{S}_i. \tag{3.10}$$

Secondly, if the non-singular modal matrices

$$\mathbf{U} = [\mathbf{u}_1, \mathbf{u}_2, \ldots, \mathbf{u}_n] \tag{3.11 a}$$

and

$$\mathbf{V} = [\mathbf{v}_1, \mathbf{v}_2, \ldots, \mathbf{v}_n] \tag{3.11 b}$$

are introduced, it can be deduced from (3.9) that

$$\sum_{i=1}^{n} \mathbf{S}_i = \mathbf{V}\mathbf{U}'. \tag{3.12}$$

Furthermore, since eqns. (3.5) can be written in the form

$$\mathbf{U}'\mathbf{V} = \mathbf{V}'\mathbf{U} = \mathbf{I},$$

so that

$$\mathbf{V} = (\mathbf{U}')^{-1}, \tag{3.13}$$

it follows from (3.12) and (3.13) that

$$\sum_{i=1}^{n} \mathbf{S}_i = \mathbf{I}. \tag{3.14}$$

3.2.3. *First-order eigenvector sensitivities*

The eigenvector changes arising from a perturbation of the element a_{kl} of $\mathbf{A}$ can be determined very simply by differentiating eqns. (3.3) and (3.4) partially with respect to a_{kl}, and by writing the resulting equations in the form

$$(\mathbf{A} - \lambda_i \mathbf{I}) \frac{\partial \mathbf{u}_i}{\partial a_{kl}} = \frac{\partial \lambda_i}{\partial a_{kl}} \mathbf{u}_i - \frac{\partial \mathbf{A}}{\partial a_{kl}} \mathbf{u}_i, \tag{3.15 a}$$

$$\frac{\partial \mathbf{v}_j'}{\partial a_{kl}} (\mathbf{A} - \lambda_j \mathbf{I}) = \frac{\partial \lambda_j}{\partial a_{kl}} \mathbf{v}_j' - \mathbf{v}_j' \frac{\partial \mathbf{A}}{\partial a_{kl}}. \tag{3.15 b}$$

Since λ_i and λ_j are eigenvalues of $\mathbf{A}$, the matrices $(\mathbf{A}-\lambda_i\mathbf{I})$ and $(\mathbf{A}-\lambda_j\mathbf{I})$ are singular, so that eqns. (3.15) cannot be solved immediately for the eigenvector sensitivity coefficients $\partial\mathbf{u}_i/\partial a_{kl}$ and $\partial\mathbf{v}_j'/\partial a_{kl}$.

Nevertheless, if eqn. (3.15 *a*) is pre-multiplied by $\mathbf{v}_j'$ $(j\neq i)$ and eqn. (3.15 *b*) is post-multiplied by $\mathbf{u}_i$ $(i\neq j)$, it follows that

$$(\lambda_j-\lambda_i)\mathbf{v}_j'\frac{\partial\mathbf{u}_i}{\partial a_{kl}}=-v_j^k u_i^l \quad (j\neq i\ ;\ j=1,2,\ldots,n) \tag{3.16 a}$$

and

$$\frac{\partial\mathbf{v}_j'}{\partial a_{kl}}\mathbf{u}_i(\lambda_i-\lambda_j)=-v_j^k u_i^l \quad (i\neq j\ ;\ i=1,2,\ldots,n) \tag{3.16 b}$$

in view of eqns. (3.5) and (3.8). Now, if eqn. (3.16 *a*) is pre-multiplied by $\mathbf{u}_j$ $(j\neq i)$ and eqn. (3.16 *b*) is post-multiplied by $\mathbf{v}_i'$, it transpires that

$$\mathbf{S}_j'\frac{\partial\mathbf{u}_i}{\partial a_{kl}}=-\left(\frac{v_j^k u_i^l}{\lambda_j-\lambda_i}\right)\mathbf{u}_j \quad (j\neq i\ ;\ j=1,2,\ldots,n) \tag{3.17 a}$$

and

$$\frac{\partial\mathbf{v}_j'}{\partial a_{kl}}\mathbf{S}_i'=\left(\frac{v_j^k u_i^l}{\lambda_j-\lambda_i}\right)\mathbf{v}_i' \quad (i\neq j\ ;\ i=1,2,\ldots,n). \tag{3.17 b}$$

Equations (3.17) can be written more compactly by introducing the matrix

$$\mathbf{H}_{ji}=[h_{ji}^{kl}]=\mathbf{v}_j\mathbf{u}_i'/(\lambda_i-\lambda_j). \tag{3.18}$$

In fact, eqns. (3.17) then become

$$\mathbf{S}_j'\frac{\partial\mathbf{u}_i}{\partial a_{kl}}=h_{ji}^{kl}\mathbf{u}_j \quad (j\neq i\ ;\ j=1,2,\ldots,n) \tag{3.19 a}$$

and

$$\frac{\partial\mathbf{v}_j'}{\partial a_{kl}}\mathbf{S}_i'=-h_{ji}^{kl}\mathbf{v}_i' \quad (i\neq j\ ;\ i=1,2,\ldots,n). \tag{3.19 b}$$

Since the eigenvectors of $\mathbf{A}$ and $\mathbf{A}'$ are linearly independent because of the assumption of distinct eigenvalues, the required vectors $\partial\mathbf{u}_i/\partial a_{kl}$ and $\partial\mathbf{v}_j'/\partial a_{kl}$ may be expanded in terms of the $\mathbf{u}_i$ and $\mathbf{v}_j'$, respectively. Thus, in the first case let

$$\frac{\partial\mathbf{u}_i}{\partial a_{kl}}=\sum_{m=1}^{n}\phi_m^{ikl}\mathbf{u}_m \tag{3.20 a}$$

so that

$$\mathbf{S}_j'\frac{\partial\mathbf{u}_i}{\partial a_{kl}}=\mathbf{u}_j\mathbf{v}_j'\sum_{m=1}^{n}\phi_m^{ikl}\mathbf{u}_m=\phi_j^{ikl}\mathbf{u}_j \quad (j\neq i\ ;\ j=1,2,\ldots,n), \tag{3.21 a}$$

and in the second case let

$$\frac{\partial\mathbf{v}_j'}{\partial a_{kl}}=\sum_{m=1}^{n}\psi_m^{jkl}\mathbf{v}_m' \tag{3.20 b}$$

so that

$$\frac{\partial\mathbf{v}_j'}{\partial a_{kl}}\mathbf{S}_i'=\left(\sum_{m=1}^{n}\psi_m^{jkl}\mathbf{v}'_m\right)\mathbf{u}_i\mathbf{v}_i'=\psi_i^{jkl}\mathbf{v}_i' \quad (i\neq j\ ;\ i=1,2,\ldots,n). \tag{3.21 b}$$

It follows by comparing (3.19 *a*) with (3.21 *a*) and (3.19 *b*) with (3.21 *b*) that

$$\phi_j^{ikl} = h_{ji}^{kl} \quad (j \neq i\ ;\ j = 1, 2, \ldots, n) \tag{3.22 a}$$

and

$$\psi_i^{jkl} = -h_{ji}^{kl} \quad (i \neq j\ ;\ i = 1, 2, \ldots, n). \tag{3.22 b}$$

If the results given in eqns. (3.22) are used in eqns. (3.20), then the latter equations become

$$\frac{\partial \mathbf{u}_i}{\partial a_{kl}} = \phi_i^{ikl}\mathbf{u}_i + \sum_{\substack{j=1\\ j\neq i}}^{n} h_{ji}^{kl}\mathbf{u}_j \quad (i, k, l = 1, 2, \ldots, n) \tag{3.23 a}$$

and

$$\frac{\partial \mathbf{v}_j'}{\partial a_{kl}} = \psi_j^{jkl}\mathbf{v}_j' + \sum_{\substack{i=1\\ i\neq j}}^{n} h_{ji}^{kl}\mathbf{v}_i' \quad (j, k, l = 1, 2, \ldots, n) \tag{3.23 b}$$

which give the required eigenvector sensitivity coefficients.

The coefficients ϕ_i^{ikl} and ψ_j^{jkl} may be left as completely arbitrary coefficients unless it is desired to constrain the perturbed eigenvectors to satisfy conditions such as (3.5). If this is the case, then the coefficients must be such that

$$(\mathbf{v}_j' + \delta\mathbf{v}_j')(\mathbf{u}_i + \delta\mathbf{u}_i) = \left(\mathbf{v}_j' + \frac{\partial \mathbf{v}_j'}{\partial a_{kl}}\,\delta a_{kl}\right)\left(\mathbf{u}_i + \frac{\partial \mathbf{u}_i}{\partial a_{kl}}\delta a_{kl}\right) = \delta_{ij}. \tag{3.24}$$

It can be readily verified by substituting from eqns. (3.23) into (3.24) that the latter equations will be satisfied if

$$\phi_i^{ikl} + \psi_i^{ikl} = 0 \quad (i = 1, 2, \ldots, n),$$

that is, if

$$\phi_i^{ikl} = -\psi_i^{ikl} = \zeta_i^{kl} \quad (i = 1, 2, \ldots, n).$$

Equations (3.23) for the eigenvector sensitivity coefficients then have the form

$$\frac{\partial \mathbf{u}_i}{\partial a_{kl}} = \zeta_i^{kl}\mathbf{u}_i + \sum_{\substack{j=1\\ j\neq i}}^{n} h_{ji}^{kl}\mathbf{u}_j \quad (i, k, l = 1, 2, \ldots, n), \tag{3.25 a}$$

and

$$\frac{\partial \mathbf{v}_j'}{\partial a_{kl}} = -\zeta_j^{kl}\mathbf{v}_j' - \sum_{\substack{i=1\\ i\neq j}}^{n} h_{ji}^{kl}\mathbf{v}_i' \quad (j, k, l = 1, 2, \ldots, n), \tag{3.25 b}$$

where the ζ_i^{kl} are arbitrary. However, the most convenient choice is perhaps

$$\zeta_i^{kl} = 0 \quad (i, k, l = 1, 2, \ldots, n),$$

in which case eqn. (3.25 *a*) assumes the form given by Laughton [5].

3.2.4. *Illustrative example: first-order sensitivities*

In the case of a system governed by an equation of the form (3.1) for which

$$\mathbf{A}=\begin{bmatrix}-2, & -1, & 1\\ 1, & 0, & 1\\ -1, & 0, & 1\end{bmatrix},$$

the eigenvalues are

$$\left.\begin{aligned}\lambda_1&=1,\\ \lambda_2&=-1+i,\\ \lambda_3&=-1-i,\end{aligned}\right\} \tag{3.26}$$

and the corresponding eigenvectors are

$$\mathbf{u}_1=\begin{bmatrix}0\\1\\1\end{bmatrix},\ \mathbf{u}_2=\begin{bmatrix}5\\-3-4i\\2+i\end{bmatrix},\ \mathbf{u}_3=\begin{bmatrix}5\\-3+4i\\2-i\end{bmatrix}, \tag{3.27 a}$$

$$\mathbf{v}_1=\begin{bmatrix}-0{\cdot}2\\0{\cdot}2\\0{\cdot}8\end{bmatrix},\ \mathbf{v}_2=\begin{bmatrix}0{\cdot}1+0{\cdot}1i\\0{\cdot}1i\\-0{\cdot}1i\end{bmatrix},\ \mathbf{v}_3=\begin{bmatrix}0{\cdot}1-0{\cdot}1i\\-0{\cdot}1i\\0{\cdot}1i\end{bmatrix}. \tag{3.27 b}$$

If the element a_{23} of $\mathbf{A}$ is perturbed by an amount $\delta a_{23}=-0{\cdot}8$, then eqns. (3.7) indicate that

$$\left.\begin{aligned}\frac{\partial\lambda_1}{\partial a_{23}}&=v_1^{\,2}u_1^{\,3}=0{\cdot}2,\\ \frac{\partial\lambda_2}{\partial a_{23}}&=v_2^{\,2}u_2^{\,3}=-0{\cdot}1+0{\cdot}2i,\\ \frac{\partial\lambda_3}{\partial a_{23}}&=v_3^{\,2}u_3^{\,3}=-0{\cdot}1-0{\cdot}2i.\end{aligned}\right\} \tag{3.28}$$

Similarly eqns. (3.25) with $\zeta_i^{23}=0$ $(i=1, 2, 3)$ indicate that

$$\left.\begin{aligned}\frac{\partial\mathbf{u}_1}{\partial a_{23}}&=h_{21}^{\ 23}\mathbf{u}_2+h_{31}^{\ 23}\mathbf{u}_3=[-1/5,\ 11/25,\ -4/25]',\\ \frac{\partial\mathbf{u}_2}{\partial a_{23}}&=h_{12}^{\ 23}\mathbf{u}_1+h_{32}^{\ 23}\mathbf{u}_3=\\ &\qquad[-1/2-i/4,\ 19/50-41i/100,\ -37/100-4i/25]',\\ \frac{\partial\mathbf{u}_3}{\partial a_{23}}&=h_{13}^{\ 23}\mathbf{u}_1+h_{23}^{\ 23}\mathbf{u}_2=\\ &\qquad[-1/2+i/4,\ 19/50+41i/100,\ -37/100+4i/25]',\end{aligned}\right\} \tag{3.29 a}$$

and

$$\left.\begin{aligned}
&\frac{\partial \mathbf{v}_1'}{\partial a_{23}} = -h_{21}{}^{23}\mathbf{v}_2' - h_{31}{}^{23}\mathbf{v}_3' = [-1/125,\ -4/125,\ 4/125)],\\
&\frac{\partial \mathbf{v}_2'}{\partial a_{23}} = -h_{12}{}^{23}\mathbf{v}_1' - h_{32}{}^{23}\mathbf{v}_3' =\\
&\qquad [1/1000 - 7i/1000,\ 1/1000 - 9i/500,\ 21/1000 - 11i/500],\\
&\frac{\partial \mathbf{v}_3'}{\partial a_{23}} = -h_{13}{}^{23}\mathbf{v}_1' - h_{23}{}^{23}\mathbf{v}_2' =\\
&\qquad [1/1000 + 7i/1000,\ -1/1000 + 9i/500,\ 21/1000 + 11i/500].
\end{aligned}\right\}\qquad (3.29\,b)$$

The numerical values of the various sensitivity coefficients given in eqns. (3.28) and (3.29) may be used to compute first-order estimates of the eigenvalues and eigenvectors of the perturbed plant matrix $\mathbf{A}$. Thus, eqns. (3.26) and (3.28) indicate that the estimated eigenvalues $\hat{\lambda}_1$, $\hat{\lambda}_2$, and $\hat{\lambda}_3$, are given by the equations

$$\left.\begin{aligned}
\hat{\lambda}_1 &= \lambda_1 + \frac{\partial \lambda_1}{\partial a_{23}}\,\delta a_{23} = 0{\cdot}840,\\
\hat{\lambda}_2 &= \lambda_2 + \frac{\partial \lambda_2}{\partial a_{23}}\,\delta a_{23} = -0{\cdot}920 + 0{\cdot}840i,\\
\hat{\lambda}_3 &= \lambda_3 + \frac{\partial \lambda_3}{\partial a_{23}}\,\delta a_{23} = -0{\cdot}920 - 0{\cdot}840i,
\end{aligned}\right\}\qquad (3.30)$$

Similarly, eqns. (3.27) and (3.29) indicate that the corresponding estimated eigenvectors $\hat{\mathbf{u}}_1$, $\hat{\mathbf{u}}_2$, $\hat{\mathbf{u}}_3$, $\hat{\mathbf{v}}_1$, $\hat{\mathbf{v}}_2$, and $\hat{\mathbf{v}}_3$, are given by the equations

$$\left.\begin{aligned}
&\hat{\mathbf{u}}_1 = \mathbf{u}_1 + \frac{\partial \mathbf{u}_1}{\partial a_{23}}\,\delta a_{23} = [0{\cdot}160,\ 0{\cdot}648,\ 1{\cdot}128]',\\
&\hat{\mathbf{u}}_2 = \mathbf{u}_2 + \frac{\partial \mathbf{u}_2}{\partial a_{23}}\,\delta a_{23} =\\
&\qquad [5{\cdot}400 + 0{\cdot}200i,\ -3{\cdot}304 - 3{\cdot}672i,\ 2{\cdot}296 + 1{\cdot}128i]',\\
&\hat{\mathbf{u}}_3 = \mathbf{u}_3 + \frac{\partial \mathbf{u}_3}{\partial a_{23}}\,\delta a_{23} =\\
&\qquad [5{\cdot}400 - 0{\cdot}200i,\ -3{\cdot}304 + 3{\cdot}672i,\ 2{\cdot}296 - 1{\cdot}128i]',
\end{aligned}\right\}\qquad (3.31\,a)$$

and

$$\left.\begin{aligned}
&\hat{\mathbf{v}}_1' = \mathbf{v}_1' + \frac{\partial \mathbf{v}_1'}{\partial a_{23}}\,\delta a_{23} = [-0{\cdot}194,\ 0{\cdot}226,\ 0{\cdot}774],\\
&\hat{\mathbf{v}}_2' = \mathbf{v}_2' + \frac{\partial \mathbf{v}_2'}{\partial a_{23}}\,\delta a_{23} =\\
&\qquad [0{\cdot}099 + 0{\cdot}106i,\ 0{\cdot}001 + 0{\cdot}114i,\ -0{\cdot}017 - 0{\cdot}082i],\\
&\hat{\mathbf{v}}_3' = \mathbf{v}_3' + \frac{\partial \mathbf{v}_3'}{\partial a_{23}}\,\delta a_{23} =\\
&\qquad [0{\cdot}099 - 0{\cdot}106i,\ 0{\cdot}001 - 0{\cdot}114i,\ -0{\cdot}017 + 0{\cdot}082i].
\end{aligned}\right\}\qquad (3.31\,b)$$

The corresponding eigenvalues and eigenvectors of the perturbed plant matrix $\mathbf{A}$ obtained by direct calculation are given by the equations

$$\left.\begin{aligned} \lambda_1 &= 0{\cdot}814, \\ \lambda_2 &= -0{\cdot}907 + 0{\cdot}808i, \\ \lambda_3 &= -0{\cdot}907 - 0{\cdot}808i, \end{aligned}\right\} \tag{3.32}$$

$$\left.\begin{aligned} \mathbf{u}_1 &= [0{\cdot}210,\ 0{\cdot}536,\ 1{\cdot}128]', \\ \mathbf{u}_2 &= [5{\cdot}489 + 0{\cdot}417i,\ -3{\cdot}304 - 3{\cdot}672i,\ 2{\cdot}362 + 1{\cdot}220i]', \\ \mathbf{u}_3 &= [5{\cdot}489 - 0{\cdot}417i,\ -3{\cdot}304 + 3{\cdot}672i,\ 2{\cdot}362 - 1{\cdot}220i]', \end{aligned}\right\} \tag{3.33 a}$$

and

$$\left.\begin{aligned} \mathbf{v}_1' &= [-0{\cdot}200,\ 0{\cdot}245,\ 0{\cdot}807], \\ \mathbf{v}_2' &= [0{\cdot}103 + 0{\cdot}107i,\ 0{\cdot}005 + 0{\cdot}122i,\ -0{\cdot}021 - 0{\cdot}078i], \\ \mathbf{v}_3' &= [0{\cdot}103 - 0{\cdot}107i,\ 0{\cdot}005 - 0{\cdot}122i,\ -0{\cdot}021 + 0{\cdot}078i], \end{aligned}\right\} \tag{3.33 b}$$

There is good agreement between the estimated and directly-calculated values of the eigenvalues and eigenvectors of the perturbed plant matrix, as can be seen by comparing (3.30) with (3.32) and (3.31) with (3.33).

3.2.5. *Second-order eigenvalue sensitivities*

The second-order eigenvalue sensitivities can be derived by extending the approach adopted in § 3.2.2. Thus, the first-order eigenvalue sensitivity coefficients are (see eqn. (3.7))

$$\frac{\partial \lambda_i}{\partial a_{kl}} = v_i^k u_i^l \quad (i, k, l = 1, 2, \ldots, n), \tag{3.34}$$

and differentiation of these expressions with respect to the element a_{st} of $\mathbf{A}$ indicates that

$$\frac{\partial^2 \lambda_i}{\partial a_{kl} \partial a_{st}} = v_i^k \frac{\partial u_i^l}{\partial a_{st}} + \frac{\partial v_i^k}{\partial a_{st}} u_i^l. \tag{3.35}$$

It was shown in § 3.2.3 that

$$\frac{\partial \mathbf{u}_i}{\partial a_{kl}} = \phi_i^{ikl} \mathbf{u}_i + \sum_{\substack{j=1 \\ j \neq i}}^{n} h_{ji}^{kl} \mathbf{u}_j \quad (i, k, l = 1, 2, \ldots, n), \tag{3.36 a}$$

and

$$\frac{\partial \mathbf{v}_j'}{\partial a_{kl}} = -\phi_j^{jkl} \mathbf{v}_j' - \sum_{\substack{i=1 \\ i \neq j}}^{n} h_{ji}^{kl} \mathbf{v}_i' \quad (j, k, l = 1, 2, \ldots, n), \tag{3.36 b}$$

where

$$h_{ji}^{kl} = v_j^k u_i^l / (\lambda_i - \lambda_j) \quad (i \neq j;\ i, j = 1, 2, \ldots, n). \tag{3.37}$$

The coefficients ϕ_i^{ikl} $(i, k, l = 1, 2, \ldots, n)$ in eqns. (3.36) are arbitrary, but the most convenient choice is usually

$$\phi_i^{ikl} = 0 \quad (i, k, l = 1, 2, \ldots, n). \tag{3.38}$$

The quantities v_j^k and u_i^l in eqn. (3.37) are respectively the kth element of $\mathbf{v}_j$ and the lth element of $\mathbf{u}_i$.

If the results given in eqns. (3.36), (3.37), and (3.38) are used in eqn. (3.35), the latter equation becomes

$$\frac{\partial^2 \lambda_i}{\partial a_{kl} \partial a_{st}} = \sum_{\substack{j=1 \\ j \neq i}}^{n} (h_{ji}{}^{st} v_i{}^k u_j{}^l - h_{ij}{}^{st} v_j{}^k u_i{}^l)$$

$$= \sum_{\substack{j=1 \\ j \neq i}}^{n} (h_{ji}{}^{st} v_i{}^k u_j{}^l + h_{ji}{}^{kl} v_i{}^s u_j{}^t) \quad (i, k, l, s, t = 1, 2, \ldots, n). \tag{3.39}$$

Equation (3.39) gives the desired second-order eigenvalue sensitivity coefficients.

In order to calculate the second-order estimate of an eigenvalue, λ_i, due to changes in the elements a_{kl} and a_{st} of the plant matrix, $\mathbf{A}$, it is necessary to substitute the various sensitivity coefficients into appropriate Taylor series expansions. Thus, if the estimates of the eigenvalues are $\hat{\hat{\lambda}}_i$ $(i = 1, 2, \ldots, n)$, then these second-order estimates are obtained by evaluating the expression

$$\hat{\hat{\lambda}}_i = \lambda_i + \frac{\partial \lambda_i}{\partial a_{kl}} \delta a_{kl} + \frac{\partial \lambda_i}{\partial a_{st}} \delta a_{st}$$

$$+ \tfrac{1}{2} \left[\frac{\partial^2 \lambda_i}{\partial a_{kl}{}^2} \delta a_{kl}{}^2 + 2 \frac{\partial^2 \lambda_i}{\partial a_{kl} \partial a_{st}} \delta a_{kl} \delta a_{st} + \frac{\partial^2 \lambda_i}{\partial a_{st}{}^2} \delta a_{st}{}^2 \right]. \tag{3.40 a}$$

The second-order estimates, $\hat{\hat{\lambda}}_i$, may be compared with the corresponding first-order estimates, $\hat{\lambda}_i$, by evaluating the expression

$$\hat{\lambda}_i = \lambda_i + \frac{\partial \lambda_i}{\partial a_{kl}} \delta a_{kl} + \frac{\partial \lambda_i}{\partial a_{st}} \delta a_{st} \tag{3.40 b}$$

in the manner of § 3.2.4. In cases where more than two elements of the plant matrix, $\mathbf{A}$, are perturbed, the first- and second-order estimates may be obtained by using the appropriately generalized form of eqns. (3.40).

3.2.6. *Second-order eigenvector sensitivities*

The corresponding second-order eigenvector coefficients can be obtained by differentiating eqns. (3.36) partially with respect to a_{st} and by making use of (3.38). It is thus found that

$$\frac{\partial^2 \mathbf{u}_i}{\partial a_{kl} \partial a_{st}} = \sum_{\substack{j=1 \\ j \neq 1}}^{n} \left[h_{ji}{}^{kl} \frac{\partial \mathbf{u}_j}{\partial a_{st}} + \frac{\partial h_{ji}{}^{kl}}{\partial a_{st}} \mathbf{u}_j \right], \tag{3.41 a}$$

and

$$\frac{\partial^2 \mathbf{v}_j{}'}{\partial a_{kl} \partial a_{st}} = - \sum_{\substack{i=1 \\ i \neq j}}^{n} \left[h_{ji}{}^{kl} \frac{\partial \mathbf{v}_i{}'}{\partial a_{st}} + \frac{\partial h_{ji}{}^{kl}}{\partial a_{st}} \mathbf{v}_i{}' \right], \tag{3.41 b}$$

where it follows from eqn. (3.37) that

$$(\lambda_i - \lambda_j) \frac{\partial h_{ji}{}^{kl}}{\partial a_{st}} = v_j{}^k \frac{\partial u_i{}^l}{\partial a_{st}} + \frac{\partial v_j{}^k}{\partial a_{st}} u_i{}^l - h_{ji}{}^{kl} \left[\frac{\partial \lambda_i}{\partial a_{st}} - \frac{\partial \lambda_j}{\partial a_{st}} \right]$$

$$(i \neq j\,;\; i, j, k, l, s, t = 1, 2, \ldots, n). \tag{3.42}$$

If the results given in eqns. (3.34) and (3.36) are used in eqn. (3.42), then the latter equation assumes the form

$$(\lambda_i-\lambda_j)\frac{\partial h_{ji}{}^{kl}}{\partial a_{st}}=\sum_{\substack{m=1\\m\neq i}}^{n} h_{mi}{}^{st}v_j{}^k u_m{}^l-\sum_{\substack{m=1\\m\neq j}}^{n} h_{jm}{}^{st}v_m{}^k u_i{}^l-h_{ji}{}^{kl}(v_i{}^s u_i{}^t-v_j{}^s u_j{}^t). \tag{3.43}$$

It then follows from eqns. (3.36), (3.41), and (3.43) that

$$\frac{\partial^2 \mathbf{u}_i}{\partial a_{kl}\partial a_{st}}=\sum_{\substack{m=1\\m\neq i}}^{n} h_{mi}{}^{kl}h_{im}{}^{st}\mathbf{u}_i+\sum_{\substack{j=1\\j\neq i}}^{n}\left\{\sum_{\substack{m=1\\m\neq i}}^{n}[(h_{mi}{}^{st}v_j{}^k u_m{}^l+h_{mi}{}^{kl}v_j{}^s u_m{}^t)\right.$$
$$\left.-h_{ji}{}^{kl}v_i{}^s u_i{}^t-h_{ji}{}^{st}v_i{}^k u_i{}^l]\mathbf{u}_j/(\lambda_i-\lambda_j)\right\}(i,k,l,s,t=1,2,\ldots,n) \tag{3.44 a}$$

and

$$\frac{\partial^2 \mathbf{v}_j{}'}{\partial a_{kl}\partial a_{st}}=\sum_{\substack{m=1\\m\neq i}}^{n} h_{jm}{}^{kl}h_{mj}{}^{st}\mathbf{v}_j{}'+\sum_{\substack{i=1\\i\neq j}}^{n}\left\{\sum_{\substack{m=1\\m\neq j}}^{n}[(h_{jm}{}^{kl}v_m{}^s u_i{}^t+h_{jm}{}^{st}v_m{}^k u_i{}^l)\right.$$
$$\left.-h_{ji}{}^{kl}v_j{}^s u_j{}^t-h_{ji}{}^{st}v_j{}^k u_j{}^l]\mathbf{v}_i{}'/(\lambda_i-\lambda_j)\right\}(j,k,l,s,t=1,2,\ldots,n). \tag{3.44 b}$$

If the elements a_{kl} and a_{st} of $\mathbf{A}$ are perturbed by increments δa_{kl} and δa_{st}, respectively, then the corresponding estimates of the eigenvectors $\mathbf{u}_i$ and $\mathbf{v}_j$ are given to a second-order of approximation by the expressions

$$\hat{\mathbf{u}}_i=\mathbf{u}_i+\frac{\partial \mathbf{u}_i}{\partial a_{kl}}\delta a_{kl}+\frac{\partial \mathbf{u}_i}{\partial a_{st}}\delta a_{st}$$
$$+\tfrac{1}{2}\left[\frac{\partial^2 \mathbf{u}_i}{\partial a_{kl}{}^2}\delta a_{kl}{}^2+2\frac{\partial^2 \mathbf{u}_i}{\partial a_{kl}\partial a_{st}}\delta a_{kl}\partial a_{st}+\frac{\partial^2 \mathbf{u}_i}{\partial a_{st}{}^2}\delta a_{st}{}^2\right], \tag{3.45 a}$$

and

$$\hat{\mathbf{v}}_j{}'=\mathbf{v}_j{}'+\frac{\partial \mathbf{v}_j{}'}{\partial a_{kl}}\delta a_{kl}+\frac{\partial \mathbf{v}_j{}'}{\partial a_{st}}\delta a_{st}$$
$$+\tfrac{1}{2}\left[\frac{\partial^2 \mathbf{v}_j{}'}{\partial a_{kl}{}^2}\delta a_{kl}{}^2+2\frac{\partial^2 \mathbf{v}_j{}'}{\partial a_{kl}\partial a_{st}}\delta a_{kl}\delta a_{st}+\frac{\partial^2 \mathbf{v}_j{}'}{\partial a_{st}{}^2}\delta a_{st}{}^2\right]$$
$$(i,j,k,l,s,t=1,2,\ldots,n). \tag{3.45 b}$$

These eigenvectors will satisfy equations of the form (3.5) if, to a second-order of approximation,

$$\hat{\mathbf{v}}_j{}'\hat{\mathbf{u}}_i=\delta_{ij}\quad (i,j=1,2,\ldots,n). \tag{3.46}$$

It can be deduced by substituting the expressions given in eqn. (3.45) into eqn. (3.46) that this will be the case if and only if the following conditions are satisfied :

$$\frac{\partial \mathbf{v}_j{}'}{\partial a_{kl}}\mathbf{u}_i+\mathbf{v}_j{}'\frac{\partial \mathbf{u}_i}{\partial a_{kl}}=0, \tag{3.47 a}$$

and

$$\mathbf{v}_j{}'\frac{\partial^2 \mathbf{u}_i}{\partial a_{kl}\partial a_{st}}+\frac{\partial \mathbf{v}_j{}'}{\partial a_{kl}}\frac{\partial \mathbf{u}_i}{\partial a_{st}}+\frac{\partial \mathbf{v}_j{}'}{\partial a_{st}}\frac{\partial \mathbf{u}_i}{\partial a_{kl}}+\frac{\partial^2 \mathbf{v}_j{}'}{\partial a_{kl}\partial a_{st}}\mathbf{u}_i=0$$
$$(i,j,k,l,s,t=1,2,\ldots,n). \tag{3.47 b}$$

In view of the assumption that $\mathbf{A}$ has n distinct eigenvalues, the eigenvectors $\mathbf{u}_i$ $(i=1, 2, \ldots, n)$ and $\mathbf{v}_j$ $(j=1, 2, \ldots, n)$ each constitute a linearly independent set. Thus, the vectors $\partial\mathbf{u}_i/\partial a_{kl}$, $\partial\mathbf{v}_j'/\partial a_{kl}$, $\partial^2\mathbf{u}_i/\partial a_{kl}\partial a_{st}$, and $\partial^2\mathbf{v}_j'/\partial a_{kl}\partial a_{st}$ may be expressed in the forms

$$\frac{\partial\mathbf{u}_i}{\partial a_{kl}}=\sum_{m=1}^{n}\phi_m{}^{ikl}\mathbf{u}_m, \tag{3.48 a}$$

$$\frac{\partial\mathbf{v}_j'}{\partial a_{kl}}=\sum_{m=1}^{n}\psi_m{}^{jkl}\mathbf{v}_m', \quad (i, j, k, l, s, t=1, 2, \ldots, n) \tag{3.48 b}$$

$$\frac{\partial^2\mathbf{u}_i}{\partial a_{kl}\partial a_{st}}=\sum_{m=1}^{n}\alpha_m{}^{iklst}\mathbf{u}_m, \tag{3.48 c}$$

$$\frac{\partial^2\mathbf{v}_j'}{\partial a_{kl}\partial a_{st}}=\sum_{m=1}^{n}\beta_m{}^{jklst}\mathbf{v}_m'. \tag{3.48 d}$$

If the expressions given in eqns. (3.48) are used in eqns. (3.47), then it follows that the conditions (3.47) will be satisfied provided that

$$\phi_j{}^{ikl}+\psi_i{}^{jkl}=0 \quad (i, j, k, l=1, 2, \ldots, n), \tag{3.49}$$

and

$$\alpha_j{}^{iklst}+\beta_i{}^{jklst}+\sum_{m=1}^{n}(\psi_m{}^{jkl}\phi_m{}^{ist}+\psi_m{}^{jst}\phi_m{}^{ikl})=0 \quad (i, j, k, l, s, t=1, 2, \ldots, n). \tag{3.50}$$

In terms of the $h_{ji}{}^{kl}$ defined in eqn. (3.37), it can be readily verified that the condition (3.50) will be satisfied provided that the coefficients $\alpha_j{}^{iklst}$ and $\beta_i{}^{jklst}$ are such that

$$\alpha_i{}^{iklst}+\beta_i{}^{iklst}-\sum_{\substack{m=1\\m\neq i}}^{n}(h_{im}{}^{kl}h_{mi}{}^{st}+h_{im}{}^{st}h_{mi}{}^{kl})=0 \quad (i, k, l, s, t=1, 2, \ldots, n), \tag{3.51 a}$$

and

$$\alpha_j{}^{iklst}+\beta_i{}^{jklst}-\sum_{\substack{m=1\\m\neq i,j}}^{n}(h_{jm}{}^{kl}h_{mi}{}^{st}+h_{jm}{}^{st}h_{mi}{}^{kl})=0 \quad (i\neq j\,;\ i, j, k, l, s, t=1, 2, \ldots, n). \tag{3.51 b}$$

Equations (3.44) clearly have the form (3.48 c, d) with

$$\alpha_i{}^{iklst}=\sum_{\substack{m=1\\m\neq i}}^{n}h_{mi}{}^{kl}h_{im}{}^{st} \quad (i, k, l, s, t=1, 2, \ldots, n), \tag{3.52 a}$$

$$\beta_i{}^{iklst}=\sum_{\substack{m=1\\m\neq i}}^{n}h_{im}{}^{kl}h_{mi}{}^{st} \quad (i, k, l, s, t=1, 2, \ldots, n), \tag{3.52 b}$$

$$\alpha_j{}^{iklst}=\sum_{\substack{m=1\\m\neq i}}^{n}[(h_{mi}{}^{st}v_j{}^k u_m{}^l+h_{mi}{}^{kl}v_j{}^s u_m{}^t)-h_{ji}{}^{kl}v_i{}^s u_i{}^t-h_{ji}{}^{st}v_i{}^k u_i{}^l]/(\lambda_i-\lambda_j) \quad (j\neq i\,;\ i, j, k, l, s, t=1, 2, \ldots, n), \tag{3.53 a}$$

$$\beta_i^{jklst} = \sum_{\substack{m=1 \\ m \neq j}}^{n} [(h_{jm}{}^{kl} v_m{}^s u_i{}^t + h_{jm}{}^{st} v_m{}^k u_i{}^l) - h_{ji}{}^{kl} v_j{}^s u_j{}^t - h_{ji}{}^{st} v_j{}^k u_j{}^l]/(\lambda_i - \lambda_j)$$

$$(i \neq j \ ; \ i, j, k, l, s, t = 1, 2, \ldots, n). \qquad (3.53\ b)$$

The coefficients given in eqns. (3.52) satisfy the condition (3.51 *a*), and it may be verified that the coefficients given in eqns. (3.53) satisfy the condition (3.51 *b*). It may therefore be concluded that, with these values for the various coefficients, the vectors $\hat{\mathbf{u}}_i$ and $\hat{\mathbf{v}}_j$ satisfy eqn. (3.46). Thus, eqns. (3.44) give the desired second-order eigenvector sensitivity coefficients and may be used to estimate eigenvectors which will be approximately orthonormal.

3.2.7. *Illustrative examples: second-order sensitivities*

Because of the form of the expressions for the eigenvector sensitivities, it is evident that all the elements of an estimated eigenvector will usually differ from all the elements of the datum eigenvector. It therefore follows that, in order to compare estimates with directly-calculated results, it is necessary to scale each estimated eigenvector by making any one of its non-zero elements equal to the corresponding element of the directly-calculated eigenvector of the perturbed plant matrix. In the two examples which follow, the eigenvectors are scaled so that the largest element in each directly-calculated eigenvector is equal to the corresponding element in the corresponding estimated eigenvector.

As a first simple illustrative example, it is instructive to investigate the sensitivity characteristics of an unforced second-order linear oscillator having the state equation

$$\dot{\mathbf{x}}(t) = \mathbf{A}\mathbf{x}(t), \qquad (3.54\ a)$$

where

$$\mathbf{A} = \begin{bmatrix} -2\zeta\omega_n, & -\omega_n{}^2 \\ 1, & 0 \end{bmatrix}. \qquad (3.54\ b)$$

In (3.54 *b*),

$$2\zeta\omega_n = c/m$$

and

$$\omega_n{}^2 = k/m,$$

where c, k, and m are respectively the damping coefficient, spring stiffness, and mass of the oscillator. The eigenvalues of the matrix, $\mathbf{A}$, given in (3.54 *b*) are

$$\left. \begin{aligned} \lambda_1 &= -\zeta\omega_n + i\omega_d, \\ \lambda_2 &= -\zeta\omega_n - i\omega_d, \end{aligned} \right\} \qquad (3.55)$$

and its corresponding normalized orthogonal eigenvectors are

$$\mathbf{u}_1 = \begin{bmatrix} -\zeta\omega_n + i\omega_d \\ 1 \end{bmatrix}, \quad \mathbf{u}_2 = \begin{bmatrix} -\zeta\omega_n - i\omega_d \\ 1 \end{bmatrix}. \qquad (3.56)$$

and

$$\mathbf{v}_1 = \tfrac{1}{2}\begin{bmatrix} -i/\omega_d \\ 1 - i(\zeta\omega_n/\omega_d) \end{bmatrix}, \quad \mathbf{v}_2 = \tfrac{1}{2}\begin{bmatrix} i/\omega_d \\ 1 + i(\zeta\omega_n/\omega_d) \end{bmatrix}. \tag{3.57}$$

In eqns. (3.55), (3.56), and (3.57),

$$\omega_d = (1 - \zeta^2)^{1/2}\omega_n.$$

In this case the effects of varying c/m and k/m can be investigated by perturbing the elements a_{11} and a_{12} in the matrix (3.54 *b*).

If the general theory developed in the previous section is applied to this system, it is readily found that, for the eigenvalue λ_1,

$$\partial\lambda_1/\partial a_{11} = [1 + i(\zeta\omega_n/\omega_d)]/2, \tag{3.58 a}$$

$$\partial\lambda_1/\partial a_{12} = -i/2\omega_d, \tag{3.58 b}$$

$$\partial^2\lambda_1/\partial a_{11}{}^2 = -i\omega_n{}^2/4\omega_d{}^3, \tag{3.59 a}$$

$$\partial^2\lambda_1/\partial a_{11}\partial a_{12} = i\zeta\omega_n/4\omega_d{}^3, \tag{3.59 b}$$

$$\partial^2\lambda_1/\partial a_{12}{}^2 = -i/4\omega_d{}^3. \tag{3.59 c}$$

Similar expressions can be obtained for the conjugate complex eigenvalue λ_2.

The expressions given in eqns. (3.40), (3.58), and (3.59) may be used to obtain first- and second-order estimates of λ_1 as a result of changes in the system parameters. Thus, consider that ω_n is fixed at the value $\omega_n = 1$ and that initially $\zeta = 0{\cdot}3$, so that $a_{11} = -0{\cdot}6$ and $a_{12} = -1$. Now consider that a_{11} is perturbed and that eqns. (3.40) are used to calculate the resulting first- and second-order estimates of λ_1. The real and imaginary parts of these estimates are plotted in fig. 3-1 (*a*) and fig. 3-1 (*b*), respectively, together with a plot of the directly-calculated values of λ_1 for the range of δa_{11} values considered. In fig. 3-1 (*a*), all three plots coincide because Re λ_1 is a linear function of ζ: in fig. 3-1 (*b*), however, it is clear that the second-order estimate is much better than the

Fig. 3-1 (*a*) Fig. 3-1 (*b*)

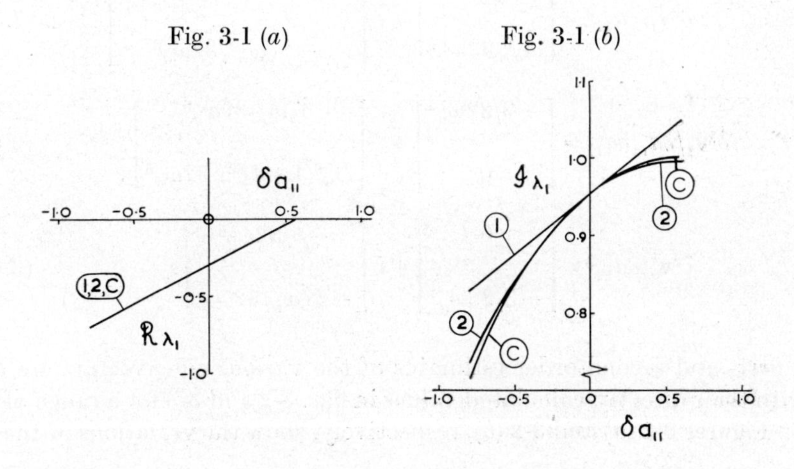

first-order estimate and that, in fact, it differs only very slightly from the directly-calculated value of λ_1.

The general theory developed in the previous section can be used to investigate the corresponding first- and second-order eigenvector estimates for perturbations in **A** by using the following partial derivatives in appropriate Taylor series expansions :

$$\partial \mathbf{u}_1/\partial a_{11} = \begin{bmatrix} \omega_n^2/4\omega_d^2 \\ -\zeta\omega_n/4\omega_d^2 \end{bmatrix} + i \begin{bmatrix} 0 \\ 1/4\omega_d \end{bmatrix}, \tag{3.60 a}$$

$$\partial \mathbf{u}_1/\partial a_{12} = \begin{bmatrix} -\zeta\omega_n/4\omega_d^2 \\ 1/4\omega_d^2 \end{bmatrix} + i \begin{bmatrix} -1/4\omega_d \\ 0 \end{bmatrix}, \tag{3.60 b}$$

$$\partial \mathbf{v}_1/\partial a_{11} = \begin{bmatrix} -1/8\omega_d^2 \\ 0 \end{bmatrix} + i \begin{bmatrix} \zeta\omega_n/8\omega_d^3 \\ \omega_n^2/8\omega_d^3 \end{bmatrix}, \tag{3.61 a}$$

$$\partial \mathbf{v}_1/\partial a_{12} = \begin{bmatrix} 0 \\ -1/8\omega_d^2 \end{bmatrix} + i \begin{bmatrix} -1/8\omega_d^3 \\ -\zeta\omega_n/8\omega_d^3 \end{bmatrix}, \tag{3.61 b}$$

$$\partial^2 \mathbf{u}_1/\partial a_{11}^2 = \begin{bmatrix} -5\zeta\omega_n^3/16\omega_d^4 \\ \omega_n^2(1+4\zeta^2)/16\omega_d^4 \end{bmatrix} + i \begin{bmatrix} \omega_n^2/16\omega_d^3 \\ -\zeta\omega_n/4\omega_d^3 \end{bmatrix}, \tag{3.62 a}$$

$$\partial^2 \mathbf{u}_1/\partial a_{11}\partial a_{12} = \begin{bmatrix} \omega_n^2(1+4\zeta^2)/16\omega_d^4 \\ -5\zeta\omega_n/16\omega_d^4 \end{bmatrix} + i \begin{bmatrix} 0 \\ 3/16\omega_d^3 \end{bmatrix}, \tag{3.62 b}$$

$$\partial^2 \mathbf{u}_1/\partial a_{12}^2 = \begin{bmatrix} -5\zeta\omega_n/16\omega_d^4 \\ 5/16\omega_d^4 \end{bmatrix} + i \begin{bmatrix} -3/16\omega_d^3 \\ 0 \end{bmatrix}, \tag{3.62 c}$$

$$\partial^2 \mathbf{v}_1/\partial a_{11}^2 = \begin{bmatrix} \zeta\omega_n/8\omega_d^4 \\ \omega_n^2/32\omega_d^4 \end{bmatrix} + i \begin{bmatrix} -\omega_n^2(1+4\zeta^2)/32\omega_d^5 \\ -5\zeta\omega_n^3/32\omega_d^5 \end{bmatrix}, \tag{3.63 a}$$

$$\partial^2 \mathbf{v}_1/\partial a_{11}\partial a_{12} = \begin{bmatrix} -3/32\omega_d^4 \\ 0 \end{bmatrix} + i \begin{bmatrix} 5\zeta\omega_n/32\omega_d^5 \\ \omega_n^2(1+4\zeta^2)/32\omega_d^5 \end{bmatrix}, \tag{3.63 b}$$

$$\partial^2 \mathbf{v}_1/\partial a_{12}^2 = \begin{bmatrix} 0 \\ -3/32\omega_d^4 \end{bmatrix} + i \begin{bmatrix} -5/32\omega_d^5 \\ -5\zeta\omega_n/32\omega_d^5 \end{bmatrix}. \tag{3.63 c}$$

The first- and second-order estimates of the various eigenvectors are compared with their directly-calculated values in figs. 3-2 and 3-3 for a range of δa_{11} values. Figures 3-2 (*a*) and 3-2 (*b*), respectively, show the variations of the real and imaginary parts of the first element of $\mathbf{u}_1$ with δa_{11}, and it is evident that the second-order estimate is better than the first-order estimate in all cases. Figure 3-3 (a) shows the variation of the imaginary part of the first element of $\mathbf{v}_1$ (there is no real part), fig. 3-3 (b) shows the variation of the real part of the second element of $\mathbf{v}_1$, and fig. 3-3 (c) shows the variation of the imaginary part of the second element of $\mathbf{v}_1$. It is again apparent that, in general, the second-order estimate is significantly closer to the directly-calculated value than the first-order estimate.

Fig. 3-2 (a) Fig. 3-2 (b)

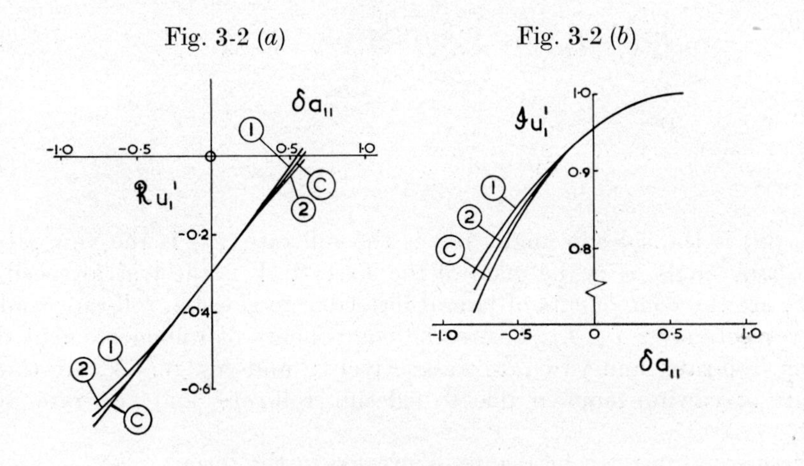

Fig. 3-3 (a)

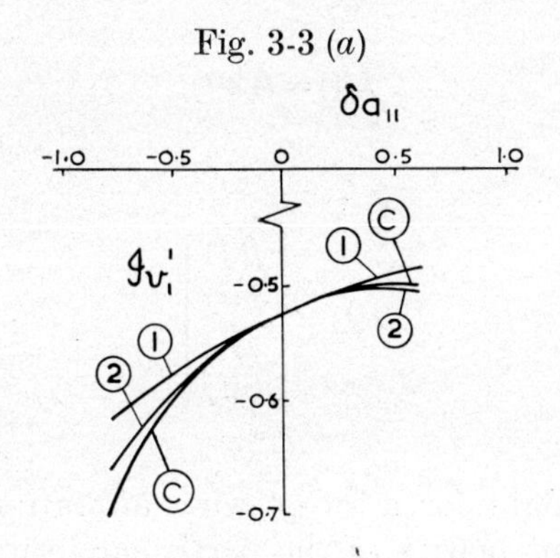

Fig. 3-3 (b) Fig. 3-3 (c)

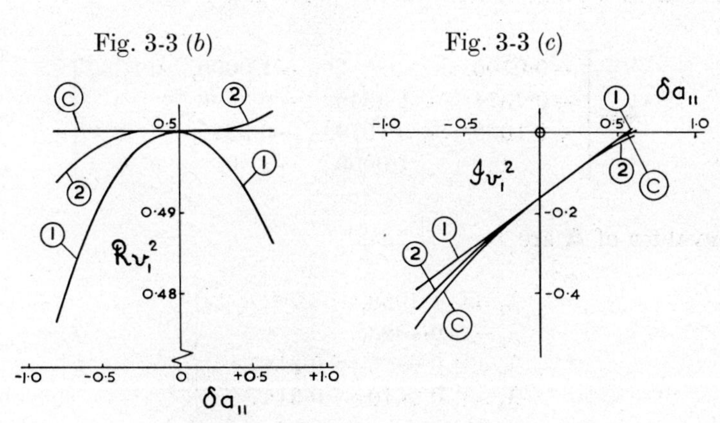

The second illustrative example concerns the sensitivity characteristics of a fourth-order representation of the lateral dynamics of an aircraft. The equations governing the lateral stability of an aircraft have the linearized form

$$mV\{\dot{\beta}(t)+r(t)\}=Y_\beta\beta(t)+Y_p p(t)+Y_r r(t)+mg\phi(t), \tag{3.64 a}$$

$$I_x\dot{p}(t)-I_{xz}\dot{r}(t)=L_\beta\beta(t)+L_p p(t)+L_r r(t), \tag{3.64 b}$$

$$I_z\dot{r}(t)-I_{xz}\dot{p}(t)=N_\beta\beta(t)+N_p p(t)+N_r r(t), \tag{3.64 c}$$

and

$$\dot{\phi}(t)=p(t), \tag{3.64 d}$$

where $\beta(t)$ is the sideslip angle, $p(t)$ is the roll rate, $r(t)$ is the yaw rate, $\phi(t)$ is the bank angle, m is the mass of the aircraft, V is the true airspeed ; Y_β, Y_p, Y_r are the components of lateral force due to sideslip, roll-rate, and yaw-rate, respectively ; L_β, L_p, L_r are the components of rolling moment due to sideslip, roll-rate, and yaw-rate, respectively ; and N_β, N_p, N_r are the components of yawing moment due to sideslip, roll-rate, and yaw-rate, respectively.

These equations can be readily expressed in the form

$$\dot{\mathbf{x}}(t)=\mathbf{A}\mathbf{x}(t), \tag{3.65 a}$$

where

$$\mathbf{x}(t)=\begin{bmatrix}\beta(t)\\ p(t)\\ r(t)\\ \phi(t)\end{bmatrix} \tag{3.65 b}$$

Typical aerodynamic data for a swept-wing aircraft flying at low speed at sea level lead to a plant matrix of the particular form

$$\mathbf{A}=\begin{bmatrix}-0{\cdot}0506, & 0, & -1{\cdot}0000, & 0{\cdot}2380\\ -0{\cdot}7374, & -1{\cdot}3345, & +0{\cdot}3696, & 0\\ +0{\cdot}1000, & +0{\cdot}1074, & -0{\cdot}3320, & 0\\ 0, & 1{\cdot}0000, & 0, & 0\end{bmatrix} \tag{3.66}$$

The eigenvalues of $\mathbf{A}$ are

$$\left.\begin{aligned}\lambda_1&=-1{\cdot}4088,\\ \lambda_2&=-0{\cdot}4023,\\ \lambda_3&=+0{\cdot}0470+i\,0{\cdot}3147,\\ \lambda_4&=+0{\cdot}0470-i\,0{\cdot}3147,\end{aligned}\right\} \tag{3.67}$$

where λ_1 and λ_2 govern the rolling and spiral modes of the aircraft, and λ_3 and λ_4 are the complex conjugate eigenvalues which govern its dutch-roll mode. The eigenvectors of **A** associated with these eigenvalues are

$$\left.\begin{aligned}
\mathbf{u}_1 &= \begin{bmatrix} +0\cdot0506 \\ +1\cdot0000 \\ -0\cdot1002 \\ -0\cdot7089 \end{bmatrix} \quad \mathbf{u}_2 = \begin{bmatrix} +0\cdot7623 \\ -0\cdot4023 \\ +0\cdot5061 \\ +1\cdot0000 \end{bmatrix} \\
\mathbf{u}_3 &= \begin{bmatrix} +0\cdot0688 - i\,0\cdot5916 \\ +0\cdot0470 + i\,0\cdot3147 \\ +0\cdot0451 + i\,0\cdot0361 \\ +1\cdot0000 \end{bmatrix} \quad \mathbf{u}_4 = \begin{bmatrix} +0\cdot0688 + i\,0\cdot5916 \\ +0\cdot0470 - i\,0\cdot3147 \\ +0\cdot0451 - i\,0\cdot0361 \\ +1\cdot0000 \end{bmatrix}
\end{aligned}\right\} \tag{3.68}$$

and the corresponding eigenvectors of **A′** are

$$\left.\begin{aligned}
\mathbf{v}_1 &= \begin{bmatrix} +0\cdot5033 \\ +0\cdot9291 \\ +0\cdot1485 \\ -0\cdot0850 \end{bmatrix} \quad \mathbf{v}_2 = \begin{bmatrix} +0\cdot1834 \\ +0\cdot1147 \\ +2\cdot0052 \\ -0\cdot1085 \end{bmatrix} \\
\mathbf{v}_3 &= \begin{bmatrix} +0\cdot0869 + i\,0\cdot6954 \\ +0\cdot2724 - i\,0\cdot1453 \\ -0\cdot9499 - i\,1\cdot1878 \\ +0\cdot5241 + i\,0\cdot0126 \end{bmatrix} \quad \mathbf{v}_4 = \begin{bmatrix} +0\cdot0869 - i\,0\cdot6954 \\ +0\cdot2724 + i\,0\cdot1453 \\ -0\cdot9499 + i\,1\cdot1878 \\ +0\cdot5241 - i\,0\cdot0126 \end{bmatrix}
\end{aligned}\right\} \tag{3.69}$$

These eigenvectors satisfy the condition (3.5).

If the element a_{21} of the matrix **A** is varied about its datum value of $-0\cdot7374$, the resulting first- and second-order sensitivities of the eigenvalues and eigenvectors of **A** may be readily calculated using the general results of the previous section and eqns. (3.40) and 3.45). The first- and second-order eigenvalue variations associated with the rolling mode, the spiral mode, and the dutch-roll mode are plotted in figs. 3-4 (*a*), 3-4 (*b*) and

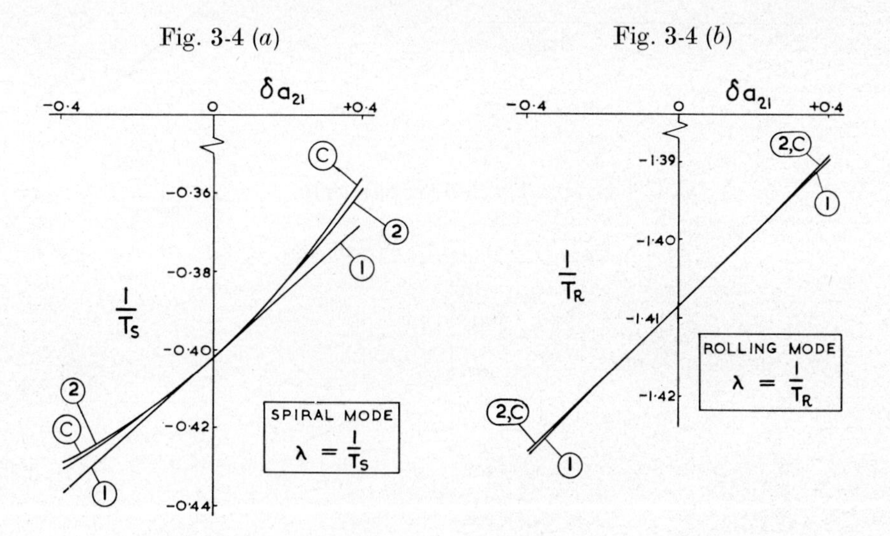

3-4 (c, d), respectively. It is apparent from these diagrams that, in general, the second-order estimates are significantly better than the first-order estimates, and are never worse. The same is true of the various eigenvector estimates plotted in figs. 3-5 and 3-6. In fact, in most cases the second order estimates are so close to the directly-calculated values that there would appear to be little point in proceeding to estimates of higher order than the second in the design of practical systems.

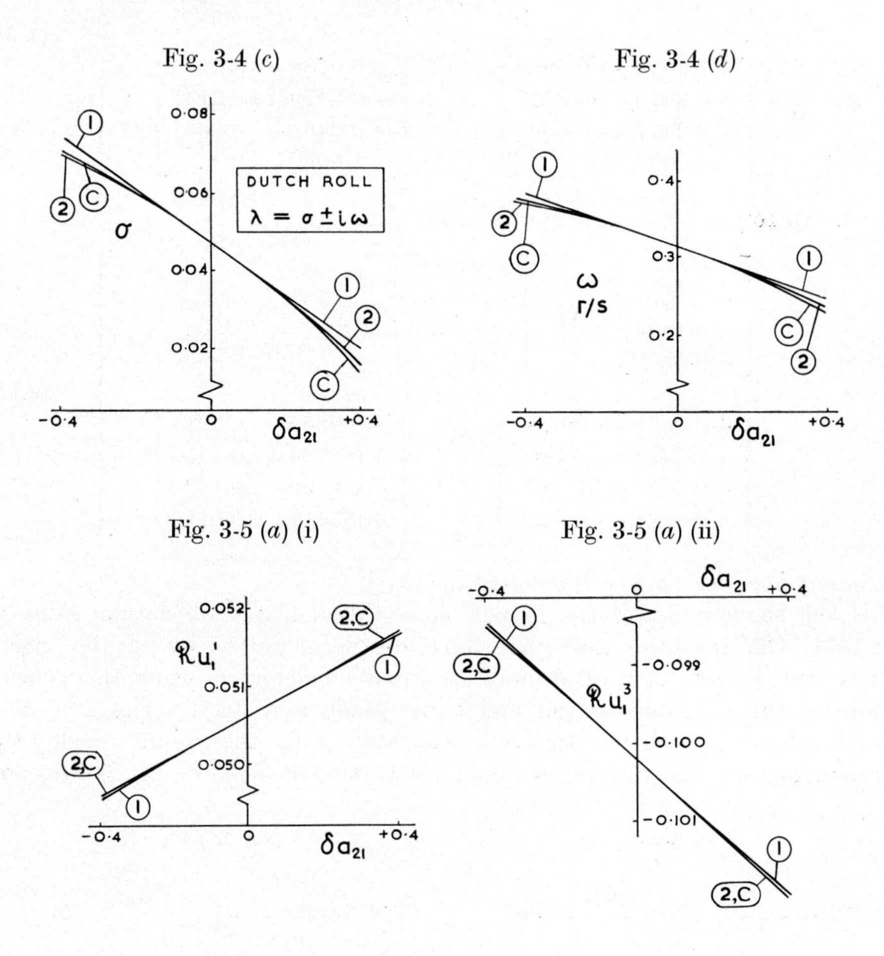

Fig. 3-4 (c)

Fig. 3-4 (d)

Fig. 3-5 (a) (i)

Fig. 3-5 (a) (ii)

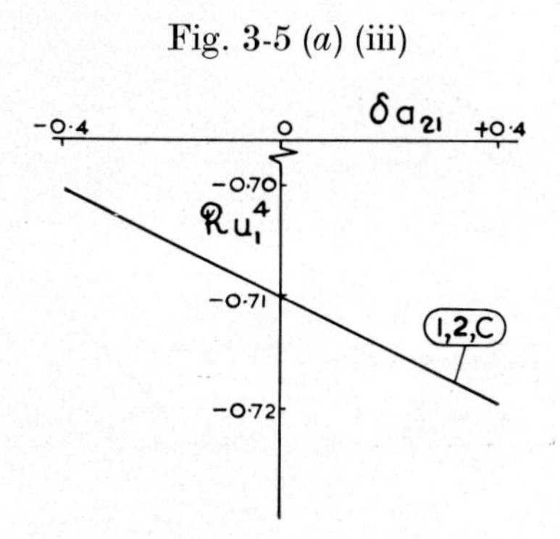

Fig. 3-5 (a) (iii)

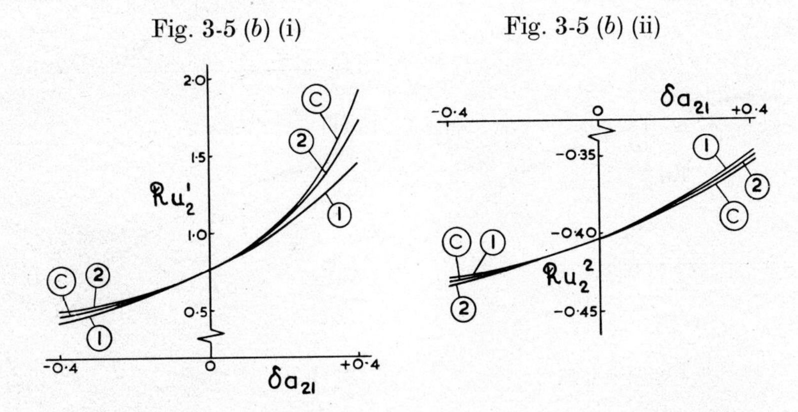

Fig. 3-5 (*b*) (i)

Fig. 3-5 (*b*) (ii)

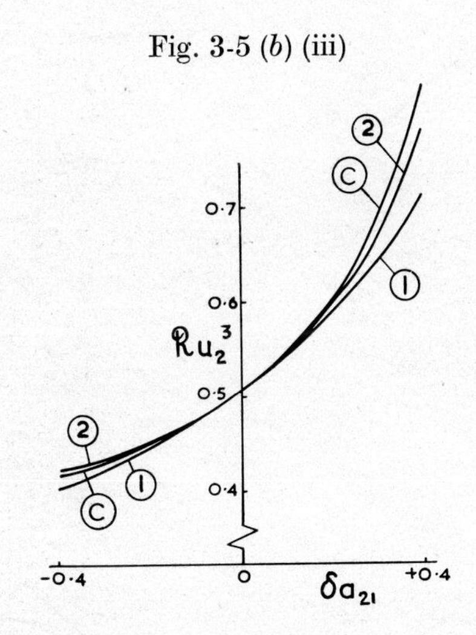

Fig. 3-5 (*b*) (iii)

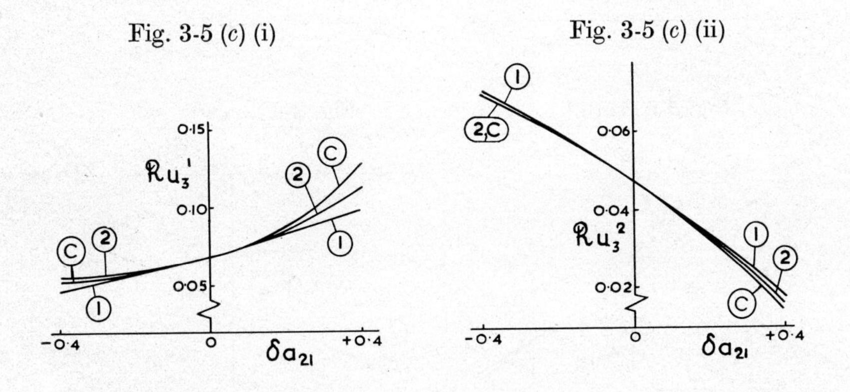

Fig. 3-5 (*c*) (i)

Fig. 3-5 (*c*) (ii)

Fig. 3-5 (*c*) (iii)

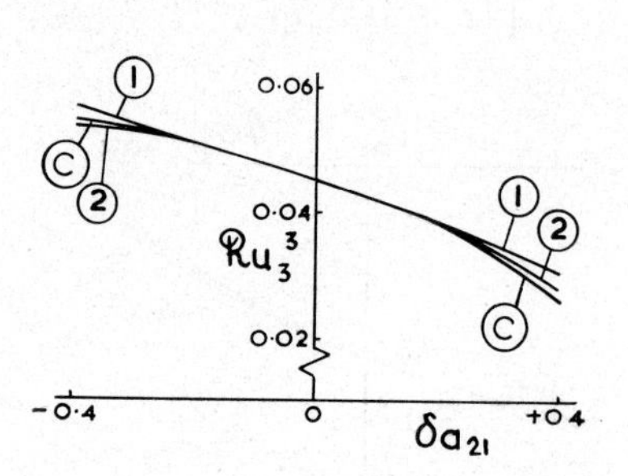

Fig. 3-5 (*c*) (iv)

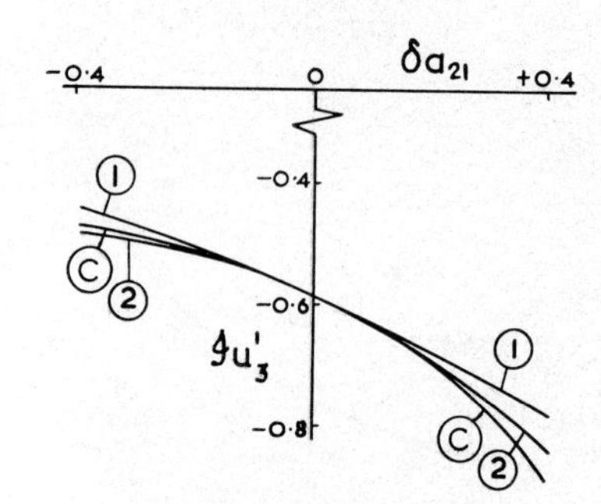

Fig. 3-5 (*c*) (v)

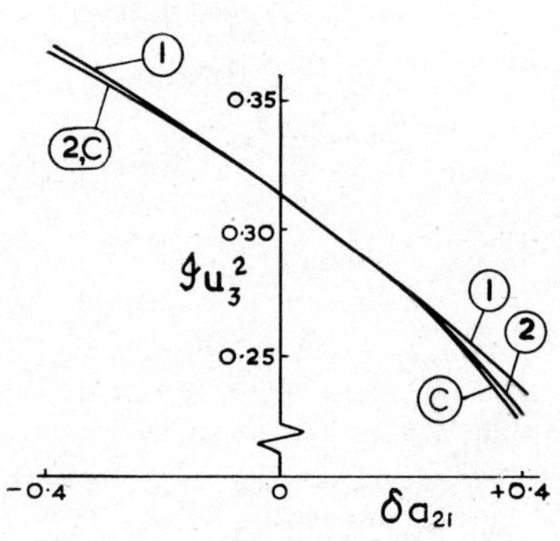

Fig. 3-5 (*c*) (vi)

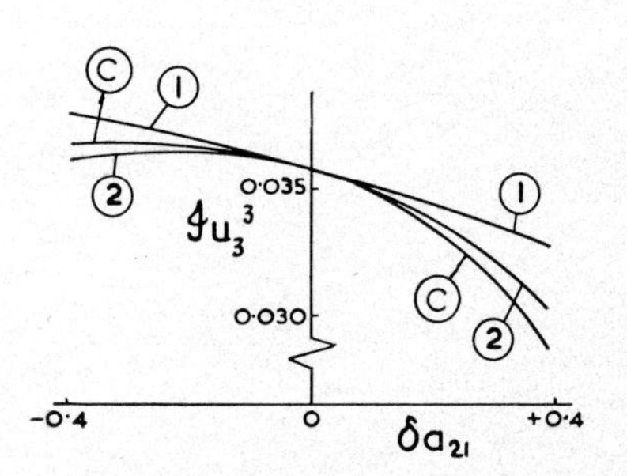

Fig. 3-6 (*a*) (i)

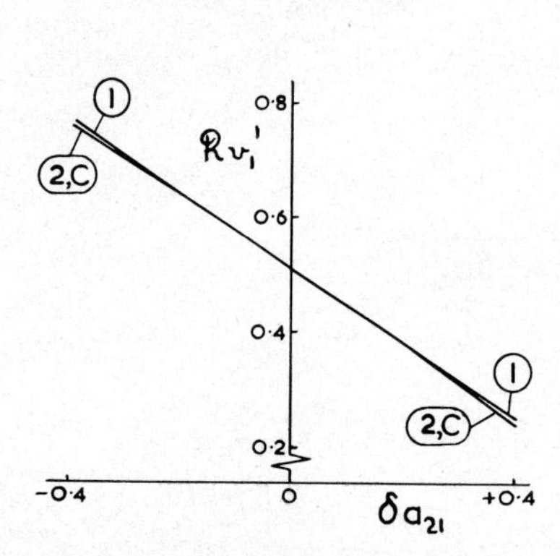

Fig. 3-6 (*a*) (ii)

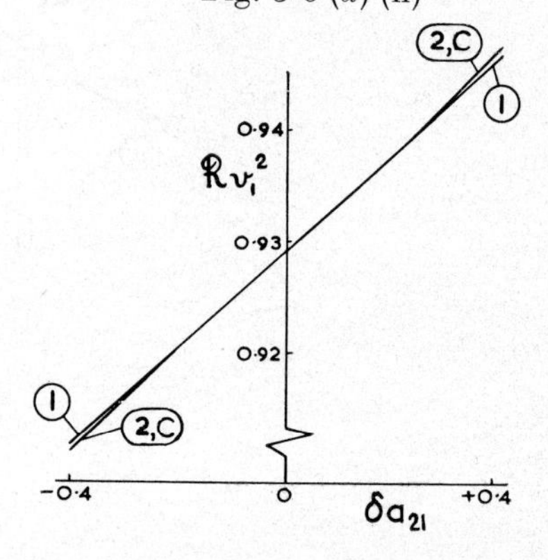

Fig. 3-6 (*a*) (iii)

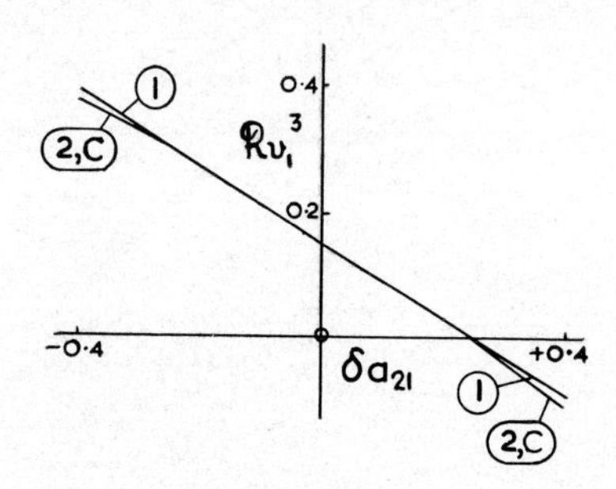

Fig. 3-6 (*a*) (iv)

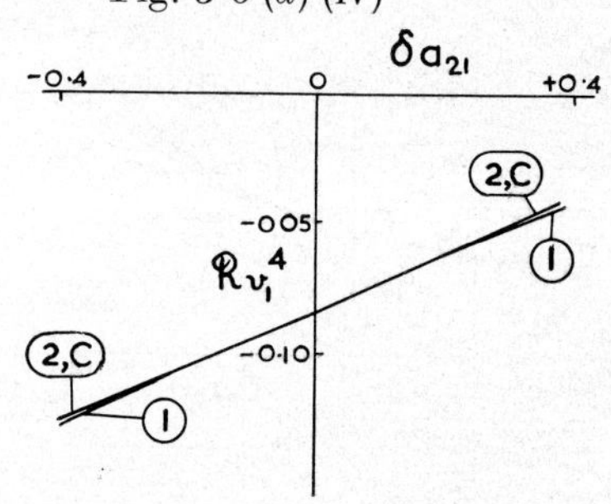

Fig. 3-6 (*b*) (i)

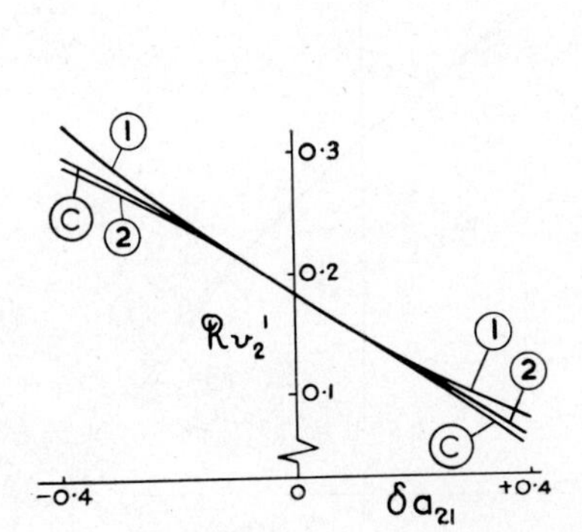

Fig. 3-6 (*b*) (ii)

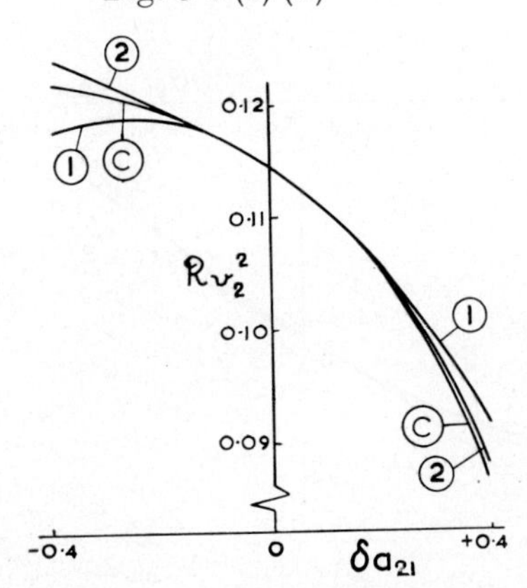

Fig. 3-6 (*b*) (iii)

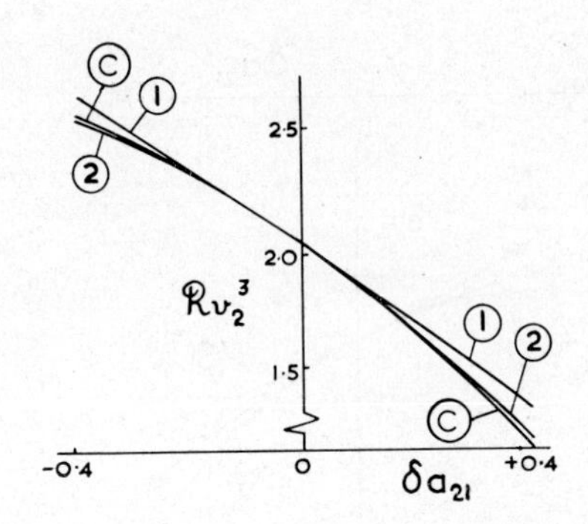

Fig. 3-6 (*b*) (iv)

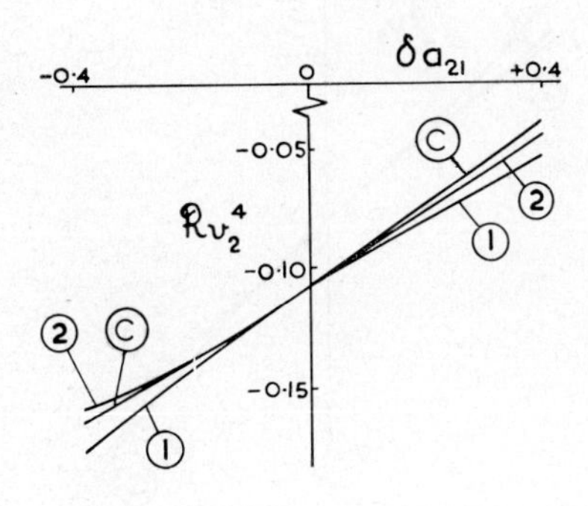

Fig. 3-6 (*c*) (i)

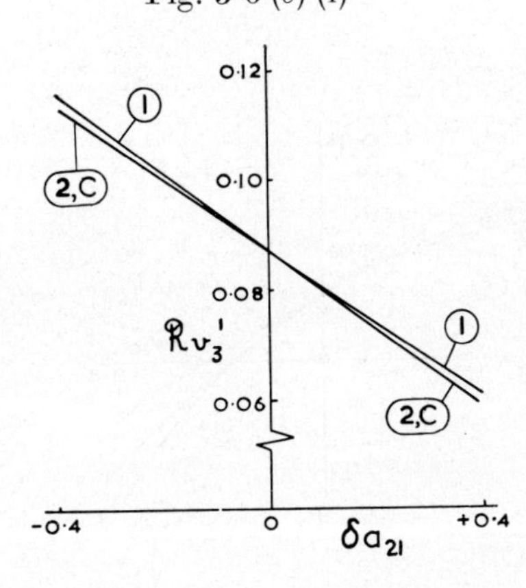

Fig. 3-6 (*c*) (ii)

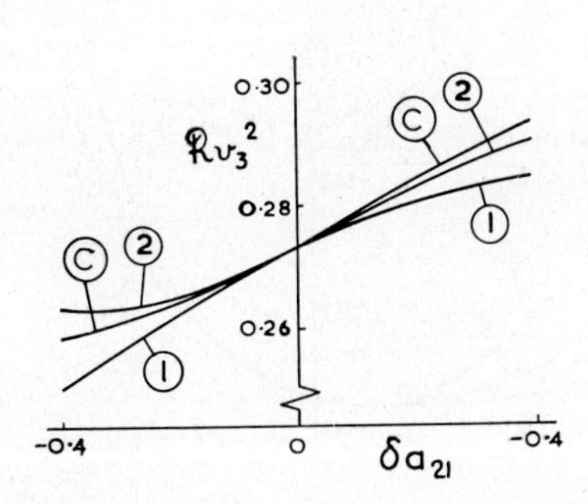

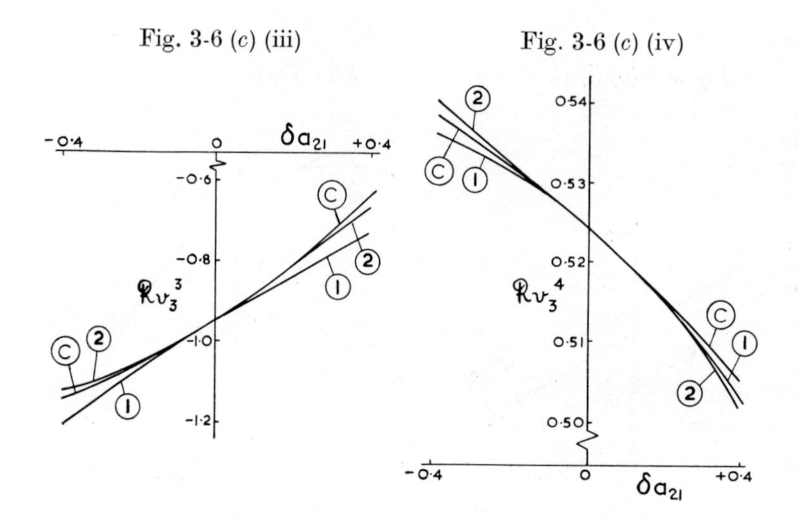

Fig. 3-6 (*c*) (iii) Fig. 3-6 (*c*) (iv)

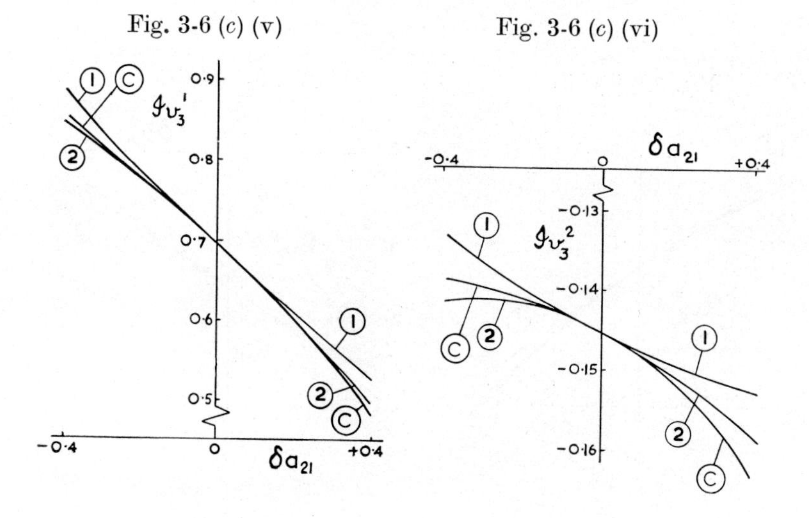

Fig. 3-6 (*c*) (v) Fig. 3-6 (*c*) (vi)

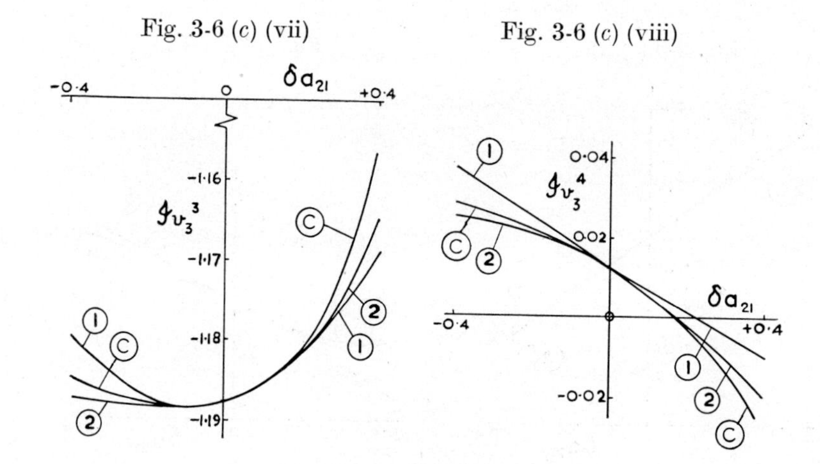

Fig. 3-6 (*c*) (vii) Fig. 3-6 (*c*) (viii)

3.3. Discrete-time systems

The results presented in the foregoing sections of this chapter concerning the sensitivity properties of continuous-time systems may be used (by analogy) in connection with discrete-time systems governed by state equations of the form

$$\mathbf{x}\{(k+1)T\} = \mathbf{\Psi}(T)\mathbf{x}(kT). \tag{3.70}$$

Indeed, since the results given in Chapter 2 indicate that the eigenvalues and eigenvectors of the plant matrix, $\mathbf{\Psi}(T)$, of the discrete-time system governed by eqn. (3.70) are respectively exp $(\lambda_1 T)$, $\mathbf{u}_i$ and $\mathbf{v}_i$ $(i=1, 2, \ldots, n)$, the results presented in this chapter for continuous-time systems may obviously be used to calculate the sensitivities of the eigenproperties of the matrix, $\mathbf{\Psi}(T)$, to changes in the elements of this matrix simply by interpreting the various sensitivity expressions given in §§ 3.2.2, 3.2.3, 3.2.5, and 3.2.6 appropriately.

References

[1] Jacobi, C. G. J., 1846, "Über ein leichtes Verfahren die in der Theorie der Säculärstörungen vorkommenden Gleichungen numerisch aufzulösen", *Crelle's J.*, **30,** 51–94.

[2] Fadeev, D. K., and Fadeeva, V. N., 1963, *Computational Methods of Linear Algebra* (W. H. Freeman & Co.).

[3] Wilkinson, J. H., 1965, *The Algebraic Eigenvalue Problem* (Oxford : Clarendon Press).

[4] Bellman, R., 1960, *Introduction to Matrix Analysis* (McGraw-Hill).

[5] Laughton, M. A., 1964, "Sensitivity in dynamical system analysis", *J. Electron. Control*, **17,** 577.

[6] Mann, J., and Marshall, S. A., 1965, "Eigenvalue trajectories in multivariable system design", *Proc. I.F.A.C. Tokyo Symposium*, p. 359.

[7] Mann, J., Marshall, S. A., and Nicholson, H., 1965, "Advances in matrix techniques in multivariable system control", *Proc. Convention on advances in Automatic Control*, Nottingham.

[8] Van Ness, J. E., Boyle, J. M., and Imad, F. P., 1965, "Sensitivities of large, multiple-loop control systems", *I.E.E.E. Trans. autom. Control*, **10,** 308.

[9] Rosenbrock, H. H., 1965, "Sensitivity of an eigenvalue to changes in the matrix", *Electron. Lett.*, **1,** 278.

[10] Morgan, B. S., 1966, "Sensitivity analysis and synthesis of multivariable systems", *I.E.E.E. Trans. autom. Control*, **11,** 506.

[11] Nicholson, H., 1967, "Eigenvalue and state-transition sensitivity of linear systems", *Proc. Instn elect. Engrs*, **114,** 1991.

Controllability and observability characteristics of linear dynamical systems

CHAPTER 4

4.1. Introduction

THE state-vector transformation

$$\mathbf{x}(t) = \mathbf{U}\boldsymbol{\xi}(t) \tag{4.1}$$

which was used in Chapter 2 in connection with the free response of the system (2.1) is also of great significance in the study of the forced modal response of continuous-time systems governed by state and output equations of the respective forms

$$\dot{\mathbf{x}}(t) = \mathbf{A}\mathbf{x}(t) + \mathbf{B}\mathbf{z}(t) \tag{4.2}$$

and

$$\mathbf{y}(t) = \mathbf{C}'\mathbf{x}(t). \tag{4.3}$$

Thus, under this transformation, eqns. (4.2) and (4.3) assume the respective forms

$$\dot{\boldsymbol{\xi}}(t) = \mathbf{J}\boldsymbol{\xi}(t) + \mathbf{U}^{-1}\mathbf{B}\mathbf{z}(t) \tag{4.4}$$

and

$$\mathbf{y}(t) = \mathbf{C}'\,\mathbf{U}\boldsymbol{\xi}(t). \tag{4.5}$$

In eqns. (4.1), (4.2), (4.3), (4.4), and (4.5), $\mathbf{x}(t)$ and $\boldsymbol{\xi}(t)$ are $n \times 1$ state vectors, $\mathbf{z}(t)$ is an $r \times 1$ input vector, $\mathbf{y}(t)$ is a $q \times 1$ output vector, $\mathbf{A}$ is an $n \times n$ plant matrix, $\mathbf{B}$ is an $n \times r$ input matrix, $\mathbf{C}$ is an $n \times q$ output matrix, $\mathbf{J}$ is the $n \times n$ Jordan canonical form of $\mathbf{A}$, and $\mathbf{U}$ is the $n \times n$ generalized modal matrix of $\mathbf{A}$.

It will be recalled from § 2.2.1 that the generalized modal matrix, $\mathbf{V}$, of $\mathbf{A}'$ can be normalized so that $\mathbf{U}^{-1} = \mathbf{V}'$: eqns. (4.4) and (4.5) can therefore be written in the respective forms

$$\dot{\boldsymbol{\xi}}(t) = \mathbf{J}\boldsymbol{\xi}(t) + \mathbf{P}\mathbf{z}(t) \tag{4.6}$$

and

$$\mathbf{y}(t) = \mathbf{R}'\boldsymbol{\xi}(t). \tag{4.7}$$

In eqn. (4.6), $\mathbf{P}$ is the $n \times r$ *mode-controllability matrix* of the system and is defined by the equation

$$\mathbf{P} = \mathbf{V}'\mathbf{B}. \tag{4.8}$$

Similarly, in eqn. (4.7), $\mathbf{R}$ is the $n \times q$ *mode-observability matrix* of the system and is defined by the equation

$$\mathbf{R} = \mathbf{U}'\mathbf{C}. \tag{4.9}$$

4.2. Mode-controllability matrix

4.2.1. *Distinct eigenvalues*

In the case when the plant matrix, $\mathbf{A}$, in eqn. (4.2) has n distinct eigenvalues $\lambda_1, \lambda_2, \ldots, \lambda_n$, then

$$\mathbf{J} = \text{diag}\ \{\lambda_1,\ \lambda_2,\ \ldots,\ \lambda_n\} \tag{4.10}$$

in eqn. (4.6). It is therefore clear that, in the case of distinct eigenvalues, the vector-matrix eqn. (4.6) is equivalent to the n uncoupled scalar equations

$$\dot{\xi}_i(t) = \lambda_i \xi_i(t) + \sum_{j=1}^{r} p_{ij} z_j(t) \quad (i = 1, 2, \ldots, n). \tag{4.11}$$

It is evident from eqn. (4.11) that the jth input variable, $z_j(t)$, can influence the element $\xi_i(t)$ of the state vector $\boldsymbol{\xi}(t)$ if and only if

$$p_{ij} = \mathbf{v}_i'\mathbf{b}_j \neq 0, \tag{4.12}$$

where $\mathbf{v}_i$ is the ith column of $\mathbf{V}$ and $\mathbf{b}_j$ is the jth column of $\mathbf{B}$. The ith mode of the system (4.11) is therefore *controllable* by the jth input variable if and only if $p_{ij} \neq 0$, and is *uncontrollable* by the jth input variable if and only if $p_{ij} = 0$: the ith mode is *controllable* if and only if it is controllable by at least one input variable. The system (4.11) is *controllable* if and only if each of the n modes is controllable.

These controllability conditions can be illustrated by considering a system for which

$$\mathbf{J} = \text{diag}\ \{1,\ 0,\ -1\} \tag{4.13 a}$$

in the notation of eqn. (4.10), and for which

$$\mathbf{P} = \begin{bmatrix} 2 & 0 \\ 7 & 1 \\ -4 & 0 \end{bmatrix}. \tag{4.13 b}$$

Since $p_{11} \neq 0$, $p_{21} \neq 0$, and $p_{31} \neq 0$, all three modes of the system are controllable by $z_1(t)$, but, since $p_{21} = 0$, $p_{22} \neq 0$, and $p_{32} = 0$, only the second mode is controllable by $z_2(t)$.

4.2.2. *Confluent eigenvalues associated with a single Jordan block*

In the case when the plant matrix, $\mathbf{A}$, in eqn. (4.2) has n repeated eigenvalues associated with a single $n \times n$ Jordan block, then

$$\mathbf{J} = \begin{bmatrix} \lambda_1, & 1, & 0, \ldots, 0, & 0 \\ 0, & \lambda_1, & 1, \ldots, 0, & 0 \\ \cdots & \cdots & \cdots & \cdots \\ \cdots & \cdots & \cdots & \cdots \\ 0, & 0, & 0, \ldots, \lambda_1, & 1 \\ 0, & 0, & 0, \ldots, 0, & \lambda_1 \end{bmatrix} \tag{4.14}$$

in eqn. (4.6). It is therefore clear that, in this case, the vector-matrix eqn. (4.6) is equivalent to the n coupled scalar equations

$$\dot{\xi}_i(t) = \lambda_1 \xi_i(t) + \xi_{i+1}(t) + \sum_{j=1}^{r} p_{ij} z_j(t) \quad (i = 1, 2, \ldots, n-1) \tag{4.15 a}$$

and

$$\dot{\xi}_n(t) = \lambda_1 \xi_n(t) + \sum_{j=1}^{r} p_{nj} z_j(t). \tag{4.15 b}$$

Furthermore, it is evident from eqns. (4.15) that the jth input variable, $z_j(t)$, can influence the elements $\xi_i(t)$ $(i = 1, 2, \ldots, m)$ of the state vector $\boldsymbol{\xi}(t)$ if and only if

$$p_{mj} = \mathbf{v}_m{}'\mathbf{b}_j \neq 0 \quad (m \leqslant n) \tag{4.16 a}$$

and

$$p_{kj} = \mathbf{v}_k{}'\mathbf{b}_j = 0 \quad (k = m+1, m+2, \ldots, n), \tag{4.16 b}$$

where $\mathbf{v}_i$ is the ith column of $\mathbf{V}$ and $\mathbf{b}_j$ is the jth column of $\mathbf{B}$. The Jordan block (4.14) is therefore *controllable* by the jth input variable if and only if $p_{nj} \neq 0$, is *partially controllable* by the jth input variable if and only if $p_{mj} \neq 0$ $(1 \leqslant m < n)$, and is *uncontrollable* by the jth input variable if and only if $p_{kj} = 0$ $(k = 1, 2, \ldots, n)$: the Jordan block (4.14) is *controllable* if and only if it is controllable by at least one input variable. The system (4.15) is *controllable* if and only if the Jordan block (4.14) is controllable.

These controllability conditions can be illustrated by considering a system for which

$$\mathbf{J} = \begin{bmatrix} -0{\cdot}5, & 1, & 0 \\ 0, & -0{\cdot}5, & 1 \\ 0, & 0, & -0{\cdot}5 \end{bmatrix} \tag{4.17 a}$$

in the notation of eqn. (4.14), and for which

$$\mathbf{P} = \begin{bmatrix} 0, & -1 \\ 2, & 1 \\ 1, & 0 \end{bmatrix}. \tag{4.17 b}$$

Since $p_{31} \neq 0$, the Jordan block (4.17 a) is controllable by $z_1(t)$, but, since $p_{22} \neq 0$ and $p_{32} = 0$, the block is only partially controllable by $z_2(t)$.

4.2.3. *Confluent eigenvalues associated with a number of distinct Jordan blocks*

The results derived in § 4.2.2 can readily be extended so as to embrace systems for which the matrix $\mathbf{J}$ in eqn. (4.6) is a block-diagonal matrix whose typical ith block ($i=1, 2, \ldots, \nu$) is the $n_i \times n_i$ Jordan block

$$\mathbf{J}_{ni}(\lambda_j)=\begin{bmatrix} \lambda_j, & 1, & 0, \ldots, 0, & 0 \\ 0, & \lambda_j, & 1, \ldots, 0, & 0 \\ \cdots & \cdots & \cdots & \cdots \\ \cdots & \cdots & \cdots & \cdots \\ 0, & 0, & 0, \ldots, \lambda_j, & 1 \\ 0, & 0, & 0, \ldots, 0, & \lambda_j \end{bmatrix}. \qquad (4.18)$$

In this section it will be assumed that each of the ν Jordan blocks is associated with a distinct eigenvalue†, that is, $\lambda_i \neq \lambda_j$ ($i \neq j$, $i, j=1, 2, \ldots, \nu$) : the matrix $\mathbf{J}$ can then be written in the form

$$\mathbf{J}=\mathbf{J}_{n_1}(\lambda_1)\oplus\mathbf{J}_{n_2}(\lambda_2)\oplus\ldots\oplus\mathbf{J}_{n_\nu}(\lambda_\nu). \qquad (4.19)$$

It is therefore clear that, in this case, the vector-matrix eqn. (4.6) is equivalent to the ν vector-matrix equations

$$\boldsymbol{\xi}_i(t)=\mathbf{J}_{ni}(\lambda_i)\boldsymbol{\xi}_i(t)+\mathbf{P}_i\mathbf{z}(t) \quad (i=1, 2, \ldots, \nu), \qquad (4.20)$$

where

$$\mathbf{P}_i=\mathbf{V}_i'\mathbf{B} \quad (i=1, 2, \ldots, \nu)$$

and the $\mathbf{V}_i$ satisfy the equations

$$\mathbf{A}'\mathbf{V}_i=\mathbf{V}_i\mathbf{J}_{ni}'(\lambda_i) \quad (i=1, 2, \ldots, \nu).$$

The ith equation of eqns. (4.20) is equivalent to the n_i scalar equations

$$\dot{\xi}_k^{(i)}(t)=\lambda_i\xi_k^{(i)}(t)+\xi_{k+1}^{(i)}(t)+\sum_{j=1}^{r} p_{kj}^{(i)}z_j(t) \quad (k=1, 2, \ldots, n_i-1) \quad (4.21\ a)$$

and

$$\dot{\xi}_{ni}^{(i)}(t)=\lambda_i\xi_{ni}^{(i)}(t)+\sum_{j=1}^{r} p_{ni,j}^{(i)}z_j(t), \qquad (4.21\ b)$$

where

$$\boldsymbol{\xi}_i(t)=[\xi_1^{(i)}(t), \xi_2^{(i)}(t), \ldots, \xi_{ni}^{(i)}(t)]', \qquad (4.22\ a)$$

and

$$\mathbf{P}_i=[p_{kj}^{(i)}] \quad (k=1, 2, \ldots, n_i\ ;\ j=1, 2, \ldots, r). \qquad (4.22\ b)$$

It is evident that the set of eqns. (4.21) is analogous to the set of eqns. (4.15) and that the ith block of the Jordan matrix (4.19) is therefore *controllable* by the jth input variable if and only if $p_{n_i,j}^{(i)} \neq 0$, is *partially controllable* by the jth input variable if and only if $p_{mj}^{(i)} \neq 0$ ($1 \leqslant m < n_i$), and is *uncontrollable* by the jth input variable if and only if $p_{kj}^{(i)}=0$ ($k=1, 2, \ldots, n_i$) : the ith Jordan block is *controllable* if and only if it is controllable by at least one input variable. The system (4.20) is *controllable* if and only if each of the ν Jordan blocks in eqn. (4.19) is controllable.

† These will be called distinct Jordan blocks.

These controllability conditions can be illustrated by considering a system for which

$$\mathbf{J} = \mathbf{J}_2(-1) \oplus \mathbf{J}_3(-2) \tag{4.23 a}$$

in the notation of eqn. (4.19), and for which

$$\mathbf{P}_1 = \begin{bmatrix} 1, & -4 \\ -2, & 0 \end{bmatrix} \tag{4.23 b}$$

and

$$\mathbf{P}_2 = \begin{bmatrix} -1, & 1 \\ 3, & -1 \\ 0, & 5 \end{bmatrix}. \tag{4.23 c}$$

Since $p_{21}^{(1)} \neq 0$, the first Jordan block in (4.23 *a*) is controllable by $z_1(t)$, but, since $p_{21}^{(2)} \neq 0$ and $p_{31}^{(2)} = 0$, the second Jordan block is only partially controllable by $z_1(t)$. In addition, since $p_{12}^{(1)} \neq 0$ and $p_{22}^{(1)} = 0$, the first Jordan block is partially controllable by $z_2(t)$, but, since $p_{32}^{(2)} \neq 0$, the second Jordan block is controllable by $z_2(t)$.

4.2.4. *Confluent eigenvalues associated with a number of non-distinct Jordan blocks*

The results derived in § 4.2.3 can be extended so as to embrace systems for which the matrix $\mathbf{J}$ in eqn. (4.6) is a block-diagonal matrix whose typical block is the $l \times l$ Jordan block

$$\mathbf{J}_l(\lambda_j) = \begin{bmatrix} \lambda_j, & 1, & 0, \ldots, & 0, & 0 \\ 0, & \lambda_j, & 1, \ldots, & 0, & 0 \\ \cdots & \cdots & \cdots & \cdots & \cdots \\ \cdots & \cdots & \cdots & \cdots & \cdots \\ 0, & 0, & 0, \ldots, & \lambda_j, & 1 \\ 0, & 0, & 0, \ldots, & 0, & \lambda_j \end{bmatrix}. \tag{4.24}$$

In this section it will be assumed initially that each of the ν Jordan blocks is associated with the same-valued eigenvalue, λ_1: the matrix $\mathbf{J}$ can then be written in the form

$$\mathbf{J} = \mathbf{J}(\lambda_1) = \mathbf{J}_{n_1}(\lambda_1) \oplus \mathbf{J}_{n_2}(\lambda_1) \oplus \ldots \oplus \mathbf{J}_{n_\nu}(\lambda_1). \tag{4.25}$$

It is therefore clear that, in this case, the vector-matrix eqn. (4.6) is equivalent to the ν vector-matrix equations

$$\boldsymbol{\xi}_i(t) = \mathbf{J}_{n_i}(\lambda_1)\boldsymbol{\xi}_i(t) + \mathbf{P}_i\mathbf{z}(t) \quad (i = 1, 2, \ldots, \nu), \tag{4.26}$$

where

$$\mathbf{P}_i = \mathbf{V}_i'\mathbf{B}$$

and the $\mathbf{V}_i$ satisfy the equations

$$\mathbf{A}'\mathbf{V}_i = \mathbf{V}_i\mathbf{J}_{ni}'(\lambda_1).$$

The ith equation of eqns. (4.26) is equivalent to the n_i scalar equations

$$\dot{\xi}_k^{(i)}(t) = \lambda_1 \xi_k^{(i)}(t) + \xi_{k+1}^{(i)}(t) + \sum_{j=1}^{r} p_{kj}^{(i)} z_j(t) \quad (k = 1, 2, \ldots, n_i - 1) \tag{4.27 a}$$

and

$$\dot{\xi}_{ni}^{(i)}(t) = \lambda_1 \xi_{ni}^{(i)}(t) + \sum_{j=1}^{r} p_{ni,j}^{(i)} z_j(t), \tag{4.27 b}$$

where the $\xi_k^{(i)}(t)$ and the $p_{kj}^{(i)}$ are given respectively by the eqns. (4.22 *a*) and (4.22 *b*).

Equation (4.27 *b*) can be written in the form

$$\dot{\xi}_{ni}^{(i)}(t) = \lambda_1 \xi_{ni}^{(i)}(t) + \mathbf{p}^{(i)\prime}\mathbf{z}(t) \quad (i = 1, 2, \ldots, \nu), \tag{4.28}$$

where

$$\mathbf{p}^{(i)\prime} = [p_{ni,1}^{(i)}, p_{ni,2}^{(i)}, \ldots, p_{ni,r}^{(i)}]. \tag{4.29}$$

If a set $\{\alpha_1, \alpha_2, \ldots, \alpha_\nu\}$ of scalars, not all zero, is introduced such that

$$\sum_{i=1}^{\nu} \alpha_i \mathbf{p}^{(i)\prime} = \boldsymbol{\pi}' = [\pi_j] \tag{4.30}$$

then eqns. (4.28) can be written in the form

$$\dot{\zeta}(t) = \lambda_1 \zeta(t) + \boldsymbol{\pi}'\mathbf{z}(t), \tag{4.31}$$

where

$$\zeta(t) = \sum_{i=1}^{\nu} \alpha_i \xi_{ni}^{(i)}(t). \tag{4.32}$$

If some, or all, of the vectors $\mathbf{p}^{(1)}, \mathbf{p}^{(2)}, \ldots, \mathbf{p}^{(\nu)}$ are linearly dependent, then $\boldsymbol{\pi}' = \mathbf{0}$ for some set of the scalars $\alpha_1, \alpha_2, \ldots, \alpha_\nu$: on the other hand, if the vectors $\mathbf{p}^{(1)}, \mathbf{p}^{(2)}, \ldots, \mathbf{p}^{(\nu)}$ are linearly independent, then $\boldsymbol{\pi}' \neq \mathbf{0}$ for all sets of the scalars $\alpha_1, \alpha_2, \ldots, \alpha_\nu$. In the former case, since eqns. (4.31) and (4.32) indicate that a transformation of state exists under which $\boldsymbol{\pi}' = \mathbf{0}$, it is evident that the control vector $\mathbf{z}(t)$ will not influence the associated transformed state variable, $\zeta(t)$: in the latter case, however, at least one of the input variables will influence the transformed state variable $\zeta(t)$ since no transformation of state exists under which $\boldsymbol{\pi}' = \mathbf{0}$. The Jordan blocks (4.25) are therefore *controllable* by the input vector $\mathbf{z}(t)$ if and only if all the rows of the mode-controllability matrix which correspond to the last rows of its constituent Jordan blocks are linearly independent : the system (4.26) is *controllable* if and only if the Jordan blocks (4.25) are controllable.

In the case of systems for which the Jordan canonical form, $\mathbf{J}$, of the plant matrix $\mathbf{A}$ in eqn. (4.2) can be written in the form

$$\mathbf{J} = \mathbf{J}(\lambda_1) \oplus \mathbf{J}(\lambda_2) \oplus \ldots \oplus \mathbf{J}(\lambda_\mu), \tag{4.33}$$

where each of the $\mathbf{J}(\lambda_i)$ is defined by an appropriate equation of the form (4.25), the controllability condition relating to the matrix $\mathbf{J}(\lambda_1)$ defined in eqn. (4.25) can be generalized [1] as follows : the constituent Jordan blocks of the matrix $\mathbf{J}$ in eqn. (4.33) are *controllable* by the input vector $\mathbf{z}(t)$ if and only if all the rows of the mode-controllability matrix which correspond to the last rows of Jordan blocks containing the same-valued eigenvalue are linearly independent. The system whose plant matrix has the Jordan canonical form (4.33) is *controllable* if and only if the constituent Jordan blocks of the matrix $\mathbf{J}$ in eqn. (4.33) are controllable.

It is interesting to note that this condition implies that it is necessary for the controllability of a system governed by a state equation of the form (4.2) that the number of input variables, r, must satisfy the inequalities

$$r \geqslant \sigma_i \quad (i=1, 2, ..., \mu), \tag{4.34}$$

where σ_i is the number of Jordan blocks associated with the eigenvalue λ_i. In particular, in the case of single-input systems, condition (4.34) can be satisfied only if $\sigma_i = 1$ $(i=1, 2, \ldots, \mu)$, that is, if all the Jordan blocks of the Jordan canonical form of the plant matrix $\mathbf{A}$ are distinct.

It is important to note that the conditions for system controllability presented in this section can be related to the well-known conditions for state-controllability [2], [3]. In fact, it can be shown in the manner of § 4.3 that systems which are controllable in the sense of §§ 4.2.1 to 4.2.4 are *state-controllable* [2], [3] if and only if the appropriate $n \times nr$ matrix

$$\mathbf{Q} = [\mathbf{B}, \mathbf{AB}, ..., \mathbf{A}^{n-1}\mathbf{B}] \tag{4.35}$$

is of rank n, in which case $(\mathbf{A}, \mathbf{B})$ is called a *controllable pair*.

The controllability conditions developed in this section can be illustrated by considering a system for which

$$\mathbf{J} = \mathbf{J}(-1) = \mathbf{J}_3(-1) \oplus \mathbf{J}_2(-1) \tag{4.36}$$

in the notation of eqn. (4.25), and for which

$$\mathbf{P}_1 = \begin{bmatrix} 1, & 0 \\ -1, & 2 \\ 1, & 3 \end{bmatrix} \tag{4.37 a}$$

and

$$\mathbf{P}_2 = \begin{bmatrix} -4, & 2 \\ 1, & -1 \end{bmatrix}. \tag{4.37 b}$$

The Jordan blocks $\mathbf{J}_3(-1)$ and $\mathbf{J}_2(-1)$ in this example are controllable since the vectors [1, 3] and [1, -1] which constitute the last rows of the respective matrices $\mathbf{P}_1$ and $\mathbf{P}_2$ are linearly independent.

These results can be further illustrated by considering a system for which

$$\mathbf{J} = \mathbf{J}(0) \oplus \mathbf{J}(-1), \tag{4.38 a}$$

where

$$\mathbf{J}(0) = \mathbf{J}_1(0) \oplus \mathbf{J}_1(0) \oplus \mathbf{J}_2(0) \tag{4.38 b}$$

and

$$\mathbf{J}(-1) = \mathbf{J}_2(-1) \oplus \mathbf{J}_3(-1) \tag{4.38 c}$$

in the notation of eqn. (4.33), and for which

$$\mathbf{P}=\begin{bmatrix} 1, & 2, & 3 \\ 1, & 0, & 0 \\ 0, & 1, & 2 \\ 2, & -1, & 1 \\ 0, & 1, & 0 \\ 2, & 7, & 6 \\ -4, & 5, & 1 \\ 2, & -3, & 4 \\ 1, & 8, & 0 \end{bmatrix} \tag{4.39}$$

The Jordan blocks in this example are controllable since the vectors [1, 2, 3], [1, 0, 0], and [2, −1, 1] associated with the last rows of the Jordan blocks $\mathbf{J}_1(0)$, $\mathbf{J}_1(0)$, and $\mathbf{J}_2(0)$ are linearly independent, and since the vectors [2, 7, 6] and [1, 8, 0] associated with the last rows of the Jordan blocks $\mathbf{J}_2(-1)$ and $\mathbf{J}_3(-1)$ are also linearly independent.

4.3. Mode-controllability structure of multivariable linear systems

4.3.1. *Introduction*

The use of the matrix $\mathbf{Q}$ defined in eqn. (4.35) as a test for state-controllability does not, of course, provide any information as to the details of the mode-controllability structure of a system governed by an equation of the form (4.2). However, it is shown in this section that it is possible to use certain properties of $\mathbf{Q}$ to determine which modes (if any) of the system (4.2) are controllable by a given component $z_i(t)$ $(i=1, 2, \ldots, r)$ of the input vector, $\mathbf{z}(t)$.

4.3.2. *Distinct eigenvalues*

Thus, let the matrix $\mathbf{B}$ in eqn. (4.2) be written in the partitioned form

$$\mathbf{B}=[\mathbf{b}_1, \mathbf{b}_2, \ldots, \mathbf{b}_r], \tag{4.40}$$

so that eqn. (4.2) has the form

$$\dot{\mathbf{x}}(t)=\mathbf{A}\mathbf{x}(t)+\sum_{i=1}^{r} \mathbf{b}_i z_i(t). \tag{4.41}$$

In order to examine the controllability characteristics of a given $z_i(t)$ $(i=1, 2, \ldots, r)$, it is only necessary to consider the properties of the $n\times n$ matrix

$$\mathbf{Q}_i=[\mathbf{b}_i, \mathbf{A}\mathbf{b}_i, \mathbf{A}^2\mathbf{b}_i, \ldots, \mathbf{A}^{n-1}\mathbf{b}_i] \tag{4.42}$$

associated with the input variable $z_i(t)$. Since it is assumed in this section that the eigenvalues of $\mathbf{A}$ are distinct, the vector $\mathbf{b}_i$ in eqn. (4.42) can be expressed in the form

$$\mathbf{b}_i=\sum_{j=1}^{n} p_{ji}\mathbf{u}_j, \tag{4.43}$$

where the $\mathbf{u}_j$ are n linearly independent eigenvectors of $\mathbf{A}$ and the p_{ji} $(j=1, 2, \ldots, n)$ are the elements of the mode-controllability matrix

$$\mathbf{p}_i=\mathbf{U}^{-1}\mathbf{b}_i, \tag{4.44}$$

where $\mathbf{U}$ is the modal matrix of $\mathbf{A}$.

Now, it can be verified that

$$\mathbf{Q}_i = \mathbf{U}\mathbf{P}_i\mathbf{G}, \tag{4.45}$$

where

$$\mathbf{P}_i = \text{diag}\{p_{1i}, p_{2i}, \ldots, p_{ni}\}, \tag{4.46}$$

and $\mathbf{G}$ is the $n \times n$ Vandermonde matrix

$$\begin{bmatrix} 1, & \lambda_1, & \lambda_1^2, \ldots, & \lambda_1^{n-1} \\ 1, & \lambda_2, & \lambda_2^2, \ldots, & \lambda_2^{n-1} \\ \cdots & \cdots & \cdots & \cdots \\ \cdots & \cdots & \cdots & \cdots \\ 1, & \lambda_n, & \lambda_n^2, \ldots, & \lambda_n^{n-1} \end{bmatrix}. \tag{4.47}$$

Since the $n \times n$ matrices $\mathbf{U}$ and $\mathbf{G}$ in eqn. (4.45) each has rank n in the case when $\mathbf{A}$ has distinct eigenvalues, it follows that

$$\text{rank } \mathbf{Q}_i = \text{rank } \mathbf{P}_i, \tag{4.48}$$

which clearly implies that rank $\mathbf{Q}_i$ is equal to the number of non-zero elements p_{ji} $(j = 1, 2, \ldots, n)$ in the $n \times n$ diagonal matrix $\mathbf{P}_i$. In § 4.2.1 it is shown that the jth mode of the system (4.38) is controllable by $z_i(t)$ if and only if $p_{ji} \neq 0$. Equation (4.48) therefore implies that the rank of $\mathbf{Q}_i$ is equal to the number of modes, n_i, of the system (4.41) which can be controlled by the ith input, $z_i(t)$. These controllable modes can be characterized without any loss of generality by the sequence of sets of eigenvalues and eigenvectors $\{\lambda_j, \mathbf{u}_j\}$ $(j = 1, 2, \ldots, n_i)$. Equation (4.43) then assumes the form

$$\mathbf{b}_i = \sum_{j=1}^{n_i} p_{ji}\mathbf{u}_j \tag{4.49}$$

so that

$$\mathbf{A}^s\mathbf{b}_i = \sum_{j=1}^{n_i} p_{ji}\lambda_j^s\mathbf{u}_j \quad (s = 1, 2, \ldots, n_i). \tag{4.50}$$

Since rank $\mathbf{Q}_i = n_i$, it follows that

$$\mathbf{A}^{n_i}\mathbf{b}_i = \sum_{k=0}^{n_i-1} \alpha_{ki}\mathbf{A}^k\mathbf{b}_i, \tag{4.51}$$

where the α_{ki} are the constants which express the linear dependence of $\mathbf{A}^{n_i}\mathbf{b}_i$ on $\mathbf{b}_i$, $\mathbf{A}\mathbf{b}_i$, ..., $\mathbf{A}^{n_i-1}\mathbf{b}_i$. It may therefore be deduced from eqns. (4.50) and (4.51) that

$$\sum_{j=1}^{n_i} p_{ji}\lambda_j^{n_i}\mathbf{u}_j = \sum_{k=0}^{n_i-1} \alpha_{ki}\left\{\sum_{j=1}^{n_i} p_{ji}\lambda_j^k\mathbf{u}_j\right\} \tag{4.52}$$

which implies that

$$p_{ji}\lambda_j^{n_i} = \sum_{k=0}^{n_i-1} \alpha_{ki}p_{ji}\lambda_j^k \quad (i = 1, 2, \ldots, n; \; j = 1, 2, \ldots, n_i). \tag{4.53}$$

Since $p_{ji} \neq 0$ $(j = 1, 2, \ldots, n_i)$, eqns. (4.53) indicate that the eigenvalues associated with the n_i modes of the system (4.41) which can be controlled by the input variable $z_i(t)$ satisfy the equation

$$\lambda^{n_i} = \sum_{k=0}^{n_i-1} \alpha_{ki}\lambda^k. \tag{4.54}$$

In order to identify the controllable modes of the system by determining the associated eigenvalues, it is thus necessary only to determine the α_{ki} in eqn. (4.51) and thus the roots of the n_ith-order polynomial eqn. (4.54).

As a simple numerical illustration of these results consider a system of the class (4.41) for which

$$\mathbf{A}=\begin{bmatrix} 0, & 1, & 0 \\ 0, & 0, & 1 \\ -6, & -11, & -6 \end{bmatrix}. \tag{4.55 a}$$

and

$$\mathbf{B}=\begin{bmatrix} 1, & 0 \\ -2, & 1 \\ 4, & -3 \end{bmatrix}. \tag{4.55 b}$$

It can be readily verified that the eigenvalues of $\mathbf{A}$ are distinct and therefore that the method described in this section is applicable. In this example,

$$\mathbf{AB}=\begin{bmatrix} -2, & 1 \\ 4, & -3 \\ -8, & 7 \end{bmatrix}, \tag{4.56 a}$$

and

$$\mathbf{A}^2\mathbf{B}=\begin{bmatrix} 4, & -3 \\ -8, & 7 \\ 16, & -15 \end{bmatrix}, \tag{4.56 b}$$

so that it can be deduced from eqns. (4.55) and (4.56) that

$$\mathbf{Q}_1=\begin{bmatrix} 1, & -2, & 4 \\ -2, & 4, & -8 \\ 4, & -8, & -16 \end{bmatrix}, \tag{4.57 a}$$

and

$$\mathbf{Q}_2=\begin{bmatrix} 0, & 1, & -3 \\ 1, & -3, & 7 \\ -3, & 7, & -15 \end{bmatrix}. \tag{4.57 b}$$

It can be verified that the matrices $\mathbf{Q}_1$ and $\mathbf{Q}_2$ are of rank one ($n_1=1$) and rank two ($n_2=2$), respectively, which implies that the system defined by eqns. (4.55) has one mode which can be controlled by the input $z_1(t)$ and two modes which can be controlled by the input $z_2(t)$, The eigenvalues of these modes can be found from eqns. (4.51) : in the present numerical example these equations are

$$\mathbf{Ab}_1=\alpha_{01}\mathbf{b}_1 \tag{4.58 a}$$

and

$$\mathbf{A}^2\mathbf{b}_2=\alpha_{02}\mathbf{b}_2+\alpha_{12}\mathbf{Ab}_2. \tag{4.58 b}$$

It is obvious by inspecting eqns. (4.55 *b*), (4.56), and (4.58) that $\alpha_{01}=-2$, $\alpha_{02}=-2$, and $\alpha_{12}=-3$: equation (4.54) thus indicates that the eigenvalue of the single mode which is controllable by the first input is -2, and that the eigenvalues of the two modes which are controllable by the second input are -1 and -2.

4.3.3. *Confluent eigenvalues associated with a single Jordan block*

If the plant matrix $\mathbf{A}$ of the system governed by equation (4.2) has n repeated eigenvalues associated with the single $n \times n$ Jordan block

$$\mathbf{J}=\begin{bmatrix} \lambda_1, & 1, & 0, \ldots, & 0, & 0 \\ 0, & \lambda_1, & 1, \ldots, & 0, & 0 \\ \cdots & \cdots & \cdots & \cdots & \cdots \\ 0, & 0, & 0, \ldots, & \lambda_1, & 1 \\ 0, & 0, & 0, \ldots, & 0, & \lambda_1 \end{bmatrix}, \tag{4.59}$$

then the generalized modal matrix $\mathbf{U}$ satisfies the equation

$$\mathbf{AU}=\mathbf{UJ}. \tag{4.60}$$

It can be verified that, since the vectors $\mathbf{b}_i$ can be expressed by an equation of the form (4.43), then the $n \times n$ matrix (4.42) can be written in the form

$$\mathbf{Q}_i=\mathbf{UP}_i\mathbf{G}, \tag{4.61}$$

where

$$\mathbf{P}_i=\begin{bmatrix} p_{1i}, & p_{2i}, \ldots, & p_{ni} \\ p_{2i}, & p_{3i}, \ldots, & 0 \\ \cdots & \cdots & \cdots \\ \cdots & \cdots & \cdots \\ p_{n-1,i}, & p_{ni}, \ldots, & 0 \\ p_{ni}, & 0, \ldots, & 0 \end{bmatrix}, \tag{4.62}$$

and

$$\mathbf{G}=\begin{bmatrix} 1, & \lambda_1, & \lambda_1^2, & \lambda_1^3, \ldots, & \lambda_1^{n-1} \\ 0, & 1, & 2\lambda_1, & 3\lambda_1^2, \ldots, & (n-1)\lambda_1^{n-2} \\ 0, & 0, & 1, & 3\lambda_1, \ldots, & (n-1)(n-2)\lambda_1^{n-3}/2 \\ \cdots & \cdots & \cdots & \cdots & \cdots \\ 0, & 0, & 0, & 0, \ldots, & 1 \end{bmatrix}. \tag{4.63}$$

This upper-triangular matrix, $\mathbf{G}$, is the so-called *alternant matrix* [4] and a typical non-zero element is given by the equation

$$\mathbf{G}=[g_{ij}]=\frac{(j-1)!\,\lambda_1^{j-i}}{(j-i)!(i-1)!} \quad (i \leqslant j\,;\ i, j=1, 2, \ldots, n). \tag{4.64}$$

Since the $n \times n$ matrices $\mathbf{U}$ and $\mathbf{G}$ in eqn. (4.61) each has rank n, it follows that

$$\text{rank } \mathbf{Q}_i=\text{rank } \mathbf{P}_i. \tag{4.65}$$

This equation implies that rank $\mathbf{Q}_i$ is equal to m_i, where

$$p_{m_i,i} \neq 0 \quad (1 \leqslant m_i \leqslant n) \tag{4.66 a}$$

and

$$p_{ji}=0 \quad (j=m_i+1, m_i+2, \ldots, n). \tag{4.66b}$$

It follows from eqns. (4.66) that

$$\mathbf{b}_i = \sum_{j=1}^{m_i} p_{ji}\mathbf{u}_j \quad (i=1, 2, \dots, r), \tag{4.67}$$

which implies that

$$\mathbf{A}^s\mathbf{b}_i = \sum_{j=1}^{m_i}\sum_{k=0}^{m_i-j} \frac{s!}{k!(s-k)!} p_{j+k,i}\lambda_1^{s-k}\mathbf{u}_k \tag{4.68}$$

in view of the fact that eqn. (4.60) implies that

$$\mathbf{A}\mathbf{u}_1 = \lambda_1\mathbf{u}_1 \tag{4.69 a}$$

and

$$\mathbf{A}\mathbf{u}_i = \lambda_1\mathbf{u}_i + \mathbf{u}_{i-1} \quad (i=2, 3, \dots, n). \tag{4.69 b}$$

Since rank $\mathbf{Q}_i = m_i$, it follows that

$$\mathbf{A}^{m_i}\mathbf{b}_i = \sum_{s=0}^{m_i-1} \alpha_{si}\mathbf{A}^s\mathbf{b}_i, \tag{4.70}$$

where the α_{si} are the constants which express the linear dependence of $\mathbf{A}^{m_i}\mathbf{b}_i$ on $\mathbf{b}_i$, $\mathbf{A}\mathbf{b}_i$, . . ., $\mathbf{A}^{m_i-1}\mathbf{b}_i$. It may now be deduced from eqns. (4.68) and (4.70) that

$$\sum_{s=0}^{m_i} \alpha_{si} \sum_{j=1}^{m_i}\sum_{k=0}^{m_i-j} \left\{\frac{s!}{k!(s-k)!}\right\} p_{j+k,i}\lambda_1^{s-k}\mathbf{u}_k = 0, \tag{4.71}$$

where

$$\alpha_{m_i,i} = -1,$$

which in turn implies that

$$p_{m_i,i}\lambda_1^{m_i} = p_{m_i,i}\sum_{j=0}^{m_i-1} \alpha_{ji}\lambda_1^j \tag{4.72 a}$$

and

$$\frac{d^k}{d\lambda^k}\left[\lambda^{m_i} - \sum_{j=0}^{m_i-1} \alpha_{ji}\lambda^j\right]_{\lambda=\lambda_1} = 0 \quad (k=1, 2, \dots, m_i-1). \tag{4.72 b}$$

Equations (4.72) indicate that the eigenvalues associated with the Jordan block (4.59) which can be controlled by the input variable $z_i(t)$ satisfy the equation

$$\lambda^{m_i} - \sum_{j=0}^{m_i} \alpha_{ji}\lambda^j = 0 \tag{4.73}$$

since

$$p_{m_i,i} \neq 0.$$

It is thus only necessary to determine the α_{si} in eqns. (4.70) and thus determine the roots of the m_ith-order polynomial eqn. (4.73). In the special case considered in this section, these m_i roots will, of course, all be equal to λ_1.

The theory developed in this section can be conveniently illustrated by considering a system of the class (4.2) for which

$$\mathbf{A}=\begin{bmatrix}1, & -1, & 3\\ 0, & -1, & 4\\ 0, & -1, & 3\end{bmatrix} \tag{4.74 a}$$

and

$$\mathbf{B}=\begin{bmatrix}1, & 5\\ 1, & 4\\ 1, & 2\end{bmatrix}. \tag{4.74 b}$$

It can readily be verified that the Jordan canonical form of $\mathbf{A}$ is given by the expression

$$\mathbf{J}=\begin{bmatrix}1, & 1, & 0\\ 0, & 1, & 1\\ 0, & 0, & 1\end{bmatrix} \tag{4.75}$$

and therefore that the method described in this section is applicable. In this example,

$$\mathbf{Q}_1=\begin{bmatrix}1, & 3, & 6\\ 1, & 3, & 5\\ 1, & 2, & 3\end{bmatrix}, \tag{4.76 a}$$

$$\mathbf{Q}_2=\begin{bmatrix}5, & 7, & 9\\ 4, & 4, & 4\\ 2, & 2, & 2\end{bmatrix}, \tag{4.76 b}$$

and

$$\mathbf{A}^3\mathbf{B}=\begin{bmatrix}10, & 11\\ 7, & 4\\ 4, & 2\end{bmatrix}. \tag{4.76 c}$$

It can be verified that the ranks of the matrices $\mathbf{Q}_1$ and $\mathbf{Q}_2$ are three ($n_1=3$) and two ($n_2=2$), respectively, and furthermore that

$$\mathbf{A}^3\mathbf{b}_1=3\mathbf{A}^2\mathbf{b}_1-3\mathbf{A}\mathbf{b}_1+\mathbf{b}_1, \tag{4.77 a}$$

and

$$\mathbf{A}^2\mathbf{b}_2=2\mathbf{A}\mathbf{b}_2-\mathbf{b}_2. \tag{4.77 b}$$

Equations (4.70), (4.73), and (4.77) thus indicate that the Jordan block (4.75) is controllable by the first input variable and partially controllable by the second input variable.

4.3.4. *Confluent eigenvalues associated with a number of distinct Jordan blocks*

The results derived in § 4.3.3 can be readily extended so as to embrace systems for which

$$\mathbf{A}\mathbf{U}=\mathbf{U}\mathbf{J} \tag{4.78 a}$$

where $\mathbf{U}$ is the non-singular generalized modal matrix of $\mathbf{A}$, but where $\mathbf{J}$ is now a block-diagonal matrix whose typical ith block ($i=1, 2, \ldots, \nu$) is the $n_i \times n_i$ Jordan block

$$\mathbf{J}_{ni}(\lambda_j)=\begin{bmatrix} \lambda_j, & 1, & 0, \ldots, 0, & 0 \\ 0, & \lambda_j, & 1, \ldots, 0, & 0 \\ \cdots & \cdots & \cdots & \cdots \\ \cdots & \cdots & \cdots & \cdots \\ 0, & 0, & 0, \ldots, \lambda_j, & 1 \\ 0, & 0, & 0, \ldots, 0, & \lambda_j \end{bmatrix}, \tag{4.78 b}$$

where λ_j is a confluent eigenvalue of the matrix $\mathbf{A}$. In this section it will be assumed that the ν Jordan blocks are each associated with a distinct eigenvalue that is, $\lambda_i \neq \lambda_j$ $(i \neq j, i, j=1, 2, \ldots, \nu)$. The Jordan matrix, $\mathbf{J}$, and the generalized modal matrix $\mathbf{U}$ in eqn. (4.78 a) can then be written in the forms

$$\mathbf{J}=\mathbf{J}_{n_1}(\lambda_1) \oplus \mathbf{J}_{n_2}(\lambda_2) \oplus \ldots \oplus \mathbf{J}_{n_\nu}(\lambda_\nu) \tag{4.79}$$

and

$$\mathbf{U}=[\mathbf{U}_1, \mathbf{U}_2, \ldots, \mathbf{U}_\nu], \tag{4.80 a}$$

where each of the $n \times n_i$ sub-matrices $\mathbf{U}_i$ has the form

$$\mathbf{U}_i=[\mathbf{u}_1{}^{(i)}, \mathbf{u}_2{}^{(i)}, \ldots, \mathbf{u}_{n_i}{}^{(i)}]. \tag{4.80 b}$$

The $n \times r$ mode-controllability matrix of the system described by eqns. (4.2), (4.78), (4.79), and (4.80) has the form

$$\mathbf{P}=\mathbf{U}^{-1}\mathbf{B}=\begin{bmatrix} \mathbf{P}_1{}^{(1)}, & \mathbf{P}_2{}^{(1)}, \ldots, & \mathbf{P}_r{}^{(1)} \\ \mathbf{P}_1{}^{(2)}, & \mathbf{P}_2{}^{(2)}, \ldots, & \mathbf{P}_r{}^{(2)} \\ \cdots & \cdots & \cdots \\ \mathbf{P}_1{}^{(\nu)}, & \mathbf{P}_2{}^{(\nu)}, \ldots, & \mathbf{P}_r{}^{(\nu)} \end{bmatrix}, \tag{4.81a}$$

where the $n_j \times 1$ mode-controllability sub-matrix $\mathbf{P}_i{}^{(j)}$ associated with the Jordan block, $\mathbf{J}_{nj}(\lambda_j)$, and with the ith input variable, $z_i(t)$, is given by

$$\mathbf{P}_i{}^{(j)}=\mathbf{V}_j{}'\mathbf{b}_i=\begin{bmatrix} p_{1i}{}^{(j)} \\ p_{2i}{}^{(j)} \\ \vdots \\ p_{nj,i}{}^{(j)} \end{bmatrix} \quad (i=1, 2, \ldots, r\,;\; j=1, 2, \ldots, \nu). \tag{4.81 b}$$

It follows from eqns. (4.81) that

$$\mathbf{b}_i=\sum_{j=1}^{\nu}\sum_{k=1}^{n_j} p_{ki}{}^{(j)}\mathbf{u}_k{}^{(j)}. \tag{4.82}$$

It can readily be shown that the $n \times n$ controllability matrix (4.42) can be expressed in the form

$$\mathbf{Q}_i=\mathbf{U}\mathbf{P}_i\mathbf{G}, \tag{4.83}$$

where

$$\mathbf{P}_i = \mathbf{P}_i^{(1)} \oplus \mathbf{P}_i^{(2)} \oplus \ldots \oplus \mathbf{P}_i^{(\nu)} \tag{4.84 a}$$

and

$$\mathbf{G} = \begin{bmatrix} \mathbf{G}_1 \\ \mathbf{G}_2 \\ \vdots \\ \mathbf{G}_\nu \end{bmatrix}. \tag{4.84 b}$$

In eqn. (4.84), the $n_j \times n_j$ sub-matrix $\mathbf{P}_i^{(j)}$ and the $n_j \times n$ sub-matrix $\mathbf{G}_j$ ($j = 1, 2, \ldots, \nu$) have the respective forms

$$\mathbf{P}_i^{(j)} = \begin{bmatrix} p_{1i}^{(j)}, & p_{2i}^{(j)}, & \ldots, & p_{n_j,i}^{(j)} \\ p_{2i}^{(j)}, & p_{3i}^{(j)}, & \ldots, & 0 \\ \cdots & \cdots & \cdots & \cdots \\ p_{n_j,i}^{(j)}, & 0, & \ldots, & 0 \end{bmatrix} \tag{4.85 a}$$

and

$$\mathbf{G}_j = \begin{bmatrix} 1, & \lambda_j, & \lambda_j^2, & \ldots, & \lambda_j^{n-1} \\ 0, & 1, & 2\lambda_j, & \ldots, & (n-1)\lambda_j^{n-2} \\ 0, & 0, & 0, & \ldots, & (n-1)(n-2)\lambda_j^{n-3} \\ \cdots & \cdots & \cdots & \cdots & \cdots \\ 0, & 0, & 0, & \ldots, & (n-1)!\lambda_j^{n-n_j}/(n-n_j)!(n_j-1)! \end{bmatrix} \tag{4.85 b}$$

The sub-matrix $\mathbf{G}_j$ is, in general, a truncated form of the alternant matrix (4.63) and the typical non-zero element is given by eqn. (4.64). Since the matrices $\mathbf{U}$ and $\mathbf{G}$ in eqn. (4.83) each has rank n in the case when $\mathbf{A}$ has distinct Jordan blocks, it follows that

$$\operatorname{rank} \mathbf{Q}_i = \sum_{j=1}^{\nu} \operatorname{rank} \mathbf{P}_i^{(j)}, \tag{4.86 a}$$

which implies that

$$\operatorname{rank} \mathbf{Q}_i = \sum_{j=1}^{\nu} m_{ji} \quad (i = 1, 2, \ldots, r), \tag{4.86 b}$$

where

$$p_{m_{ji},i}^{(j)} \neq 0 \quad (1 \leqslant m_{ji} \leqslant n_j\,;\; j = 1, 2, \ldots, \nu) \tag{4.87 a}$$

and

$$p_{ki}^{(j)} = 0 \quad (k = m_{ji}+1, m_{ji}+2, \ldots, n_j;\; j = 1, 2 \ldots, \nu). \tag{4.87 b}$$

It follows from eqns. (4.87) that eqn. (4.82) can be written in the truncated form

$$\mathbf{b}_i = \sum_{j=1}^{\nu} \sum_{k=1}^{m_{ji}} p_{ki}^{(j)} \mathbf{u}_k^{(j)} \quad (i = 1, 2, \ldots, r). \tag{4.88}$$

It may therefore be inferred that

$$\mathbf{A}^s \mathbf{b}_i = \sum_{l=1}^{\nu} \sum_{j=1}^{m_{li}} \sum_{k=1}^{m_{li}-j} \left\{ \frac{s!}{(s-k)!k!} \right\} p_{j+k,i}^{(l)} \lambda_j^{s-k} \mathbf{u}_j^{(l)} \tag{4.89}$$

in view of the fact that

$$\left.\begin{aligned} \mathbf{A}\mathbf{u}_1^{(l)} &= \lambda_l \mathbf{u}_1^{(l)} \\ \text{and} \qquad \mathbf{A}\mathbf{u}_j^{(l)} &= \lambda_l \mathbf{u}_j^{(l)} + \mathbf{u}_{j-1}^{(l)} \end{aligned}\right\} (l=1, 2, \ldots, \nu). \tag{4.90}$$

Since it has been shown that the rank of $\mathbf{Q}_i = \sum_{j=1}^{\nu} m_{ji} = m_i$(say), it follows that

$$\mathbf{A}^{m_i}\mathbf{b}_i = \sum_{s=0}^{m_i-1} \alpha_{si}\mathbf{A}_s\mathbf{b}_i, \tag{4.91}$$

where the terms $\mathbf{A}^s\mathbf{b}_i$ ($s=0, 1, \ldots, m_i$) are given by the expression (4.89). It may readily be shown that

$$p_{m_{ji},i}^{(j)}\lambda_j^{m_i} = p_{m_{ji},i}^{(j)} \sum_{s=0}^{m_i-1} \alpha_{si}\lambda_j^s \tag{4.92 a}$$

and

$$\frac{d^k}{d\lambda^k}\left[\lambda^{m_i} - \sum_{s=0}^{m_i-1} \alpha_{si}\lambda^s\right]_{\lambda=\lambda_j} = 0 \tag{4.92 b}$$

$$(j=1, 2, \ldots, \nu\,;\ k=1, 2, \ldots, m_{ji}-1).$$

Equations (4.92) indicate that the eigenvalues associated with the Jordan blocks (4.79) which can be controlled by the input variable $z_i(t)$ satisfy the equation

$$\lambda^{m_i} - \sum_{s=0}^{m_i-1} \alpha_{si}\lambda^s = 0 \tag{4.93}$$

since

$$p_{m_{ji},i}^{(j)} \neq 0 \quad (j=1, 2, \ldots, \nu\,;\ i=1, 2, \ldots, r).$$

It is therefore only necessary to determine the α_{si} in eqn. (4.91) and thus determine the roots of the m_ith-order polynomial eqn. (4.93).

The theory described in this section can be illustrated by considering a system of the class described by eqns. (4.2) and (4.78) for which

$$\mathbf{A} = \begin{bmatrix} 1, & 1, & -2, & 5, & -21 \\ 0, & 1, & -1, & 1, & -4 \\ 1, & 1, & -2, & 6, & -27 \\ 0, & 0, & 0, & -3, & 16 \\ 0, & 0, & 0, & 1, & 5 \end{bmatrix} \tag{4.94 a}$$

and

$$\mathbf{B} = \begin{bmatrix} 1, & 1 \\ 1, & 0 \\ 1, & 1 \\ 4, & 2 \\ 1, & 1 \end{bmatrix}. \tag{4.94 b}$$

It can be verified that the eigenvalues of $\mathbf{A}$ are $\{0, 0, 0, 1, 1\}$, and that there are two distinct Jordan blocks: the Jordan canonical form of the matrix (4.94 a) is accordingly

$$\mathbf{J}=\mathbf{J}_3(0)\oplus\mathbf{J}_2(1)=\begin{bmatrix}0, & 1, & 0, & 0, & 0\\ 0, & 0, & 1, & 0, & 0\\ 0, & 0, & 0, & 0, & 0\\ 0, & 0, & 0, & 1, & 1\\ 0, & 0, & 0, & 0, & 1\end{bmatrix}, \tag{4.95}$$

which indicates that the method described in this section is applicable. In this example,

$$\mathbf{Q}_1=\begin{bmatrix}1, & -1, & 4, & 2, & 2\\ 1, & 0, & 3, & 1, & 1\\ 1, & -3, & 2, & 0, & 0\\ 4, & 4, & 4, & 4, & 4\\ 1, & 1, & 1, & 1, & 1\end{bmatrix} \tag{4.96 a}$$

and

$$\mathbf{Q}_2=\begin{bmatrix}1, & -12, & 4, & 8, & 12\\ 0, & -3, & 11, & 13, & 15\\ 1, & -16, & -4, & -4, & -4\\ 2, & 10, & 18, & 26, & 34\\ 1, & 3, & 5, & 7, & 9\end{bmatrix}. \tag{4.96 b}$$

It can be verified that the ranks of the matrices $\mathbf{Q}_1$ and $\mathbf{Q}_2$ are both four ($m_1=m_2=4$), and furthermore that

$$\mathbf{A}^4\mathbf{b}_1=\mathbf{A}^3\mathbf{b}_1,$$

and

$$\mathbf{A}^4\mathbf{b}_2=2\mathbf{A}^3\mathbf{b}_2-\mathbf{A}^2\mathbf{b}_2.$$

Equations (4.93) therefore assume the forms

$$\lambda^3(\lambda-1)=0$$

and

$$\lambda^2(\lambda-1)^2=0.$$

These equations indicate that the Jordan block, $\mathbf{J}_3(0)$, is controllable by the first input and partially controllable by the second input, and that the Jordan block, $\mathbf{J}_2(1)$, is partially controllable by the first input and controllable by the second input.

4.4. Mode-observability matrices

4.4.1. *Distinct eigenvalues*

In the case when the plant matrix, $\mathbf{A}$, in eqn. (4.2) has n distinct eigenvalues, the vector-matrix eqn. (4.6) is equivalent to the n uncoupled scalar eqns. (4.11). In addition, it follows from eqn. (4.7) that

$$y_j(t)=\sum_{i=1}^{n} r_{ij}\xi_i(t), \quad (j=1, 2, \ldots, q) \tag{4.97}$$

and it is evident from this latter equation that the ith element, $\xi_i(t)$, of the modal state vector will be present in the jth output variable, $y_j(t)$, if and only if

$$r_{ij}=\mathbf{u}_i{}'\mathbf{c}_j\neq 0,$$

where $\mathbf{u}_i$ is the ith column of $\mathbf{U}$ and $\mathbf{c}_j$ is the jth column of $\mathbf{C}$. The ith mode of the system (4.2) and (4.3) is therefore *observable* in the jth output variable if and only if $r_{ij} \neq 0$, and is *unobservable* in the jth output variable if and only if $r_{ij}=0$: the ith mode is *observable* if and only if it is observable in at least one output variable. The system (4.11) and (4.97) is *observable* if and only if each of the n modes is observable.

These observability conditions can be illustrated by considering a system for which

$$\mathbf{J}=\text{diag}\{2, \; -1{\cdot}5, \; 0{\cdot}4\} \tag{4.98 a}$$

and

$$\mathbf{R}=\begin{bmatrix} 0, & 1 \\ -1, & 2 \\ 4, & -5 \end{bmatrix}. \tag{4.98 b}$$

Since $r_{11}=0$, $r_{21}\neq 0$, and $r_{31}\neq 0$, only the second and third modes are observable in $y_1(t)$ but, since $r_{12}\neq 0$, $r_{22}\neq 0$, and $r_{32}\neq 0$, all three modes are observable in $y_2(t)$.

4.4.2. *Confluent eigenvalues*

In the manner of § 4.4.1, the theoretical results and definitions presented in §§ 4.2.2 to 4.2.4 concerning the mode- and system-controllability conditions for systems having plant matrices with confluent eigenvalues can also be dualized to obtain the corresponding mode- and system-observability characteristics of such systems.

It is important to note that these conditions for system observability can be related to the well-known conditions for state-observability [2], [3]. In fact, it can be shown in the manner of § 4.5 that systems which are observable in the sense of §§ 4.4.1 and 4.4.2 are *state-observable* [2], [3] if and only if the appropriate $n \times nq$ matrix

$$\mathbf{S}=[\mathbf{C}, \mathbf{A}'\mathbf{C}, \ldots, (\mathbf{A}')^{n-1}\mathbf{C}] \tag{4.99}$$

is of rank n, in which case $(\mathbf{A}, \mathbf{C})$ is called an *observable pair*.

4.5. Mode-observability structure of multivariable linear systems

4.5.1 *Introduction*

The use of the matrix $\mathbf{S}$ defined in eqn. (4.99) as a test for state-observability does not, of course, provide any information as to the details of the mode-observability structure of a system governed by state and output equations of the respective forms (4.2) and (4.3). However, it is shown in this section that it is possible to use certain properties of $\mathbf{S}$ to determine which modes (if any) of the system (4.2) and (4.3) are observable in a given component $y_i(t)$ $(i=1, 2, \ldots, q)$ of the output vector, $\mathbf{y}(t)$.

4.5.2. *Distinct eigenvalues*

Thus, let the matrix $\mathbf{C}$ in eqn. (4.3) be written in the partitioned form

$$\mathbf{C}=[\mathbf{c}_1, \mathbf{c}_2, \ldots, \mathbf{c}_q] \tag{4.100}$$

so that eqn. (4.2) can be written in the form

$$y_i(t) = \mathbf{c}_i{}'\mathbf{x}(t) \quad (i=1, 2, ..., q). \tag{4.101}$$

The observability characteristics of a given $y_i(t)$ $(i=1, 2, \ldots, q)$, can, of course, be determined by considering the rank of the $n \times n$ matrix

$$\mathbf{S}_i = [\mathbf{c}_i, \mathbf{A}'\mathbf{c}_i, ..., (\mathbf{A}')^{n-1}\mathbf{c}_i] \tag{4.102}$$

associated with the output variable $y_i(t)$. Since it is assumed that the eigenvalues of $\mathbf{A}$ are distinct, the vector $\mathbf{c}_i$ in eqn. (4.102) can be expressed in the form

$$\mathbf{c}_i = \sum_{j=1}^{n} r_{ji}\mathbf{v}_j, \tag{4.103}$$

where the $\mathbf{v}_j$ are n linearly independent eigenvectors of $\mathbf{A}'$ and the r_{ji} $(j=1, 2, \ldots, n)$ are the elements of the mode-observability matrix

$$\mathbf{r}_i = \mathbf{U}'\mathbf{c}_i, \tag{4.104}$$

where $\mathbf{U}$ is the modal matrix of $\mathbf{A}$.

Now, it can be verified that

$$\mathbf{S}_i = \mathbf{V}\mathbf{R}_i\mathbf{G}, \tag{4.105}$$

where

$$\mathbf{R}_i = \text{diag}\,\{r_{1i}, r_{2i}, ..., r_{ni}\}, \tag{4.106}$$

and $\mathbf{G}$ is the $n \times n$ Vandermonde matrix

$$\begin{bmatrix} 1, & \lambda_1, & \lambda_1{}^2, ..., & \lambda_1{}^{n-1} \\ 1, & \lambda_2, & \lambda_2{}^2, ..., & \lambda_2{}^{n-1} \\ \cdots & \cdots & \cdots & \cdots \\ \cdots & \cdots & \cdots & \cdots \\ 1, & \lambda_n, & \lambda_n{}^2, ..., & \lambda_n{}^{n-1} \end{bmatrix}. \tag{4.107}$$

Since the $n \times n$ matrices $\mathbf{V}$ and $\mathbf{G}$ in eqn. (4.105) each has rank n in the case when $\mathbf{A}$ has distinct eigenvalues, it follows that

$$\text{rank}\ \mathbf{S}_i = \text{rank}\ \mathbf{R}_i, \tag{4.108}$$

which clearly implies that rank $\mathbf{S}_i$ is equal to the number of non-zero elements r_{ji} $(j=1, 2, \ldots, n)$ in the $n \times n$ diagonal matrix $\mathbf{R}_i$. It will be recalled from § 4.4.1 that the jth mode of the system (4.2) and (4.3) is observable in $y_i(t)$ if and only if $r_{ji} \neq 0$ $(j=1, 2, \ldots, n)$. Thus, the rank of $\mathbf{S}_i$ is equal to the number of modes, n_i, of the system (4.2) which can be observed in the ith output $y_i(t)$. These modes can be characterized without any loss of generality by the sequence of sets of eigenvalues and eigenvectors $\{\lambda_j, \mathbf{u}_j\}$ $(j=1, 2, ..., n_i)$. Equation (4.103) then assumes the form

$$\mathbf{c}_i = \sum_{j=1}^{n_i} r_{ji}\mathbf{v}_j \tag{4.109}$$

so that

$$(\mathbf{A}')^s \mathbf{c}_i = \sum_{j=1}^{n_i} r_{ji} \lambda^s \mathbf{v}_j \quad (s = 1, 2, \ldots, n_i). \tag{4.110}$$

Since rank $\mathbf{S}_i = n_i$, it follows that

$$(\mathbf{A}')^{n_i} \mathbf{c}_i = \sum_{k=0}^{n_i - 1} \beta_{ki} (\mathbf{A}')^k \mathbf{c}_i, \tag{4.111}$$

where the β_{ki} are the constants which express the linear dependence of $(\mathbf{A}')^{n_i}\mathbf{c}_i$ on $\mathbf{c}_i$, $\mathbf{A}'\mathbf{c}_i, \ldots, (\mathbf{A}')^{n_i - 1}\mathbf{c}_i$. It can now be deduced from eqns. (4.110) and (4.111) that

$$\sum_{j=1}^{n_i} r_{ji} \lambda_j^{n_i} \mathbf{v}_j = \sum_{k=0}^{n_i-1} \beta_{ki} \left\{ \sum_{j=1}^{n_i} r_{ji} \lambda_j^k \mathbf{v}_j \right\} \tag{4.112}$$

which implies that

$$r_{ji} \lambda_j^{n_i} = \sum_{k=0}^{n_i - 1} \beta_{ki} r_{ji} \lambda_j^k \quad (i = 1, 2, \ldots, n\,;\ j = 1, 2, \ldots, n_i). \tag{4.113}$$

Since $r_{ji} \neq 0$ $(j = 1, 2, \ldots, n_i)$, eqns. (4.113) indicate that the eigenvalues associated with the n_i modes of the system (4.2) which can be observed in the output variable $y_i(t)$ satisfy the equation

$$\lambda^{n_i} = \sum_{k=0}^{n_i - 1} \beta_{ki} \lambda^k. \tag{4.114}$$

In order to identify the observable modes of the system by determining the associated eigenvalues, it is thus necessary only to determine the β_{ki} in eqn. (4.111) and thus the roots of the n_ith-order polynomial eqn. (4.114).

As a simple numerical illustration of this result consider a system of the class (4.2) and (4.3) for which

$$\mathbf{A} = \begin{bmatrix} 0, & 1, & 0 \\ 0, & 0, & 1 \\ -8, & -14, & -7 \end{bmatrix} \tag{4.115 a}$$

and

$$\mathbf{C} = \begin{bmatrix} 2, & 2 \\ -3, & 3 \\ -2, & 1 \end{bmatrix}. \tag{4.115 b}$$

It can be readily verified that the eigenvalues of **A** are distinct and therefore that the method described in this section is applicable. In this example,

$$\mathbf{A}'\mathbf{C} = \begin{bmatrix} 16, & -8 \\ 30, & -12 \\ 11, & -4 \end{bmatrix} \tag{4.116 a}$$

and

$$(\mathbf{A}')^2\mathbf{C} = \begin{bmatrix} -88, & 32 \\ -138, & 48 \\ -47, & 16 \end{bmatrix} \tag{4.116 b}$$

so that it may be deduced from eqns. (4.115) and (4.116) that

$$\mathbf{S}_1 = \begin{bmatrix} 2, & 16, & -88 \\ -3, & 30, & -138 \\ -2, & 11, & -47 \end{bmatrix} \qquad (4.117\,a)$$

and

$$\mathbf{S}_2 = \begin{bmatrix} 2, & -8, & 32 \\ 3, & -12, & 48 \\ 1, & -4, & 16 \end{bmatrix}. \qquad (4.117\,b)$$

It can be verified that the matrices $\mathbf{S}_1$ and $\mathbf{S}_2$ are of rank two ($n_1 = 2$) and rank one ($n_2 = 1$) respectively, which implies that the system defined by eqns. (4.110) has two modes which can be observed in the output $y_1(t)$ and one mode which can be observed in the output $y_2(t)$. The eigenvalues of these modes can be found from eqns. (4.111) : in the present numerical example these equations are

$$(\mathbf{A}')^2\mathbf{c}_1 = \beta_{11}\mathbf{A}'\mathbf{c}_1 + \beta_{01}\mathbf{c}_1 \qquad (4.118\,a)$$

and

$$\mathbf{A}'\mathbf{c}_2 = \beta_{02}\mathbf{c}_2. \qquad (4.118\,b)$$

It is obvious by inspecting eqns. (4.115 *b*), (4.116) and (4.118) that $\beta_{11}\ -5$, $\beta_{01} = -4$, and $\beta_{02} = -4$.

Equation (4.114) thus indicates that the eigenvalues of the two modes which are observable in the first output, $y_1(t)$, are -1 and -4, and that the eigenvalue of the mode which is observable in the second output, $y_2(t)$, is -4.

4.5.3. *Confluent eigenvalues*

The theoretical results presented in §§ 4.3.3 and 4.3.4 concerning the mode-controllability characteristics for systems having plant matrices with confluent eigenvalues may readily be dualized to obtain the corresponding mode observability characteristics of such systems.

4.6. Discrete-time systems

The results presented in the foregoing sections of this chapter concerning the controllability and observability characteristics of continuous-time systems may be used (by analogy) in connection with discrete-time systems governed by state and output equations of the respective forms

$$\mathbf{x}\{(k+1)T\} = \mathbf{\Psi}(T)\mathbf{x}(kT) + \mathbf{\Delta}(T)\mathbf{z}(kt) \qquad (4.119)$$

and

$$\mathbf{y}(kT) = \mathbf{\Gamma}'(T)\mathbf{x}(kT). \qquad (4.120)$$

Indeed, in the discrete-time case, eqns. (4.119) and (4.120) indicate that the appropriate plant matrix is $\mathbf{\Psi}(T)$ instead of $\mathbf{A}$, the input matrix is $\mathbf{\Delta}(T)$ instead of $\mathbf{B}$, and that the output matrix is $\mathbf{\Gamma}'(T)(=\mathbf{C}')$. Furthermore, the results presented in Chapter 2 indicate that, in the discrete-time case, the eigenvalues of the plant matrix are exp $(\lambda_i T)$ $(i=1, 2, \ldots, n)$ instead of λ_i $(i=1, 2, \ldots, n)$ whilst the corresponding eigenvectors $\mathbf{u}_i$ and $\mathbf{v}_i$ $(i=1, 2, \ldots, n)$ are unchanged, Thus, the mode-controllability matrix of the discrete-time system (4.119) corresponding to $\mathbf{P}$ is $\mathbf{P}(T)=\mathbf{V}'\mathbf{\Delta}(T)$, and the system-controllability matrix corresponding to $\mathbf{Q}$ is

$$\mathbf{Q}(T)=[\mathbf{\Delta}(T), \mathbf{\Psi}(T)\mathbf{\Delta}(T), \ldots, \{\mathbf{\Psi}(T)\}^{n-1}\mathbf{\Delta}(T)]. \tag{4.121}$$

Similarly, the mode-observability matrix corresponding to $\mathbf{R}$ is $\mathbf{R}(T)=\mathbf{U}'\mathbf{\Gamma}(T)$, and the system-observability matrix corresponding to $\mathbf{S}$ is

$$\mathbf{S}(T)=[\mathbf{\Gamma}(T), \mathbf{\Psi}'(T)\mathbf{\Gamma}(T), \ldots, \{\mathbf{\Psi}'(T)\}^{n-1}\mathbf{\Gamma}(T)]. \tag{4.122}$$

The controllability and observability characteristics of the class of discrete-time systems governed by eqns. (4.119) and (4.120) may accordingly be inferred from the properties of the matrices $\mathbf{P}(T)$, $\mathbf{Q}(T)$, $\mathbf{R}(T)$, and $\mathbf{S}(T)$ in precisely the same way in which the corresponding characteristics of the class of continuous-time systems governed by eqns. (4.2) and (4.3) are inferred from the properties of the matrices $\mathbf{P}$, $\mathbf{Q}$, $\mathbf{R}$, and $\mathbf{S}$.

In this connection it is important to note that in the case of the governing equations of discrete-time systems obtained from continuous-time systems in the manner of Chapter 2, the eigenvalues exp $(\lambda_i T)$ of the plant matrix $\mathbf{\Psi}(T)$ will only be distinct provided that one of the following conditions obtains :

$$(a)\ \sigma_i \neq \sigma_j \quad (i, j=1, 2, \ldots, n) \tag{4.123 a}$$

$$(b)\ T \neq 2\pi k/(\omega_i-\omega_j) \quad (k=0, \pm 1, \pm 2, \ldots), \tag{4.123 b}$$

$$\text{if } \sigma_i=\sigma_j \quad (i \neq j\ ;\ i, j=1, 2, \ldots, n).$$

In conditions (4.123), $\lambda_l=\sigma_l+\omega_l\sqrt{-1}$.

References

[1] Simon, J. D., and Mitter, S. K., 1968, " A theory of modal control ", *Inf. Control*, **13,** 316.

[2] Kalman, R. E., 1961, " On the general theory of control systems ", *Proc. 1st IFAC Congress*, Moscow, (London : Butterworths), **1,** 481.

[3] Gilbert, E. G., 1963, " Controllability and observability in multivariable control systems ", *J. Soc. ind. appl. Math. Control*, A, **1,** 128.

[4] Aitken, A. C., 1939, *Determinants and Matrices* (Oliver & Boyd).

Single-input modal control systems

CHAPTER 5

5.1. Introduction

EQUATIONS (1.1.) and (1.3) indicate that the state equation of a single-input linear plant will have the form

$$\dot{\mathbf{x}}(t) = \mathbf{A}\mathbf{x}(t) + \mathbf{b}z(t) \tag{5.1 a}$$

in the continuous-time domain, and the corresponding form

$$\mathbf{x}\{(k+1)T\} = \mathbf{\Psi}(T)\mathbf{x}(kT) + \boldsymbol{\delta}(T)z(kT) \tag{5.1 b}$$

in the discrete-time domain. It will be recalled from Chapter 2 that the free response of an uncontrolled plant of either class is given by a linear combination of the dynamical modes of the system, where the mode shapes are determined by the eigenvectors and the time-domain characteristics by the eigenvalues of the appropriate plant matrix, $\mathbf{A}$ or $\mathbf{\Psi}(T)$.

However, if control loops are introduced which generate the input vector by linear feedback of the state vector of the plant, then the response characteristics of the resulting closed-loop system will no longer be determined by the eigenproperties of $\mathbf{A}$ or $\mathbf{\Psi}(T)$, but by those of some new closed-loop plant matrix whose eigenvectors and eigenvalues will depend upon the precise nature of the feedback loops. It transpires that, by introducing appropriate feedback loops, it is possible to design a closed-loop system whose plant matrix is such that those of its eigenvalues which correspond to the controllable modes of the uncontrolled system can be assigned new values which lead to closed-loop response characteristics that are superior to the corresponding characteristics of the original uncontrolled plant.

5.2. Continuous-time systems : plant matrices with distinct eigenvalues

5.2.1. *Single-mode control*

If all the elements of the state vector, $\mathbf{x}(t)$, can be measured by appropriate transducers, it will be possible to combine these transducer outputs to generate a signal given by an equation of the form

$$s(t)=\sum_{k=1}^{n} \mu_k x_k(t)=\boldsymbol{\mu}'\mathbf{x}(t), \tag{5.2}$$

where $\boldsymbol{\mu}$ is a measurement vector and the prime denotes matrix transposition. This signal, $s(t)$, can then be amplified by a single proportional controller having a gain K, thus yielding an input variable, $z(t)$, given by

$$z(t)=Ks(t)=K\boldsymbol{\mu}'\mathbf{x}(t). \tag{5.3}$$

It follows by substituting the expression for $z(t)$ given in eqn. (5.3) into eqn. (5.1 *a*) that the governing equation of the resulting closed-loop system is

$$\dot{\mathbf{x}}(t)=\mathbf{A}\mathbf{x}(t)+K\mathbf{b}\boldsymbol{\mu}'\mathbf{x}(t)=\mathbf{C}\mathbf{x}(t). \tag{5.4}$$

Equation (5.4) indicates that the effect of the input variable defined by eqn. (5.3) is to change the plant matrix $\mathbf{A}$ to a new matrix $\mathbf{C}$ given by

$$\mathbf{C}=\mathbf{A}+K\mathbf{b}\boldsymbol{\mu}'. \tag{5.5}$$

If $\boldsymbol{\mu}$ is now chosen to be equal to $\mathbf{v}_j$, the jth eigenvector of the matrix $\mathbf{A}'$, then the plant matrix of the controlled system given by eqn. (5.5) will have the form

$$\mathbf{C}=\mathbf{A}+K\mathbf{b}\mathbf{v}_j'. \tag{5.6}$$

It then follows from eqns. (2.14) and (5.6) that

$$\mathbf{C}\mathbf{u}_k=\mathbf{A}\mathbf{u}_k=\lambda_k\mathbf{u}_k, \tag{5.7}$$

indicating that $\mathbf{u}_k$ and λ_k $(k\neq j,\ k=1, 2, \ldots, n)$ are eigenvectors and eigenvalues of $\mathbf{C}$ as well as of $\mathbf{A}$. It is also evident from eqns. (2.16) and (5.6) that

$$\mathbf{C}\mathbf{u}_j=\mathbf{A}\mathbf{u}_j+K\mathbf{b}=\lambda_j\mathbf{u}_j+K\mathbf{b}, \tag{5.8}$$

which implies that λ_j is not an eigenvalue of $\mathbf{C}$ and also that $\mathbf{u}_j$ is no longer the corresponding eigenvector if $K\neq 0$. The effect of using the measurement vector $\mathbf{v}_j$ is therefore to change the eigenvalue λ_j to some new value ρ_j and the eigenvector $\mathbf{u}_j$ to some corresponding new vector $\mathbf{w}_j$, leaving the remaining $(n-1)$ pairs of eigenvalues and eigenvectors of the plant matrix of the uncontrolled system unchanged.

Since the n eigenvectors of $\mathbf{A}$ are linearly independent, it follows that the input vector can be expressed in the form

$$\mathbf{b}=\sum_{k=1}^{n} p_k\mathbf{u}_k, \tag{5.9}$$

where it will be recalled from Chapter 4 that the scalars p_k $(k=1, 2, \ldots, n)$ are the elements of the mode-controllability matrix of the single-input system (5.1 *a*). It can be deduced from eqns. (2.6), (2.17), (5.6), and (5.9) that

$$\mathbf{C}'\mathbf{v}_k=\lambda_k\mathbf{v}_k+Kp_k\mathbf{v}_j \quad (k\neq j,\ k=1, 2, \ldots, n) \tag{5.10}$$

and

$$\mathbf{C}'\mathbf{v}_j=(\lambda_j+Kp_j)\mathbf{v}_j, \tag{5.11}$$

which indicates that

$$\rho_j = \lambda_j + K p_j. \tag{5.12}$$

If λ_j is real, then the vector $\mathbf{v}_j$ and the scalar p_j will also be real, and the feedback loop incorporating $\mathbf{v}_j$ as a measurement vector will therefore by physically realizable (in principle, at least) : eqn. (5.12) implies that the proportional-controller gain necessary to alter the jth eigenvalue to any desired real value is given by the expression

$$K = (\rho_j - \lambda_j)/p_j, \tag{5.13}$$

if $p_j \neq 0$, that is, if the jth mode is controllable.

5.2.2. *Multi-mode control*

The analysis presented in § 5.2.1 can be readily extended in order to determine a control law for the single-input variable, $z(t)$, which will alter more than one eigenvalue of the plant matrix of the uncontrolled system, provided, of course, that the appropriate modes are controllable.

Thus, if the control signals given by the expressions

$$s_j(t) = \sum_{k=1}^{n} \mu_{jk} x_k = \boldsymbol{\mu}_j' \mathbf{x}(t) \quad (j = 1, 2, \ldots, m) \tag{5.14}$$

are amplified by m proportional controllers having gains, K_j, yielding an input variable, $z(t)$, given by

$$z(t) = \sum_{j=1}^{m} K_j \boldsymbol{\mu}_j' \mathbf{x}(t), \tag{5.15}$$

then it follows by substituting the expression for $z(t)$ given in eqn. (5.15) into eqn. (5.1 *a*) that the effect of the input variable (5.15) is to change the plant matrix $\mathbf{A}$ to a new plant matrix $\mathbf{C}$ given by

$$\mathbf{C} = \mathbf{A} + \mathbf{b} \sum_{j=1}^{m} K_j \boldsymbol{\mu}_j'. \tag{5.16}$$

If $\boldsymbol{\mu}_j$ is now chosen to be equal to $\mathbf{v}_j$, the jth eigenvector of the matrix $\mathbf{A}'$, then the plant matrix of the controlled system given by eqn. (5.16) will have the form

$$\mathbf{C} = \mathbf{A} + \mathbf{b} \sum_{j=1}^{m} K_j \mathbf{v}_j', \tag{5.17}$$

where the set $\{\mathbf{v}_1, \mathbf{v}_2, \ldots, \mathbf{v}_m\}$ consists of an appropriate selection of m eigenvectors of $\mathbf{A}'$.

It is thus evident from eqns. (2.14) and (5.17) that if

$$m + 1 \leqslant k \leqslant n,$$

then

$$\mathbf{C}\mathbf{u}_k = \mathbf{A}\mathbf{u}_k = \lambda_k \mathbf{u}_k, \tag{5.18}$$

indicating that $\mathbf{u}_k$ and λ_k $(k = m+1, m+2, \ldots, n)$ are eigenvectors and eigenvalues of $\mathbf{C}$ as well as of $\mathbf{A}$. It is also evident from eqns. (2.16) and (5.17) that if

$$1 \leqslant j \leqslant m,$$

then

$$\mathbf{Cu}_j = \mathbf{Au}_j + \mathbf{b}K_j = \lambda_j \mathbf{u}_j + \mathbf{b}K_j, \tag{5.19}$$

which implies that λ_j is not an eigenvalue of $\mathbf{C}$ and also that $\mathbf{u}_j$ is no longer the corresponding eigenvector if $K_j \neq 0$. The effect of using m measurement vectors $\mathbf{v}_j$ $(j=1, 2, \ldots, m)$ is therefore to change the eigenvalues λ_j to some new values ρ_j and the eigenvectors $\mathbf{u}_j$ to some corresponding new vectors $\mathbf{w}_j$, leaving the remaining $(n-m)$ pairs of eigenvalues and eigenvectors of the plant matrix of the uncontrolled system unchanged.

In order to calculate values of the proportional-controller gains, K_j, in terms of the eigenvalues of $\mathbf{A}$ and the required eigenvalues of $\mathbf{C}$, let

$$\mathbf{w}_j = \sum_{k=1}^{n} q_{jk} \mathbf{u}_k \quad (j=1, 2, \ldots, m). \tag{5.20}$$

It follows from eqns. (2.14), (2.16), and (5.9) that the appropriate elements of the mode-controllability matrix are

$$p_j = \mathbf{v}_j' \mathbf{b} \quad (j=1, 2, \ldots, n). \tag{5.21}$$

Since, by definition, ρ_j and $\mathbf{w}_j$ satisfy the equation

$$\mathbf{Cw}_j = \rho_j \mathbf{w}_j \quad (j=1, 2, \ldots, m), \tag{5.22}$$

substituting from eqns. (5.17) and (5.20) into eqn. (5.22) yields the expression

$$\left(\mathbf{A} + \mathbf{b} \sum_{l=1}^{m} K_l \mathbf{v}_l'\right) \sum_{k=1}^{n} q_{jk} \mathbf{u}_k = \rho_j \sum_{k=1}^{n} q_{jk} \mathbf{u}_k \quad (j=1, 2, \ldots, m). \tag{5.23}$$

In eqn. (5.23),

$$\mathbf{A} \sum_{k=1}^{n} q_{jk} \mathbf{u}_k = \sum_{k=1}^{n} q_{jk} \lambda_k \mathbf{u}_k \quad (j=1, 2, \ldots, m) \tag{5.24}$$

in view of eqns. (2.2) : also, in eqn. (5.23),

$$\mathbf{b} \sum_{l=1}^{m} K_l \mathbf{v}_l' \sum_{k=1}^{n} q_{jk} \mathbf{u}_k = \mathbf{b} \sum_{l=1}^{m} K_l q_{jl} = \sum_{k=1}^{n} p_k \mathbf{u}_k \sum_{l=1}^{m} K_l q_{jl} \quad (j=1, 2, \ldots, m) \tag{5.25}$$

in view of eqns. (2.14), (2.16), and (5.9). Equations (5.24) and (5.25) therefore indicate that eqn. (5.23) has the form

$$\sum_{k=1}^{n} q_{jk} \lambda_k \mathbf{u}_k + \sum_{k=1}^{n} p_k \mathbf{u}_k \sum_{l=1}^{m} K_l q_{jl} = \rho_j \sum_{k=1}^{n} q_{jk} \mathbf{u}_k \quad (j=1, 2, \ldots, m). \tag{5.26}$$

Equation (5.26) is a vector equation in the $\mathbf{u}_k$ and is equivalent to the $n \times m$ scalar equations

$$(\rho_j - \lambda_k) q_{jk} - p_k \sum_{l=1}^{m} K_l q_{jl} = 0 \quad (j=1, 2, \ldots, m\,;\ k=1, 2, \ldots, n). \tag{5.27}$$

It is evident that, for a given value of j, the m equations involving $\lambda_1, \lambda_2, \ldots,$ and λ_m can be written in the matrix form

$$\mathbf{F}_j \mathbf{q}_j = \mathbf{0}, \tag{5.28 a}$$

where

$$\mathbf{F}_j = [f_{kl}^{(j)}], \tag{5.28 b}$$

$$f_{kl}^{(j)} = (\rho_j - \lambda_k)\delta_{kl} - p_k K_l, \tag{5.28 c}$$

$$\mathbf{q}_j = [q_{jl}], \tag{5.28 d}$$

and δ_{kl} is the Kronecker delta. Since $\mathbf{q}_j \neq \mathbf{0}$, it follows from eqn. (5.28 *a*) that

$$|\mathbf{F}_j| = 0, \tag{5.29}$$

which, in view of (5.28 *b*) and (5.28 *c*) has the explicit form

$$\begin{vmatrix} (\rho_j - \lambda_1) - p_1 K_1, & -p_1 K_2, \ldots, & -p_1 K_m \\ -p_2 K_1, & (\rho_j - \lambda_2) - p_2 K_2, \ldots, & -p_2 K_m \\ \vdots & \vdots & \vdots \\ -p_m K_1, & -p_m K_2, \ldots, (\rho_j - \lambda_m) - p_m K_m \end{vmatrix} = 0 \quad (j = 1, 2, \ldots, m). \tag{5.30}$$

Expansion of the determinant in (5.30) yields the equation

$$\prod_{k=1}^{m} (\rho_j - \lambda_k) = \sum_{l=1}^{m} K_l p_l \prod_{\substack{k=1 \\ k \neq l}}^{m} (\rho_j - \lambda_k) \quad (j = 1, 2, \ldots, m) \tag{5.31}$$

which implies that

$$\sum_{k=1}^{m} \left(\frac{K_k p_k}{\rho_j - \lambda_k} \right) = 1 \quad (j = 1, 2, \ldots, m) \tag{5.32}$$

provided that

$$\rho_j - \lambda_k \neq 0 \quad (j, k = 1, 2, \ldots, m).$$

Equations (5.32) can be solved for the K_j to give

$$K_j = \prod_{k=1}^{m} (\rho_k - \lambda_j) \Big/ \Bigg(p_j \prod_{\substack{k \neq j \\ k=1}}^{m} (\lambda_k - \lambda_j) \Bigg) \quad (j = 1, 2, \ldots, m), \tag{5.33}$$

which indicates that the gains, K_j, will be calculable if

$$p_j \neq 0 \quad (j = 1, 2, \ldots, m).$$

In view of eqn. (5.21), this condition is equivalent to the requirement that the eigenvalues λ_1, λ_2, . . . and λ_m are associated with controllable modes.

The control law obtained by substituting from eqn. (5.33) into eqn. (5.15) and by putting $\boldsymbol{\mu}_j = \mathbf{v}_j$ in the latter equation has the form

$$z(t) = \sum_{j=1}^{m} \left[\frac{\prod_{k=1}^{m} (\rho_k - \lambda_j) \mathbf{v}_j'}{p_j \prod_{\substack{k=1 \\ k \neq j}}^{m} (\lambda_k - \lambda_j)} \right] \mathbf{x}(t) = \mathbf{g}' \mathbf{x}(t). \tag{5.34}$$

This control law will alter the eigenvalues $\lambda_1, \lambda_2, \ldots, \lambda_m$ of the plant matrix, $\mathbf{A}$, of the uncontrolled system to prescribed new values $\rho_1, \rho_2, \ldots, \rho_m$, leaving the remaining $(n-m)$ eigenvalues $\lambda_{m+1}, \ldots, \lambda_n$ unaltered. It is important to note that, in the case of real systems, the control law defined in (5.34) will be real even in the case of complex eigenvalues. This follows from the fact that if $\lambda_k = \lambda_j^*$ then $\mathbf{v}_k = \mathbf{v}_j^*$ and $p_k = p_j^*$, together with the fact that the required new eigenvalues, ρ_k, will either occur in conjugate complex pairs or be real.

5.2.3. *Illustrative examples*

The theory of single-input modal control developed in the preceding section can be illustrated by synthesizing appropriate feedback loops for the third-order system for which the state eqn. (5.1 *a*) has the particular form

$$\dot{\mathbf{x}}(t) = \begin{bmatrix} -2, & -1, & 1 \\ 1, & 0, & 1 \\ -1, & 0, & 1 \end{bmatrix} \mathbf{x}(t) + \begin{bmatrix} 1 \\ 1 \\ 1 \end{bmatrix} z(t). \tag{5.35}$$

The eigenstructure of this system is given by the following matrices :

$$\mathbf{\Lambda} = \begin{bmatrix} 1, & 0, & 0 \\ 0, & -1+i, & 0 \\ 0, & 0, & -1-i \end{bmatrix}, \tag{5.36 a}$$

$$\mathbf{U} = \begin{bmatrix} 0, & 5, & 5 \\ 1, & -3-4i, & -3+4i \\ 1, & 2+i, & 2-i \end{bmatrix}, \tag{5.36 b}$$

and

$$\mathbf{V} = \tfrac{1}{10} \begin{bmatrix} -2, & 1+i, & 1-i \\ 2, & i, & -i \\ 8, & -i, & i \end{bmatrix}. \tag{5.36 c}$$

The corresponding mode-controllability matrix is

$$\mathbf{p} = \mathbf{V}'\mathbf{b} = \tfrac{1}{10} \begin{bmatrix} 8 \\ 1+i \\ 1-i \end{bmatrix}. \tag{5.37}$$

It is evident that, in the absence of control, the system (5.35) has eigenvalues $\lambda_1 = 1$, $\lambda_2 = -1+i$, and $\lambda_3 = -1-i$. The first mode is therefore unstable but controllable by the input $z(t)$ since $p_1 \neq 0$. The second and third modes are asymptotically stable and are also both controllable by $z(t)$ since $p_2 \neq 0$ and $p_3 \neq 0$: it will be noted that p_2 and p_3, corresponding to the conjugate complex eigenvalues λ_2 and λ_3, are themselves conjugate complex numbers.

If it is desired to change the unstable eigenvalue $\lambda_1 = 1$ to the value $\rho_1 = -3$, then the proportional-controller gain associated with the appropriate measurement vector, $\mathbf{v}_1'$, is given by

$$K = (\rho_1 - \lambda_1)/p_1 = 5(-3-1)/4 = -5, \tag{5.38}$$

using eqn. (5.13). Thus, eqn. (5.3) indicates that the required feedback control law is

$$z(t) = K\mathbf{v}_1'\mathbf{x}(t) = x_1(t) - x_2(t) - 4x_3(t). \tag{5.39}$$

It may readily be verified that the plant matrix of the closed-loop system defined by eqns. (5.35) and (5.39) is

$$\mathbf{C}=\begin{bmatrix}-1, & -2, & -3\\ 2, & -1, & -3\\ 0, & -1, & -3\end{bmatrix}$$

and that the eigenvalues of this matrix are $\rho_i=-3$, $\lambda_2=-1+i$, and $\lambda_3=-1-i$, as required.

If, more generally, it is now desired to change the unstable eigenvalue $\lambda_1=1$ to the value $\rho_1=-1$ and to change the eigenvalues $\lambda_2=-1+i$ and $\lambda_3=-1-i$ to the values $\rho_2=-2+2i$ and $\rho_3=-2-2i$, then the proportional-controller gains associated with the appropriate measurement vectors $\mathbf{v}_1'$, $\mathbf{v}_2'$ and $\mathbf{v}_3'$ are given by

$$K_1=\frac{(\rho_1-\lambda_1)(\rho_2-\lambda_1)(\rho_3-\lambda_1)}{p_1(\lambda_2-\lambda_1)(\lambda_3-\lambda_1)}=-6{\cdot}5, \tag{5.40 a}$$

$$K_2=\frac{(\rho_1-\lambda_2)(\rho_2-\lambda_2)(\rho_3-\lambda_2)}{p_2(\lambda_1-\lambda_2)(\lambda_3-\lambda_2)}=7+i, \tag{5.40 b}$$

and

$$K_3=\frac{(\rho_1-\lambda_3)(\rho_2-\lambda_3)(\rho_3-\lambda_3)}{p_3(\lambda_1-\lambda_3)(\lambda_2-\lambda_3)}=7-i, \tag{5.40 c}$$

using eqn. (5.33). Thus, eqns. (5.33) and (5.34) indicate that the required feedback control law is given by the expression

$$z(t)=K_1\mathbf{v}_1'\mathbf{x}(t)+K_2\mathbf{v}_2'\mathbf{x}(t)+K_3\mathbf{v}_3'\mathbf{x}(t),$$

which, in this case, assumes the form

$$z(t)=2{\cdot}5x_1(t)-1{\cdot}5x_2(t)-5x_3(t). \tag{5.41}$$

It may be readily verified that the plant matrix of the closed-loop system defined by eqns. (5.35) and (5.41) is

$$\mathbf{C}=\begin{bmatrix}0{\cdot}5, & -2{\cdot}5, & -4\\ 3{\cdot}5, & -1{\cdot}5, & -4\\ 1{\cdot}5, & -1{\cdot}5, & -4\end{bmatrix},$$

and that the eigenvalues of this matrix are -1, $-2+2i$, and $-2-2i$, as required.

5.3. Continuous-time systems : plant matrices with confluent eigenvalues

5.3.1. *Multi-Jordan block control*

In this section it is shown that the theory presented in § 5.2 can be extended to embrace single-input systems whose plant matrices have a number of sets of confluent eigenvalues. Thus, if $\mathbf{U}$ is a generalized modal matrix of the plant matrix $\mathbf{A}$ and $\mathbf{V}$ is the corresponding generalized modal matrix of $\mathbf{A}'$ for which

$$\mathbf{V}'\mathbf{U}=\mathbf{I}, \tag{5.42}$$

then

$$\mathbf{AU}=\mathbf{UJ}, \tag{5.43 a}$$

and

$$\mathbf{A'V}=\mathbf{VJ'}, \tag{5.43 b}$$

where $\mathbf{J}$ is a block-diagonal matrix whose typical ith block ($i=1, 2, \ldots, \nu$) is the $n_i \times n_i$ Jordan block

$$\mathbf{J}_{n_i}(\lambda_j)=\begin{bmatrix} \lambda_j, & 1, & 0, \ldots, & 0, & 0 \\ 0, & \lambda_j, & 1, \ldots, & 0, & 0 \\ \cdots & \cdots & \cdots & \cdots & \cdots \\ 0, & 0, & 0, \ldots, & \lambda_j, & 1 \\ 0, & 0, & 0, \ldots, & 0, & \lambda_j \end{bmatrix}, \tag{5.44}$$

where λ_j is a confluent eigenvalue of the matrix $\mathbf{A}$. In this section it will be assumed that the ν Jordan blocks are each associated with a distinct eigenvalue, that is, $\lambda_i \neq \lambda_j$; $i \neq j, i, j=1, 2, \ldots, \nu$.

The Jordan matrix $\mathbf{J}$ and the generalized modal matrices $\mathbf{U}$ and $\mathbf{V}$ in eqns. (5.42) and (5.43) can therefore be written in the respective forms

$$\mathbf{J}=\mathbf{J}_{n_1}(\lambda_1)\oplus\mathbf{J}_{n_2}(\lambda_2)\oplus\ldots\oplus\mathbf{J}_{n_\nu}(\lambda_\nu); \tag{5.45}$$

$$\mathbf{U}=[\mathbf{U}_1, \mathbf{U}_2, \ldots, \mathbf{U}_\nu], \tag{5.46}$$

and

$$\mathbf{V}=[\mathbf{V}_1, \mathbf{V}_2, \ldots, \mathbf{V}_\nu], \tag{5.47}$$

where the $n \times n_i$ sub-matrices $\mathbf{U}_i$ and $\mathbf{V}_i$ respectively have the forms

$$\mathbf{U}_i=[\mathbf{u}_1{}^{(i)}, \mathbf{u}_2{}^{(i)}, \ldots, \mathbf{u}_{n_i}{}^{(i)}] \tag{5.48}$$

and

$$\mathbf{V}_i=[\mathbf{v}_1{}^{(i)}, \mathbf{v}_2{}^{(i)}, \ldots, \mathbf{v}_{n_i}{}^{(i)}]. \tag{5.49}$$

The $n \times 1$ mode-controllability matrix of the system described by eqns. (5.1 a), (5.45), and (5.49) has the form

$$\mathbf{p}=\mathbf{V'b}=\begin{bmatrix} \mathbf{p}^{(1)} \\ \mathbf{p}^{(2)} \\ \cdot \\ \cdot \\ \mathbf{p}^{(\nu)} \end{bmatrix}, \tag{5.50}$$

where the $n_i \times 1$ mode-controllability sub-matrix $\mathbf{p}^{(i)}$ associated with the Jordan block, $\mathbf{J}_{n_i}(\lambda_i)$, is given by the expression

$$\mathbf{p}^{(i)}=\mathbf{V}_i{}'\mathbf{b}=\begin{bmatrix} p_1{}^{(i)} \\ p_2{}^{(i)} \\ \cdot \\ \cdot \\ p_{n_i}{}^{(i)} \end{bmatrix}. \tag{5.51}$$

It follows from eqns. (5.42), (5.50), and (5.51) that

$$\mathbf{b}=\sum_{i=1}^{\nu}\sum_{j=1}^{n_i} p_j^{(i)}\mathbf{u}_j^{(i)}. \tag{5.52}$$

If $p_{m_i}^{(i)} \neq 0$ and $p_j^{(i)}=0\,(j=m_i+1, m_i+2, \ldots, n_i)$ then the mode-controllability sub-matrix of a partially controllable Jordan block has the structure

$$\mathbf{p}^{(i)\prime}=[p_1^{(i)}, p_2^{(i)}, \ldots, p_{m_i}^{(i)}, 0, 0, \ldots, 0]$$

and eqn. (5.52) can therefore be reduced to the truncated form

$$\mathbf{b}=\sum_{i=1}^{\nu}\sum_{j=1}^{m_i} p_j^{(i)}\mathbf{u}_j^{(i)}. \tag{5.53}$$

This truncated form can be used to investigate the effect of the control law

$$z(t)=\sum_{i=1}^{\nu}\sum_{j=1}^{m_i} K_j^{(i)}\mathbf{v}_j^{(i)\prime}\mathbf{x}(t) \tag{5.54}$$

on the eigenstructure of the system (5.1 a). Thus, it follows by substituting the expression for $z(t)$ given in eqn. (5.54) into (5.1 a) that the state equation of the resulting closed-loop system is

$$\dot{\mathbf{x}}(t)=\mathbf{A}\mathbf{x}(t)+\mathbf{b}\sum_{i=1}^{\nu}\sum_{j=1}^{m_i} K_j^{(i)}\mathbf{v}_j^{(i)\prime}\mathbf{x}(t)=\mathbf{C}\mathbf{x}(t). \tag{5.55}$$

Equation (5.55) indicates that the effect of the input variable defined by eqn. (5.54) is to change the plant matrix $\mathbf{A}$ to a new matrix $\mathbf{C}$ given by the expression

$$\mathbf{C}=\mathbf{A}+\mathbf{b}\sum_{i=1}^{\nu}\sum_{j=1}^{m_i} K_j^{(i)}\mathbf{v}_j^{(i)\prime}. \tag{5.56}$$

It follows from eqns. (5.44), (5.46), and (5.47) that the eigenvectors of the system (5.1 a) satisfy the equations

$$\left.\begin{aligned}
&\mathbf{A}\mathbf{u}_1^{(i)}=\lambda_i\mathbf{u}_1^{(i)},\\
&\mathbf{A}\mathbf{u}_j^{(i)}=\lambda_i\mathbf{u}_j^{(i)}+\mathbf{u}_{j-1}^{(i)} \quad (j=2, 3, \ldots, n_i),\\
&\mathbf{A}'\mathbf{v}_j^{(i)}=\lambda_i\mathbf{v}_j^{(i)}+\mathbf{v}_{j+1}^{(i)} \quad (j=1, 2, \ldots, n_{i-1}),\\
&\mathbf{A}'\mathbf{v}_{n_i}^{(i)}=\lambda_i\mathbf{v}_{n_i}^{(i)},
\end{aligned}\right\} \quad (i=1, 2, \ldots, n_i), \tag{5.57}$$

where, in view of eqn. (5.42),

$$\left.\begin{aligned}
&\mathbf{v}_i^{(j)\prime}\mathbf{u}_k^{(l)}=0 \quad (j\neq l, j, l=1, 2, \ldots, \nu;\; i=1, 2, \ldots, n_j;\\
&\qquad\qquad\qquad k=1, 2, \ldots, n_l),\\
&\mathbf{v}_i^{(j)\prime}\mathbf{u}_k^{(j)}=0 \quad (j=1, 2, \ldots, \nu;\; i\neq k, i, k=1, 2, \ldots, n_j),\\
&\text{and}\\
&\mathbf{v}_i^{(j)\prime}\mathbf{u}_i^{(j)}=1 \quad (j=1, 2, \ldots, \nu;\; i=1, 2, \ldots, n_j).
\end{aligned}\right\} \tag{5.58}$$

It can now be deduced from eqns. (5.56), (5.57), and (5.58) that

$$\mathbf{C}\mathbf{u}_k^{(i)} = \mathbf{A}\mathbf{u}_k^{(i)} + \mathbf{b}\sum_{l=1}^{\nu}\sum_{j=1}^{m_l} K_j^{(l)}\mathbf{v}_j^{(l)\prime}\mathbf{u}_k^{(l)},$$

so that

$$\mathbf{C}\mathbf{u}_k^{(i)} = \lambda_i\mathbf{u}_k^{(i)} + \mathbf{u}_{k-1}^{(i)} \quad (m_i+1 \leqslant k \leqslant n_i). \tag{5.59}$$

Similarly,

$$\mathbf{C}'\mathbf{v}_k^{(i)} = \mathbf{A}'\mathbf{v}_k^{(i)} + \sum_{l=1}^{\nu}\sum_{j=1}^{m_l} K_j^{(l)}\mathbf{v}_j^{(l)}\mathbf{b}'\mathbf{v}_k^{(i)},$$

so that

$$\mathbf{C}'\mathbf{v}_k^{(i)} = \lambda_i\mathbf{v}_k^{(i)} + \mathbf{v}_{k+1}^{(i)} \quad (m_i+1 \leqslant k \leqslant n_i - 1) \tag{5.60 a}$$

and

$$\mathbf{C}'\mathbf{v}_{ni}^{(i)} = \lambda_i\mathbf{v}_{ni}^{(i)}. \tag{5.60 b}$$

Equations (5.59) and (5.60) indicate that the $\mathbf{u}_k^{(i)}$ $(m_i+1 \leqslant k \leqslant n_i;\ i=1, 2, \ldots, \nu)$ are eigenvectors of $\mathbf{C}$ as well as of $\mathbf{A}$, and that the $\mathbf{v}_k^{(i)}$ $(m_i+1 \leqslant k \leqslant n_i;\ i=1, 2, \ldots, \nu)$ are eigenvectors of $\mathbf{C}'$ as well as of $\mathbf{A}'$. In addition, the matrix $\mathbf{C}$ has ν Jordan blocks $\mathbf{J}_{ni-mi}(\lambda_i)$ associated with the eigenvalues λ_i $(i=1, 2, \ldots, \nu)$.

It is also evident from eqns. (5.56), (5.57), and (5.58) that

$$\mathbf{C}\mathbf{u}_k^{(i)} = \mathbf{A}\mathbf{u}_k^{(i)} + \mathbf{b}\sum_{l=1}^{\nu}\sum_{j=1}^{m_l} K_j^{(l)}\mathbf{v}_j^{(l)\prime}\mathbf{u}_k^{(i)},$$

so that

$$\mathbf{C}\mathbf{u}_1^{(i)} = \lambda_1\mathbf{u}_1^{(i)} + K_1^{(i)}\sum_{j=1}^{m_i} p_j^{(i)}\mathbf{u}_j^{(i)}, \tag{5.61 a}$$

and

$$\mathbf{C}\mathbf{u}_k^{(i)} = \lambda_k\mathbf{u}_k^{(i)} + \mathbf{u}_{k-1}^{(i)} + K_k^{(i)}\sum_{j=1}^{m_i} p_j^{(i)}\mathbf{u}_j^{(i)} \quad (2 \leqslant k \leqslant m_i). \tag{5.61 b}$$

Similarly,

$$\mathbf{C}'\mathbf{v}_k^{(i)} = \mathbf{A}'\mathbf{v}_k^{(i)} + \sum_{l=1}^{\nu}\sum_{j=1}^{m_l} K_j^{(l)}\mathbf{v}_j^{(l)}\mathbf{b}'\mathbf{v}_k^{(i)},$$

so that

$$\mathbf{C}'\mathbf{v}_k^{(i)} = \lambda_k\mathbf{v}_k^{(i)} + \mathbf{v}_{k+1}^{(i)} + p_k^{(i)}\sum_{l=1}^{\nu}\sum_{j=1}^{m_l} K_j^{(l)}\mathbf{v}_j^{(l)} \quad (k=1, 2, \ldots, m_i). \tag{5.62}$$

Equations (5.61) and (5.62) imply that λ_k $(k=1, 2, \ldots, m_i)$ is not an eigenvalue of $\mathbf{C}$ and that $\mathbf{u}_k$ and $\mathbf{v}_k$ are no longer the corresponding eigenvectors if $K_j^{(i)} \neq 0$ $(i=1, 2, \ldots, \nu;\ j=1, 2, \ldots, m_j)$. The effect of the control law (5.54) is therefore to change the eigenvalues λ_j $(j=1, 2, \ldots, \nu)$ to new distinct values $\rho_j^{(i)}$ $(i=1, 2, \ldots, \nu;\ j=1, 2, \ldots, m_i)$ and the eigenvectors $\mathbf{u}_j^{(i)}$ to corresponding new vectors $\mathbf{w}_j^{(i)}$ $(i=1, 2, \ldots, \nu;\ j=1, 2, \ldots, m_i)$.

In order to calculate values of the proportional-controller gains, $K_j^{(i)}$, in terms of the eigenvalues of $\mathbf{A}$ and the required eigenvalues of $\mathbf{C}$, let

$$\mathbf{w}_j^{(i)} = \sum_{k=1}^{\nu} \sum_{l=1}^{n_k} q_{jl}^{(k)} \mathbf{u}_l^{(k)} \quad (i=1, 2, \ldots, \nu\, ;\; j=1, 2, \ldots, m_i). \tag{5.63}$$

Since, by definition, $\rho_j^{(i)}$ and $\mathbf{w}_j^{(i)}$ satisfy the equation

$$\mathbf{C}\mathbf{w}_j^{(i)} = \rho_j^{(i)} \mathbf{w}_j^{(i)} \quad (i=1, 2, \ldots, \nu\, ;\; j=1, 2, \ldots, m_i) \tag{5.64}$$

substituting from eqns. (5.56) and (5.63) into eqn. (5.64) yields the equation

$$\left(\mathbf{A} + \mathbf{b} \sum_{s=1}^{\nu} \sum_{t=1}^{m_s} K_t^{(s)} \mathbf{v}_t^{(s)\prime}\right) \sum_{k=1}^{\nu} \sum_{l=1}^{n_k} q_{jl}^{(k)} \mathbf{u}_l^{(k)} = \rho_j^{(i)} \sum_{k=1}^{\nu} \sum_{l=1}^{n_k} q_{jl}^{(k)} \mathbf{u}_l^{(k)}. \tag{5.65}$$

This equation may be simplified, using eqns. (5.53) and (5.58), to the form

$$\sum_{k=1}^{\nu} \left\{ \sum_{l=2}^{n_k} q_{jl}^{(k)} (\lambda_k \mathbf{u}_l^{(k)} + \mathbf{u}_{l-1}^{(k)}) + q_{j1}^{(k)} \lambda_k \mathbf{u}_1^{(k)} \right\}$$
$$+ \left\{ \sum_{k=1}^{\nu} \sum_{l=1}^{m_k} p_l^{(k)} \mathbf{u}_l^{(k)} \right\} \left\{ \sum_{s=1}^{\nu} \sum_{t=1}^{m_s} q_{jt}^{(s)} K_t^{(s)} \right\} = \rho_j^{(i)} \sum_{k=1}^{\nu} \sum_{l=1}^{n_k} q_{jl}^{(k)} \mathbf{u}_l^{(k)}$$
$$(i=1, 2, \ldots, \nu\, ;\; j=1, 2, \ldots, m_i). \tag{5.66}$$

Equation (5.66) is a vector equation in the $\mathbf{u}_l^{(k)}$ and is thus equivalent to the scalar equations

$$(\rho_j^{(i)} - \lambda_k) q_{jl}^{(k)} - q_{j,\,l+1}^{(k)} - p_l^{(k)} \sum_{s=1}^{\nu} \sum_{t=1}^{m_s} q_{jt}^{(s)} K_t^{(s)} = 0$$
$$(i, k=1, 2, \ldots, \nu\, ;\; j=1, 2, \ldots, m_i\, ;\; l=1, 2, \ldots, m_k), \tag{5.67 a}$$

and

$$(\rho_j^{(i)} - \lambda_k) q_{jl}^{(k)} - q_{j,\,l+1}^{(k)} = 0 \quad (i, k=1, 2, \ldots, \nu\, ;\; j=1, 2, \ldots, m_i\, ;$$
$$l = m_k+1, m_k+2, \ldots, n_k). \tag{5.67 b}$$

In order to reduce the number of equations in the set (5.67) it has been assumed that the coefficient $q_{jl}^{(k)}$ is zero if the l-subscript is outside the permissible range.

It is evident that, for given values of i and j, eqns. (5.67) can be written in the form

$$\mathbf{F}^{(ij)} \mathbf{q}^{(j)} = \mathbf{0} \quad (i=1, 2, \ldots, \nu\, ;\; j=1, 2, \ldots, m_i), \tag{5.68}$$

or, in the equivalent partitioned form

$$\begin{bmatrix} \mathbf{F}_{11}^{(ij)}, & \mathbf{F}_{12}^{(ij)} \\ \mathbf{F}_{21}^{(ij)}, & \mathbf{F}_{22}^{(ij)} \end{bmatrix} \begin{bmatrix} \mathbf{q}_1^{(j)} \\ \mathbf{q}_2^{(j)} \end{bmatrix} = \begin{bmatrix} \mathbf{0} \\ \mathbf{0} \end{bmatrix}. \tag{5.69}$$

In eqn. (5.59), the sub-matrices $\mathbf{F}_{st}^{(ij)}$ $(s, t=1,\ 2)$ and the sub-vectors $\mathbf{q}_s^{(j)}$ $(s=1, 2)$ may be partitioned further into $\nu \times \nu$ sub-matrices $\mathbf{F}_{st}^{(ij)}(k, l)$ and $\nu \times 1$ sub-vectors $\mathbf{q}_s^{(j)}(k)$, respectively, using the notation

$$\mathbf{F}_{st}^{(ij)} = [\mathbf{F}_{st}^{(ij)}(k, l)] \quad (k, l=1, 2, \ldots, \nu). \tag{5.70}$$

and

$$\mathbf{q}_s^{(j)} = [\mathbf{q}_s^{(j)}(k)] \quad (k=1, 2, \ldots, \nu). \tag{5.71}$$

In eqn. (5.69), $\mathbf{F}_{11}^{(ij)}(k, k)$ is the $m_k \times m_k$ matrix

$$\begin{bmatrix} \rho_j^{(i)} - \lambda_k - K_1^{(k)} p_1^{(k)}, & -1 - K_2^{(k)} p_1^{(k)}, & \ldots, & -K_{m_k}^{(k)} p_1^{(k)} \\ -K_1^{(k)} p_2^{(k)}, & \rho_j^{(i)} - \lambda_k - K_2^{(k)} p_2^{(k)}, & \ldots, & -K_{m_k}^{(k)} p_2^{(k)} \\ \cdots & \cdots & \cdots & \cdots \\ -K_1^{(k)} p_{m_k - 1}^{(k)}, & -K_2^{(k)} p_{m_k-1}^{(k)}, & \ldots, & -1 - K_{m_k}^{(k)} p_{m_k-1}^{(k)} \\ -K_1^{(k)} p_{m_k}^{(k)}, & -K_2^{(k)} p_{m_k}^{(k)}, & \ldots, & \rho_j^{(i)} - \lambda_k - K_{m_k}^{(k)} p_{m_k}^{(k)} \end{bmatrix}$$

$$(k=1, 2, \ldots, \nu), \tag{5.72}$$

$\mathbf{F}_{11}^{(ij)}(k, l)$ is the $m_k \times m_l$ matrix

$$\begin{bmatrix} -K_1^{(l)} p_1^{(k)}, & -K_2^{(l)} p_1^{(k)}, & \ldots, & -K_{m_l}^{(l)} p_1^{(k)} \\ -K_1^{(l)} p_2^{(k)}, & -K_2^{(l)} p_2^{(k)}, & \ldots, & -K_{m_l}^{(l)} p_2^{(k)} \\ \cdots & \cdots & \cdots & \cdots \\ -K_1^{(l)} p_{m_k}^{(k)}, & -K_2^{(l)} p_{m_k}^{(k)}, & \ldots, & -K_{m_l}^{(l)} p_{m_k}^{(k)} \end{bmatrix}$$

$$(k \neq l, k, l=1, 2, \ldots, \nu), \tag{5.73}$$

$\mathbf{F}_{12}^{(ij)}(k, k)$ is the $m_k \times (n_k - m_k)$ matrix

$$\begin{bmatrix} 0, & 0, \ldots, & 0 \\ 0, & 0, \ldots, & 0 \\ -1, & 0, \ldots, & 0 \end{bmatrix} \quad (k=1, 2, \ldots, \nu), \tag{5.74}$$

$\mathbf{F}_{12}^{(ij)}(k, l)$ is an $m_k \times (n_l - m_l)$ null matrix

$$\mathbf{0}_{m_k, (n_l - m_l)} \quad (k \neq l, k, l=1, 2, \ldots, \nu), \tag{5.75}$$

$\mathbf{F}_{21}^{(ij)}(k, l)$ is the $(n_k - m_k) \times m_l$ null matrix

$$\mathbf{0}_{(n_k - m_k), m_l} \quad (k, l=1, 2, \ldots, \nu), \tag{5.76}$$

$\mathbf{F}_{22}^{(ij)}(k, k)$ is the $(n_k - m_k) \times (n_k - m_k)$ matrix

$$\begin{bmatrix} \rho_j^{(i)} - \lambda_k, & -1, & 0, \ldots, & 0, & 0 \\ 0, & \rho_j^{(i)} - \lambda_k, & -1, \ldots, & 0, & 0 \\ \cdots & \cdots & \cdots & \cdots & \cdots \\ 0, & 0, & 0, \ldots, & \rho_j^{(i)} - \lambda_k, & -1 \\ 0, & 0, & 0, \ldots, & 0, & \rho_j^{(i)} - \lambda_k \end{bmatrix}$$

$$(k=1, 2, \ldots, \nu), \tag{5.77}$$

$\mathbf{F}_{22}^{(ij)}(k, l)$ is the $(n_k - m_k) \times (n_l - m_l)$ null matrix

$$\mathbf{0}_{(n_k - m_k)\ (n_l - m_l)} \quad (k \neq l, k, l=1, 2, \ldots, \nu), \tag{5.78}$$

$\mathbf{q}_1^{(i)}(k)'$ is the $m_k \times 1$ vector

$$[q_{j1}^{(k)}, q_{j2}^{(k)}, \ldots, q_{j,m_k}^{(k)}] \quad (k=1, 2, \ldots, \nu), \tag{5.79}$$

and $\mathbf{q}_2^{(j)}(k)'$ is the $(n_k - m_k) \times 1$ vector

$$[q_{j,m_k+1}^{(k)}, q_{j,m_k+2}^{(k)}, \ldots, q_{j,n_k}^{(k)}] \quad (k=1, 2, \ldots, \nu). \tag{5.80}$$

Since $\mathbf{q}^{(j)} \neq \mathbf{0}$, it follows from eqn. (5.68) that

$$|\mathbf{F}^{(ij)}| = 0, \tag{5.81}$$

which, in view of eqns. (5.69), (5.72), and (5.78), implies that

$$|\mathbf{F}_{11}^{(ij)}| \prod_{k=1}^{\nu} (\rho_j^{(i)} - \lambda_k)^{n_k - m_k} = 0 \quad (i = 1, 2, \ldots, \nu;\ j = 1, 2, \ldots, m_i). \tag{5.82}$$

If the closed-loop eigenvalues, $\rho_j^{(i)}$, are chosen so that

$$\rho_j^{(i)} \neq \lambda_k \quad (i, k = 1, 2, \ldots, \nu;\ j = 1, 2, \ldots, m_i), \tag{5.83}$$

then it follows from eqns. (5.82) and (5.83) that

$$|\mathbf{F}_{11}^{(ij)}| = 0. \tag{5.84}$$

It may be shown that the foregoing results imply that

$$|\mathbf{F}_{11}^{(ij)}| = \prod_{k=1}^{\nu} (\rho_j^{(i)} - \lambda_k)^{m_k} \left[1 - \sum_{t=1}^{\nu} \sum_{l=0}^{m_t - 1} \sum_{s=1}^{m_t - l} \frac{K_s^{(t)} p_{l+s}^{(t)}}{(\rho_j^{(i)} - \lambda_t)^{l+1}}\right] \tag{5.85}$$

which, in view of (5.83) and (5.84), is equivalent to the equations

$$\sum_{t=1}^{\nu} \sum_{l=0}^{m_t - 1} \sum_{s=1}^{m_t - l} \left[\frac{K_s^{(t)} p_{l+s}^{(t)}}{(\rho_j^{(i)} - \lambda_t)^{l+1}}\right] = 1 \quad (i = 1, 2, \ldots, \nu;\ j = 1, 2, \ldots, m_i). \tag{5.86}$$

It is evident that, for a given value of i and j, this equation may be written in the form

$$\mathbf{R}^{(ij)} \mathbf{P} \mathbf{K} = 1, \tag{5.87 a}$$

where

$$\mathbf{R}^{(ij)} = [\mathbf{R}_1^{(ij)}, \mathbf{R}_2^{(ij)}, \ldots, \mathbf{R}_\nu^{(ij)}], \tag{5.87 b}$$

$$\mathbf{P} = \mathbf{P}_1 \oplus \mathbf{P}_2 \oplus \ldots \oplus \mathbf{P}_\nu, \tag{5.87 c}$$

and

$$\mathbf{K}' = [\mathbf{K}_1', \mathbf{K}_2', \ldots, \mathbf{K}_\nu']. \tag{5.87 d}$$

In eqns. (5.87),

$$\mathbf{R}_k^{(ij)} = \left[\frac{1}{(\rho_j^{(i)} - \lambda_k)}, \frac{1}{(\rho_j^{(i)} - \lambda_k)^2}, \ldots, \frac{1}{(\rho_j^{(i)} - \lambda_k)^{m_k}}\right], \tag{5.88 a}$$

$$\mathbf{P}_k = \begin{bmatrix} p_1^{(k)}, & p_2^{(k)}, & \ldots, & p_{m_k}^{(k)} \\ p_2^{(k)}, & p_3^{(k)}, & \ldots, & 0 \\ \cdots & \cdots & \cdots & \cdots \\ p_{m_k}^{(k)}, & 0, & \ldots, & 0 \end{bmatrix}, \tag{5.88 b}$$

and

$$\mathbf{K}_k' = [K_1^{(k)}, K_2^{(k)}, \ldots, K_{m_k}^{(k)}]. \tag{5.88 c}$$

In eqns. (5.87 a), the matrices $\mathbf{P}$ and $\mathbf{K}$ are independent of the integers i and j: these equations can therefore be written as the single matrix equation

$$\mathbf{RPK} = \mathbf{e} \tag{5.89}$$

where

$$\mathbf{R}=\begin{bmatrix}\mathbf{R}^{(1)}\\ \mathbf{R}^{(2)}\\ \vdots\\ \mathbf{R}^{(\nu)}\end{bmatrix}, \tag{5.90 a}$$

$$\mathbf{R}^{(i)}=\begin{bmatrix}\mathbf{R}^{(i1)}\\ \mathbf{R}^{(i2)}\\ \vdots\\ \mathbf{R}^{(i,m_i)}\end{bmatrix}, \tag{5.90 b}$$

and $\mathbf{e}$ is the $n\times 1$ column vector given by

$$\mathbf{e}'=[1, 1, \ldots, 1]. \tag{5.91}$$

The matrix $\mathbf{P}$ has a block-diagonal form and is non-singular since each block $\mathbf{P}_k$ is non-singular in view of the fact that $p_{m_k}{}^{(k)}\neq 0$ $(k=1, 2, \ldots, \nu)$. It follows that it is possible to solve eqn. (5.90) for the proportional-controller gains if

$$\rho_j{}^{(i)}\neq\rho_l{}^{(k)} \quad (i\neq k,\ i, k=1, 2, \ldots, \nu;\ j=1, 2, \ldots, m_i;\ l=1, 2, \ldots, m_k) \tag{5.92}$$

since, if this latter condition is satisfied by an appropriate choice of the closed-loop eigenvalues, then $\mathbf{R}$ is non-singular : in this case

$$\mathbf{K}=\mathbf{P}^{-1}\mathbf{R}^{-1}\mathbf{e} \tag{5.93}$$

which is the required closed-form solution for the vector of proportional-controller gains for a single-input system with confluent eigenvalues. The required control law is obtained by substituting into eqn. (5.54) the values of $K_j{}^{(i)}$ determined using eqn. (5.93).

5.3.2. *Illustrative Examples*

The theory of modal control developed in the preceding section can be conveniently illustrated by synthesizing appropriate feedback loops for two systems. In the first example, the system is partially controllable and has a plant matrix which is similar to a single Jordan block : in the second example, the system is controllable and has a plant matrix which is similar to the direct sum of two distinct Jordan blocks.

Thus, as the first example, consider a system governed by the state equation

$$\dot{\mathbf{x}}(t)=\begin{bmatrix}1, & -1, & 3\\ 0, & -1, & 4\\ 0, & -1, & 3\end{bmatrix}\mathbf{x}(t)+\begin{bmatrix}5\\ 4\\ 2\end{bmatrix}z(t). \tag{5.94}$$

The eigenstructure of this system is given by the following matrices :

$$\mathbf{U}=\begin{bmatrix}1, & 2, & 3\\ 0, & 2, & 1\\ 0, & 1, & 1\end{bmatrix}, \tag{5.95 a}$$

$$\mathbf{V}=\begin{bmatrix}1, & 0, & 0\\ 1, & 1, & -1\\ -4, & -1, & 2\end{bmatrix}, \tag{5.95 b}$$

and

$$\mathbf{J}=\mathbf{J}_3(\lambda_1)=\begin{bmatrix}1, & 1, & 0\\ 0, & 1, & 1\\ 0, & 0, & 1\end{bmatrix}. \tag{5.95 c}$$

The matrix given in eqn. (5.95 c) is the Jordan canonical form of the plant matrix of the system governed by eqn. (5.94) : this matrix clearly corresponds to an eigenvalue $\lambda_1=1$ of multiplicity three. The corresponding mode-controllability matrix is

$$\mathbf{p}^{(1)}=\mathbf{V}^{(1)\prime}\mathbf{b}=\begin{bmatrix}1\\ 2\\ 0\end{bmatrix} \tag{5.96}$$

which indicates that only the first two eigenvalues of the system may be changed since $p_3^{(1)}=0$ and $p_2^{(1)}\neq 0$. If the required eigenvalues of the closed-loop plant matrix are $\rho_1^{(1)}=-1$, $\rho_2^{(1)}=-2$, and $\lambda_1=1$, then eqn. (5.93) assumes the form

$$\begin{bmatrix}K_1^{(1)}\\ K_2^{(1)}\end{bmatrix}=\begin{bmatrix}p_1^{(1)}, & p_2^{(1)}\\ p_2^{(1)}, & 0\end{bmatrix}^{-1}\begin{bmatrix}1/(\rho_1^{(1)}-\lambda_1), & 1/(\rho_1^{(1)}-\lambda_1)^2\\ 1/(\rho_2^{(1)}-\lambda_1), & 1/(\rho_2^{(1)}-\lambda_1)^2\end{bmatrix}^{-1}\begin{bmatrix}1\\ 1\end{bmatrix} \tag{5.97}$$

which, on substitution of the appropriate values, may be reduced to the equation

$$\begin{bmatrix}K_1^{(1)}\\ K_2^{(1)}\end{bmatrix}=\begin{bmatrix}-3\\ -1\end{bmatrix}. \tag{5.98}$$

The required control law is determined by substituting the values of the proportional-controller gains given in eqn. (5.98) into eqn. (5.54) : this control law is given by the expression

$$\begin{aligned} z(t) &= K_1^{(1)}\mathbf{v}_1^{(1)\prime}\mathbf{x}(t)+K_2^{(1)}\mathbf{v}_2^{(1)\prime}\mathbf{x}(t)\\ &=[-3, \ -4, \ 13]\mathbf{x}(t). \end{aligned} \tag{5.99}$$

It may readily be verified that the plant matrix of the closed-loop system defined by eqns. (5.94) and (5.99) is

$$\mathbf{C}=\begin{bmatrix}-14, & -19, & 64\\ -12, & -15, & 52\\ -6, & -8, & 27\end{bmatrix}, \tag{5.100}$$

and that the eigenvalues of this matrix are -1, -2, and 1, as required.

Similarly, as the second example, consider a system governed by the state equation

$$\dot{\mathbf{x}}(t)=\begin{bmatrix}-4, & 1, & 2\\ -11, & 4, & 4\\ -7, & 1, & 4\end{bmatrix}\mathbf{x}(t)+\begin{bmatrix}1\\ -3\\ 5\end{bmatrix}z(t). \tag{5.101}$$

The eigenstructure for this sytem is given by the following matrices :

$$\mathbf{U}=\begin{bmatrix}1, & 0, & 0\\ 1, & -1, & -2\\ 2, & 1, & 1\end{bmatrix}, \tag{5.102 a}$$

$$\mathbf{V}=\begin{bmatrix}1, & -5, & 3\\ 0, & 1, & -1\\ 0, & 2, & -1\end{bmatrix}, \tag{5.102 b}$$

and

$$\mathbf{J}=\mathbf{J}_2(\lambda_1)\oplus\mathbf{J}_1(\lambda_2)=\begin{bmatrix}1, & 1, & 0\\ 0, & 1, & 0\\ 0, & 0, & 2\end{bmatrix}. \tag{5.102 c}$$

The matrix given in eqn. (5.102 c) is the Jordan canonical form of the plant matrix of the system governed by eqn. (5.101) : this matrix clearly corresponds to an eigenvalue $\lambda_1=1$ of multiplicity two and to a single eigenvalue $\lambda_2=2$. The corresponding mode-controllability sub-matrices are

$$\mathbf{p}^{(1)}=\mathbf{V}^{(1)\prime}\mathbf{b}=\begin{bmatrix}1\\ 2\end{bmatrix} \tag{5.103 a}$$

and

$$\mathbf{p}^{(2)}=\mathbf{V}^{(2)\prime}\mathbf{b}=[1]. \tag{5.103 b}$$

Since $p_2^{(1)}$ and $p_1^{(2)}$ are both non-zero, it is possible to change all the eigenvalues of the closed-loop plant matrix. Thus, if the required eigenvalues are $\rho_1^{(1)}=-2$, $\rho_2^{(1)}=-3$, and $\rho_1^{(2)}=-1$, then eqn. (5.93) assumes the form

$$\begin{bmatrix}K_1^{(1)}\\ K_2^{(1)}\\ K_1^{(2)}\end{bmatrix}=\begin{bmatrix}p_1^{(1)}, & p_2^{(1)}, & 0\\ p_2^{(1)}, & 0, & 0\\ 0, & 0, & p_1^{(2)}\end{bmatrix}^{-1}$$
$$\times\begin{bmatrix}1/(\rho_1^{(1)}-\lambda_1), & 1/(\rho_1^{(1)}-\lambda_1)^2, & 1/(\rho_1^{(1)}-\lambda_2)\\ 1/(\rho_2^{(1)}-\lambda_1), & 1/(\rho_2^{(1)}-\lambda_1)^2, & 1/(\rho_2^{(1)}-\lambda_2)\\ 1/(\rho_1^{(2)}-\lambda_1), & 1/(\rho_1^{(2)}-\lambda_1)^2, & 1/(\rho_2^{(2)}-\lambda_2)\end{bmatrix}^{-1}\begin{bmatrix}1\\ 1\\ 1\end{bmatrix} \tag{5.104}$$

which, on substitution of the appropriate values, may be reduced to the equation

$$\begin{bmatrix}K_1^{(1)}\\ K_2^{(1)}\\ K_1^{(2)}\end{bmatrix}=\begin{bmatrix}12\\ 19\\ -60\end{bmatrix}. \tag{5.105}$$

The required control law is determined by substituting the values of the proportional-controller gains given in eqn. (5.105) into eqn. (5.54) : this control law is given by the expression

$$\begin{aligned} z(t) &= K_1^{(1)}\mathbf{v}_1^{(1)\prime}\mathbf{x}(t)+K_2^{(1)}\mathbf{v}_2^{(1)\prime}\mathbf{x}(t)+K_1^{(2)}\mathbf{v}_1^{(2)\prime}\mathbf{x}(t)\\ &= [-263,\ 79,\ 98]\mathbf{x}(t). \end{aligned} \tag{5.106}$$

It may readily be verified that the plant matrix of the closed-loop system defined by eqns. (5.101) and (5.106) is

$$\mathbf{C}=\begin{bmatrix} -267, & 80, & 100 \\ 778, & -233, & -290 \\ -1322, & 396, & 494 \end{bmatrix} \tag{5.107}$$

and that the eigenvalues of this matrix are -1, -2, and -3, as required.

5.4. Discrete-time systems

The results presented in the foregoing sections of this chapter concerning the modal control of single-input continuous-time systems may be used (by analogy) in connection with discrete-time systems governed by state equations of the form

$$\mathbf{x}\{(k+1)T\}=\mathbf{\Psi}(T)\mathbf{x}(kT)+\mathbf{\delta}(T)z(kT). \tag{5.108}$$

Indeed, in the discrete-time case, eqn. (5.108) indicates that the appropriate plant matrix is $\mathbf{\Psi}(T)$ instead of $\mathbf{A}$, and that the input vector is $\mathbf{\delta}(T)$ instead of $\mathbf{b}$. Furthermore, the results presented in Chapter 2 indicate that, in the discrete-time case, the eigenvalues of the plant matrix are $\exp(\lambda_i T)$ $(i=1, 2, \ldots, n)$, whilst the corresponding eigenvectors are $\mathbf{u}_i$ and $\mathbf{v}_i$ $(i=1, 2, \ldots, n)$: the mode-controllability matrix of the discrete-time system (5.108) corresponding to $\mathbf{p}$ is therefore $\mathbf{p}(T)=\mathbf{V}'\mathbf{\delta}(T)$.

Thus, in the case of discrete-time systems of the class (5.108) whose plant matrices have distinct eigenvalues, it may be inferred from eqn. (5.34) that the control law

$$z(kT)=\sum_{j=1}^{m}\left[\frac{\prod_{l=1}^{m}(\rho_l-\exp\lambda_j T)\,\mathbf{v}_j'}{p_j(T)\sum_{\substack{l=1\\ l\neq j}}^{m}(\exp\lambda_l T\text{-}\exp\lambda_j T)}\right]\mathbf{x}(kT) \tag{5.109}$$

will alter the eigenvalues $\exp\lambda_1 T$, $\exp\lambda_2 T$, . . ., $\exp\lambda_m T$ of the plant matrix, $\mathbf{\Psi}(T)$, of the uncontrolled system to prescribed new values $\rho_1, \rho_2, \ldots, \rho_m$, leaving the remaining $(n-m)$ eigenvalues $\exp\lambda_{m+1}T$, . . ., $\exp\lambda_n T$ unchanged : in eqn. (5.109), $p_j(T)$ is the jth element of the mode-controllability matrix $\mathbf{p}(T)$. In cases when the plant matrix $\mathbf{\Psi}(T)$ has confluent eigenvalues, analogous results to those presented in § 5.3.1 may be used to determine the appropriate laws.

Multi-Input Modal Control Systems

CHAPTER 6

6.1. Introduction

IN this chapter, the *single*-input modal control theory presented in Chapter 5 is extended so as to embrace both continuous-time and discrete-time *multi*-input systems governed by state equations of the respective forms (1.1) and (1.3). It is shown that a number of alternative methods for designing multi-input modal control systems exist since, in general, the proportional-controller gains satisfy a set of underdetermined non-linear algebraic equations.

6.2. Continuous-time systems : plant matrices with distinct eigenvalues

6.2.1. *General theory*

This section is concerned with the design of modal controllers of multi-input continuous-time systems governed by state equations of the form

$$\dot{\mathbf{x}}(t)=\mathbf{A}\mathbf{x}(t)+\mathbf{B}\mathbf{z}(t), \tag{6.1}$$

where $\mathbf{x}(t)$ is the $n \times 1$ state vector of the system, $\mathbf{A}$ is the $n \times n$ plant matrix of the uncontrolled system, $\mathbf{B}$ is the $n \times r$ input matrix and $\mathbf{z}(t)$ is the $r \times 1$ input vector. In this section, it is assumed that $\mathbf{A}$ has n distinct eigenvalues, λ_j $(j=1, 2, \ldots, n)$, and that the corresponding eigenvectors of $\mathbf{A}$ and $\mathbf{A}'$ are $\mathbf{u}_j$ and $\mathbf{v}_j$ $(j=1, 2, \ldots, n)$, respectively.

If all the elements of the state vector $\mathbf{x}(t)$ can be measured by appropriate transducers, it will be possible to combine the transducer outputs to generate $m(m \leqslant n)$ signals given by equations of the form

$$s_j(t)=\sum_{k=1}^{n} \mu_{jk} x_k(t)=\boldsymbol{\mu}_j{}'\mathbf{x}(t) \quad (j=1, 2, \ldots, m), \tag{6.2}$$

where $\boldsymbol{\mu}_j$ is a measurement vector and the prime denotes matrix transposition. These new signals, $s_j(t)$, can then be amplified by $r \times m$ proportional controllers having gains K_{ij} ($i=1, 2, \ldots, r$; $j=1, 2, \ldots, m$) thus yielding input variables given by

$$z_i(t)=\sum_{j=1}^{m} K_{ij}s_j(t)=\sum_{j=1}^{m} K_{ij}\boldsymbol{\mu}_j{}'\mathbf{x}(t) \quad (i=1, 2, \ldots, r). \tag{6.3}$$

It follows by substituting the expression for $z_i(t)$ given in eqn. (6.3) into eqn. (6.1) that the governing equation of the resulting closed-loop system is

$$\dot{\mathbf{x}}(t)=\mathbf{A}\mathbf{x}(t)+\sum_{i=1}^{r}\mathbf{b}_i\sum_{j=1}^{m}K_{ij}\boldsymbol{\mu}_j{}'\mathbf{x}(t)=\mathbf{C}\mathbf{x}(t), \tag{6.4}$$

where the $\mathbf{b}_i$ are the columns of $\mathbf{B}$, and can be expressed in the form

$$\mathbf{b}_i=\sum_{j=1}^{n}p_{ji}\mathbf{u}_j \quad (i=1, 2, \ldots, r). \tag{6.5}$$

Equation (6.4) indicates that the effect of the input variables defined by eqns. (6.3) is to change the plant matrix, $\mathbf{A}$, to a new matrix, $\mathbf{C}$, given by

$$\mathbf{C}=\mathbf{A}+\sum_{i=1}^{r}\mathbf{b}_i\sum_{j=1}^{m}K_{ij}\boldsymbol{\mu}_j{}'. \tag{6.6}$$

If the $\boldsymbol{\mu}_j$ are now chosen to be equal to the $\mathbf{v}_j$ ($j=1, 2, \ldots, m$), then the plant matrix of the controlled system given by eqn. (6.6) will have the form

$$\mathbf{C}=\mathbf{A}+\sum_{i=1}^{r}\mathbf{b}_i\sum_{j=1}^{m}K_{ij}\mathbf{v}_j{}', \tag{6.7}$$

where the set $\{\mathbf{v}_1, \mathbf{v}_2, \ldots, \mathbf{v}_m\}$ consists of an appropriate selection of m eigenvectors of $\mathbf{A}'$.

It is thus evident from eqns. (2.14) and (6.7) that if

$$m+1\leqslant k\leqslant n,$$

then

$$\mathbf{C}\mathbf{u}_k=\mathbf{A}\mathbf{u}_k=\lambda_k\mathbf{u}_k, \tag{6.8}$$

thus indicating that $\mathbf{u}_k$ and λ_k ($k=m+1, m+2, \ldots, n$) are eigenvectors and eigenvalues of $\mathbf{C}$ as well as of $\mathbf{A}$. It is also evident from eqns. (2.16) and (6.7) that if

$$1\leqslant j\leqslant m,$$

then

$$\mathbf{C}\mathbf{u}_j=\mathbf{A}\mathbf{u}_j+\sum_{i=1}^{r}K_{ij}\mathbf{b}_i=\lambda_j\mathbf{u}_j+\sum_{i=1}^{r}K_{ij}\mathbf{b}_i \tag{6.9}$$

which implies that, in general, λ_j is not an eigenvalue of $\mathbf{C}$ and also that $\mathbf{u}_j$ is no longer the corresponding eigenvector if $K_{ij}\neq 0$. The effect of using m measurement vectors, $\mathbf{v}_j$ ($j=1, 2, \ldots, m$), is therefore to change the eigenvalues, λ_j, to some new values, ρ_j, and the eigenvectors, $\mathbf{u}_j$, to some corresponding new vectors, $\mathbf{w}_j$ $(j=1, 2, \ldots, m)$, leaving the remaining $(n-m)$ pairs of eigenvalues and eigenvectors of the plant matrix of the uncontrolled system unchanged.

In order to calculate values of the proportional-controller gains, K_{ij}, in terms of the eigenvalues of $\mathbf{A}$ and the required eigenvalues of $\mathbf{C}$, let

$$\mathbf{w}_j = \sum_{k=1}^{n} q_{jk}\mathbf{u}_k \quad (j=1, 2, \ldots, m). \tag{6.10}$$

It follows from eqns. (2.14), (2.16) and (6.5) that the appropriate elements of the mode-controllability matrix are

$$p_{ji} = \mathbf{v}_j{}'\mathbf{b}_i \quad (i=1, 2, \ldots, r\,;\; j=1, 2, \ldots, n). \tag{6.11}$$

Since, by definition ρ_j and $\mathbf{w}_j$ satisfy the equations

$$\mathbf{C}\mathbf{w}_j = \rho_j\mathbf{w}_j \quad (j=1, 2, \ldots, m), \tag{6.12}$$

substituting from eqns. (6.7) and (6.10) into eqn. (6.12) yields the equations

$$\left(\mathbf{A} + \sum_{i=1}^{r} \mathbf{b}_i \sum_{l=1}^{m} K_{il}\mathbf{v}_l{}'\right) \sum_{k=1}^{n} q_{jk}\mathbf{u}_k = \rho_j \sum_{k=1}^{n} q_{jk}\mathbf{u}_k \quad (j=1, 2, \ldots, m). \tag{6.13}$$

Equations (2.14), (2.16), (6.5), and (6.7) imply that eqn. (6.13) can be expressed in the form

$$\sum_{k=1}^{n} q_{jk}\lambda_k\mathbf{u}_k + \sum_{i=1}^{r}\sum_{k=1}^{n} p_{ki}\mathbf{u}_k \sum_{l=1}^{m} K_{il}q_{jl} = \rho_j \sum_{k=1}^{n} q_{jk}\mathbf{u}_k \quad (j=1, 2, \ldots, m). \tag{6.14}$$

Equation (6.14) is a vector equation in the $\mathbf{u}_k$ and is equivalent to the $n \times m$ scalar equations

$$(\rho_j - \lambda_k)q_{jk} - \sum_{i=1}^{r} p_{ki} \sum_{l=1}^{m} K_{il}q_{jl} = 0 \quad (j=1, 2, \ldots, m\,;\; k=1, 2, \ldots, n). \tag{6.15}$$

It is evident that for a given value of j, the m equations involving $\lambda_1, \lambda_2, \ldots, \lambda_m$ can be written in the matrix form

$$\mathbf{F}_j\mathbf{q}_j = \mathbf{0} \quad (j=1, 2, \ldots, m), \tag{6.16 a}$$

where

$$\mathbf{F}_j = [f_{kl}{}^{(j)}], \tag{6.16 b}$$

$$f_{kl}{}^{(j)} = (\rho_j - \lambda_k)\delta_{kl} - \sum_{i=1}^{r} p_{ki}K_{il}, \tag{6.16 c}$$

$$\mathbf{q}_j = [q_{jl}], \tag{6.16 d}$$

and δ_{kl} is the Kronecker delta. Since $\mathbf{q}_j \neq \mathbf{0}$, it follows from eqn. (6.16 *a*) that

$$|\mathbf{F}_j| = 0, \tag{6.17}$$

which, in view of (6.16 *b*) and (6.16 *c*), has the explicit form

$$\begin{vmatrix} \rho_j - \lambda_1 - \sum_{i=1}^{r} p_{1i}K_{i1}, & -\sum_{i=1}^{r} p_{1i}K_{i2}, \ldots, & -\sum_{i=1}^{r} p_{1i}K_{im} \\ -\sum_{i=1}^{r} p_{2i}K_{i1}, & \rho_j - \lambda_2 - \sum_{i=1}^{r} p_{2i}K_{i2}, \ldots, & -\sum_{i=1}^{r} p_{2i}K_{im} \\ \cdots & \cdots & \cdots \\ \cdots & \cdots & \cdots \\ -\sum_{i=1}^{r} p_{mi}K_{i1}, & -\sum_{i=1}^{r} p_{mi}K_{i2}, \ldots, & \rho_j - \lambda_m - \sum_{i=1}^{r} p_{mi}K_{im} \end{vmatrix} = 0$$

$$(j = 1, 2, \ldots, m). \qquad (6.18)$$

The expansion of the determinants in eqn. (6.18) yields a set of m algebraic equations in the $m \times r$ proportional-controller gains, K_{ij}. It is evident that, in general, these equations will be non-linear in the K_{ij} when $r > 1$, and that there will be no unique solution for the K_{ij} since eqns. (6.18) are underdetermined when $r > 1$. The special case, $r = 1$, is, of course, that discussed in detail in § 5.2.2.

6.2.2. *Minimum-gain modal controllers*

If, in the case $r > 1$, the m underdetermined eqns. (6.18) are augmented by a set of $m \times (r-1)$ algebraic equations involving the K_{ji}, then the resulting $m \times r$ equations will, in general, determine a set of values for the $m \times r$ proportional-controller gains, K_{ji}. Two types of augmentation are discussed in this section : the first corresponds to a design procedure in which the sum of the squares of the feedback gains is minimized, and the second to a design procedure in which the sum of the moduli of the feedback gains is minimized. Each of these procedures is applied to the design of a controller which controls only one mode of a system. The class of system considered is characterized by the fact that the mode to be controlled is controllable by *every* input variable : thus, if the mode to be controlled is characterized by the eigenvalue λ_i, then

$$p_{ij} \neq 0 \quad (j = 1, 2, \ldots, r). \qquad (6.19)$$

In the case of the first design procedure applied to the control of the ith mode, the special form of eqn. (6.18) is

$$\rho_i - \lambda_i = \sum_{j=1}^{r} p_{ij}K_{ji} \qquad (6.20)$$

if the input variable, $z_j(t)$, is given by an expression of the form

$$z_j(t) = K_{ji}\mathbf{v}_i'\mathbf{x}(t) \quad (j = 1, 2, \ldots, r), \qquad (6.21\ a)$$

where $\mathbf{v}_i$ is the ith eigenvector of the transposed plant matrix, $\mathbf{A}'$. Equation (6.21 *a*) can be expressed in the equivalent form

$$z_j(t) = \mathbf{g}_j'\mathbf{x}(t) \quad (j = 1, 2, \ldots, r), \qquad (6.21\ b)$$

where

$$\mathbf{g}_j = [g_{jk}] = [K_{ji} v_i^k] \tag{6.21 c}$$

and v_i^k is the kth element of the eigenvector, $\mathbf{v}_i$. Thus, in terms of this nomenclature, it follows directly from eqn. (6.21 c) that the sum of the squares, S_i, of the $n \times r$ feedback gains, g_{jk}, is given by the equation

$$S_i = \sum_{j=1}^{r} \sum_{k=1}^{n} (K_{ji} v_i^k)^2 = N_i \sum_{j=1}^{r} K_{ji}^2, \tag{6.22}$$

where N_i is the square of the euclidean norm of the eigenvector $\mathbf{v}_i$. The values of the $m \times r$ proportional-controller gains, K_{ji}, can be determined in this case by minimizing S_i/N_i, subject to the constraint (6.20). These values can be readily obtained by forming the function, s_i, given by the equation

$$s_i = (S_i/N_i) + L_i \left[\rho_i - \lambda_i - \sum_{j=1}^{r} p_{ij} K_{ji} \right], \tag{6.23}$$

where L_i is a Lagrange multiplier. It follows from eqn. (6.23) that

$$\left.\begin{aligned} \frac{\partial s_i}{\partial K_{ji}} &= 2K_{ji} - L_i p_{ij} \\ \text{and} \qquad \frac{\partial s_i}{\partial L_i} &= \rho_i - \lambda_i - \sum_{j=1}^{r} p_{ij} K_{ji} \end{aligned}\right\} \quad (j = 1, 2, \ldots, r). \tag{6.24}$$

Since the condition (6.19) is assumed to hold, it may readily be verified, using eqns. (6.24), that S_i/N_i assumes an absolute minimum value if, and only if,

$$\frac{K_{1i}}{p_{i1}} = \frac{K_{2i}}{p_{i2}} = \ldots = \frac{K_{ri}}{p_{ir}}. \tag{6.25}$$

Equations (6.20) and (6.25) imply that the required proportional-controller gains, $\hat{K}_{ji}$, and the corresponding minimum value, $\hat{S}_i$, of S_i are given by the equations

$$\hat{K}_{ji} = p_{ij}(\rho_i - \lambda_i) / \sum_{j=1}^{r} (p_{ij})^2 \quad (j = 1, 2, \ldots, r), \tag{6.26}$$

and

$$\hat{S}_i = N_i(\rho_i - \lambda_i)^2 / \sum_{j=1}^{r} (p_{ij})^2. \tag{6.27}$$

The contribution, $\hat{S}_{ik}$, of the feedback gains associated with the kth input to $\hat{S}_i$ is given by the equation

$$\hat{S}_{ik} = N_i(\rho_i - \lambda_i)^2 (p_{ik})^2 / \left\{ \sum_{j=1}^{r} (p_{ik})^2 \right\}^2. \tag{6.28}$$

Similarly, S_{ik}, the value of S_i when only the kth input is used to control the ith mode, is given by the equation

$$S_{ik}=N_i(\rho_i-\lambda_i)^2/(p_{ik})^2. \tag{6.29}$$

Equations (6.28) and (6.29) imply that

$$\hat{S}_{ik}/S_{ik}=1\Big/\Big\{1+\sum_{\substack{j=1\\ j\neq k}}^{r}(p_{ij}/p_{ik})^2\Big\}. \tag{6.30}$$

Since the condition (6.19) is assumed to hold, eqn. (6.30) implies that

$$\hat{S}_{ik}<S_{ik} \tag{6.31}$$

in *all* cases. However, unless

$$|p_{ij}|\gg|p_{ik}|, \tag{6.32}$$

eqn. (6.30) indicates that $\hat{S}_{ik}$ will not be significantly less than S_{ik}: in view of this fact, the implications of controlling the ith mode with a proper subset of the complete set, $\{z_1(t), z_2(t), \ldots, z_r(t)\}$, of inputs should be carefully examined. Indeed, in many cases this subset will contain only the single input associated with the largest absolute value of p_{ij} $(1\leqslant j\leqslant r)$.

In the case of the second design procedure applied to the control of the ith mode of a system characterized by the property (6.19), eqns. (6.20) and (6.21) imply that the sum of moduli, M_i, of the $n\times r$ feedback gains, g_{jk}, is given by the equation

$$M_i=\sum_{j=1}^{r}\sum_{k=1}^{n}|K_{ji}v_i^k|=H_i\sum_{j=1}^{r}|K_{ji}|, \tag{6.33}$$

where H_i is the sum of the moduli of the elements of the eigenvector $\mathbf{v}_i$. The values of the $m\times r$ proportional-controller gains, K_{ji}, can be determined in this case by minimizing M_i/H_i, subject to the constraint (6.20). These values can be readily obtained by forming the function, m_i, given by the equation

$$m_i=(M_i/H_i)+L_i\left[\rho_i-\lambda_i-\sum_{j=1}^{r}p_{ij}K_{ji}\right], \tag{6.34}$$

where L_i is a Lagrange multiplier. It follows from eqn. (6.34) that

$$\left.\begin{aligned}\frac{\partial m_i}{\partial K_{ji}}&=\operatorname{sgn}K_{ji}-L_ip_{ij}\\ &\text{and}\\ \frac{\partial m_i}{\partial L_i}&=\rho_i-\lambda_i-\sum_{i=1}^{r}p_{ij}K_{ji}\end{aligned}\right\}\quad(j=1,2,\ldots,r). \tag{6.35}$$

Equation (6.35) indicates that, if

$$L_i=\frac{\operatorname{sgn}K_{ji}}{p_{ij}}\quad(j=1,2,\ldots,r), \tag{6.36}$$

then

$$\frac{\partial m_i}{\partial K_{ji}} = 0 \quad (j=1, 2, \ldots, r), \tag{6.37}$$

which implies that a necessary condition for m_i to be a minimum is that

$$\frac{\operatorname{sgn} K_{1i}}{p_{i1}} = \frac{\operatorname{sgn} K_{2i}}{p_{i2}} = \ldots = \frac{\operatorname{sgn} K_{ri}}{p_{ir}}. \tag{6.38}$$

Equations (6.38) can clearly be satisfied only if

$$|p_{ij}| = c_i \quad (j=1, 2, \ldots, r), \tag{6.39}$$

where c_i is a real number. If the conditions (6.39) are satisfied, then it is evident that an infinite number of sets $\{K_{1i}, K_{2i}, \ldots, K_{ri}\}$ exist which satisfy eqns. (6.20) and (6.38) and thus minimize M_i. However, if it is desired to choose the contribution, M_{ij}, of the sum of the moduli of the feedback gains associated with the jth input to M_i such that

$$M_{ij} = M_i / r, \tag{6.40}$$

then a unique solution set $\{\hat{K}_{1i}, \hat{K}_{2i}, \ldots, \hat{K}_{ri}\}$ of eqns. (6.20) and (6.38) exists. It is evident from these equations that the elements of the set and the corresponding minimum value, $\hat{M}_i$, of M_i are given by the expressions

$$\hat{K}_{ji} p_{ij} = (\rho_i - \lambda_i)/r \quad (j=1, 2, \ldots, r) \tag{6.41}$$

and

$$\hat{M}_i = H_i |\rho_i - \lambda_i| / c_i. \tag{6.42}$$

The contribution, $\hat{M}_{ik}$, of the feedback gains associated with the kth input to $\hat{M}_i$ is given by the equation

$$\hat{M}_{ik} = H_i |\rho_i - \lambda_i| / (r c_i). \tag{6.43}$$

Similarly, M_{ik}, the value of M_i when only the kth input is used to control the ith mode, is given by the equation

$$M_{ik} = H_i |\rho_i - \lambda_i| / c_i. \tag{6.44}$$

Equations (6.43) and (6.44) imply that

$$\hat{M}_{ik} / M_{ik} = 1/r \tag{6.45}$$

which implies that $\hat{M}_{ik} < M_{ik}$ in all cases for which the conditions (6.39) obtain.

The theory of multi-input modal control developed in this section can be illustrated by designing appropriate feedback loops for the second-order system for which the state eqn. (6.1) has the particular form

$$\dot{\mathbf{x}}(t) = \begin{bmatrix} -5, & 12 \\ -1, & 2 \end{bmatrix} \mathbf{x}(t) + \begin{bmatrix} 10, & 6 \\ 3, & 1 \end{bmatrix} \mathbf{z}(t). \tag{6.46}$$

The eigenstructure of this system is given by the following matrices :

$$\boldsymbol{\Lambda}=\begin{bmatrix}-2, & 0\\ 0, & -1\end{bmatrix}, \tag{6.47 a}$$

$$\mathbf{U}=\begin{bmatrix}4, & 3\\ 1, & 1\end{bmatrix} \tag{6.47 b}$$

and

$$\mathbf{V}=\begin{bmatrix}1, & -1\\ -3, & 4\end{bmatrix}. \tag{6.47 c}$$

The corresponding mode-controllability matrix is

$$\mathbf{P}=\mathbf{V}'\mathbf{B}=\begin{bmatrix}1, & 3\\ 2, & -2\end{bmatrix}. \tag{6.48}$$

In the case of the first design procedure described in this section, eqns. (6.21) and (6.26) indicate that

$$z_1(t)=-\frac{1}{5}x_1(t)+\frac{3}{5}x_2(t) \tag{6.49 a}$$

and

$$z_2(t)=-\frac{3}{5}x_1(t)+\frac{9}{5}x_2(t) \tag{6.49 b}$$

if it is desired to control the first mode of the system (6.46) such that $\rho_1=-4$. The corresponding single-input control laws are

$$z_1(t)=-2x_1(t)+6x_2(t), \tag{6.50 a}$$

and

$$z_2(t)=-\frac{2}{3}x_1(t)+2x_2(t). \tag{6.50 b}$$

It is evident from eqns. (6.27, (6.28), (6.29), (6.49), and (6.50) that

$$\hat{S}_1=4, \tag{6.51 a}$$

$$\hat{S}_{11}=\frac{2}{5}, \tag{6.51 b}$$

$$\hat{S}_{12}=\frac{18}{5}, \tag{6.51 c}$$

$$S_{11}=40, \tag{6.51 d}$$

and

$$S_{12}=\frac{40}{9}. \tag{6.51 e}$$

These results clearly indicate the benefits of using multi-input control in this case.

In the case of the second design procedure described in this section, eqns. (6.21) and (6.41) indicate that

$$z_1(t)=\tfrac{1}{2}x_1(t)-2x_2(t) \tag{6.52 a}$$

and

$$z_2(t)=-\tfrac{1}{2}x_1(t)+2x_2(t) \tag{6.52 b}$$

if it is desired to control the second mode of the system (6.46) such that $\rho_2=-3$. The corresponding single-input control laws are

$$z_1(t)=x_1(t)-4x_2(t) \tag{6.53 a}$$

or

$$z_2(t)=-x_1(t)+4x_2(t). \tag{6.53 b}$$

It is evident from eqns. (6.42), (6.43), (6.44), (6.52) and (6.53) that

$$\hat{M}_2=5, \tag{6.54 a}$$

$$\hat{M}_{21}=\frac{5}{2}, \tag{6.54 b}$$

$$\hat{M}_{22}=\frac{5}{2}, \tag{6.54 c}$$

$$M_{21}=5, \tag{6.54 d}$$

and

$$M_{22}=5. \tag{6.54 e}$$

6.2.3. *Prescribed-gain modal controllers*

An alternative method of augmenting the underdetermined eqns. (6.18) can be effected by prescribing $m\times(r-1)$ of the $n\times r$ feedback gains, g_{ik} ($j=1, 2, \ldots, r$; $k=1, 2, \ldots, n$), defined in eqns. (6.21 *c*). Equation (6.21 *c*) implies that even if the kth state variable is not available for measurement, then it may nevertheless be possible, in certain circumstances, to choose any set of desired closed-loop eigenvalues by ensuring that

$$g_{jk}=0 \quad (j=1, 2, \ldots, r).$$

This design approach thus obviates the need for incorporating state observers† to estimate the state variables that are not available for feedback.

---

† State observers are discussed in Chapter 7.

6.2.4. *Multi-stage design procedure for modal controllers*

The theory for the modal control of single-input systems presented in §§5.2 and 5.3 can be used sequentially to design feedback loops for multi-input systems governed by state equations of the form (6.1). Thus, if the input matrix, $\mathbf{B}$, has the partitioned form

$$\mathbf{B} = [\mathbf{b}_1, \mathbf{b}_2, \ldots, \mathbf{b}_r], \tag{6.55}$$

then eqn. (6.1) can be written as

$$\dot{\mathbf{x}}(t) = \mathbf{A}\mathbf{x}(t) + \sum_{i=1}^{r} \mathbf{b}_i z_i(t). \tag{6.56}$$

If eqn. (6.56) is compared with eqn. (5.1 *a*), it is evident that each of the $z_i(t)$ in eqn. (6.56) can be generated in the same manner as $z(t)$ in eqn. (5.1 *a*) to effect the desired eigenvalue changes. Each $z_i(t)$ in eqn. (6.56) will therefore have the form

$$z_i(t) = \sum_{j=1}^{n} K_j^{(i)} \mathbf{v}_j^{(i)\prime} \mathbf{x}(t) = \mathbf{g}_i' \mathbf{x}(t) \quad (i = 1, 2, \ldots, r), \tag{6.57}$$

where the $K_j^{(i)}$ are the proportional-controller gains associated with the ith stage of the design procedure, and the $\mathbf{v}_j^{(i)}$ are the corresponding measurement vectors. Also it may be inferred from eqn. (5.34) that

$$\mathbf{g}_i' = \sum_{j=1}^{n} \frac{\prod_{k=1}^{n} (\rho_k - \lambda_j) \mathbf{v}_j^{(i)\prime}}{\mathbf{v}_j^{(i)\prime} \mathbf{b}_i \prod_{\substack{k=1 \\ k \neq j}}^{n} (\lambda_k - \lambda_j)}. \tag{6.58}$$

In eqns. (6.57) and (6.58), the measurement vectors, $\mathbf{v}_j^{(i)}$ ($j = 1, 2, \ldots, n$), are the columns of the modal matrix, $\mathbf{V}_i$, of the matrix $\mathbf{C}_i$: the latter matrix is given by the recurrence relation

$$\mathbf{C}_{i+1} = \mathbf{C}_i + \mathbf{b}_i \mathbf{g}_i' \quad (i = 1, 2, \ldots, r-1), \tag{6.59 a}$$

where

$$\mathbf{C}_1 = \mathbf{A}. \tag{6.59 b}$$

The mode-controllability matrix, $\mathbf{P}_i$, associated with the ith stage of the design procedure is given by the equation

$$\mathbf{P}_i = \mathbf{V}_i' \mathbf{B}_i \quad (i = 1, 2, \ldots, r), \tag{6.60 a}$$

where

$$\mathbf{B}_i = [\mathbf{b}_i, \mathbf{b}_{i+1}, \ldots, \mathbf{b}_r]. \tag{6.60 b}$$

In applying the multi-stage design procedure, the modes to be controlled at the ith stage are chosen on the basis of the magnitudes of the elements of the appropriate mode-controllability matrix, $\mathbf{P}_i$, in view of the form of eqn. (5.34). Thus, for example, consider that at the second stage of the design of a four-input modal controller for a plant modelled by a sixth-order state equation, the appropriate mode-controllability matrix is

$$\mathbf{P}_2 = \begin{bmatrix} 1, & 4, & -1 \\ 6, & 2, & 2 \\ -4, & 1, & 1 \\ 7, & -3, & -2 \\ -2, & 2, & 1 \\ -1, & 6, & 3 \end{bmatrix}, \tag{6.61}$$

and that the input $z_1(t)$ has already been used to control the second and fifth modes of the system. In choosing the modes to be controlled by $z_2(t)$, only the first, third, fourth, and sixth rows of $\mathbf{P}_2$ are relevant : examination of these rows clearly indicates that the third and fourth modes would be chosen since

$$|-4| > |1|, \quad |-4| > |1|, \tag{6.62 a}$$

and

$$|7| > |-3|, \quad |7| > |-2|. \tag{6.62 b}$$

6.2.5. *Dyadic modal controllers*

In this section the formula (5.33) for the loop gains of a single-input modal control system is applied to obtain closed-form expressions for the feedback gains of a class of multi-input modal control systems.

Thus, consider systems governed by state equations of the form (6.1) with state feedback given by the equation

$$\mathbf{z}(t) = \mathbf{G}\mathbf{x}(t) \tag{6.63}$$

where $\mathbf{G}$ is an $r \times n$ matrix. There is, in general, of course, an infinity of matrices, $\mathbf{G}$, which will result in the controllable modes of the closed-loop system defined by eqns. (6.1) and (6.63) having arbitrary eigenvalues. It will be recalled from § 6.2.1, however, that the calculation of $\mathbf{G}$ usually involves the solution of a set of non-linear algebraic equations.

The computational difficulties associated with this non-linearity may be circumvented by considering the special class of system characterized by the fact that the feedback matrix, $\mathbf{G}$, in eqn. (6.63) is chosen to have the dyadic form

$$\mathbf{G} = \mathbf{f}\mathbf{d}', \tag{6.64}$$

where $\mathbf{f}$ and $\mathbf{d}$ are respectively $r \times 1$ and $n \times 1$ vectors. This class of system has been investigated by Simon and Mitter [1], Gould *et al.* [2], and Retallack and Macfarlane [3] : the results presented in this section, however, are derived on the basis of modal control theory and have a simple closed-form structure which facilitates their direct application to the design of control loops for both partially-controllable and controllable systems.

Thus, under the action of the control law (6.63), the system governed by eqn. (6.1) has the equivalent single-input form

$$\dot{\mathbf{x}}(t) = \mathbf{A}\mathbf{x}(t) + \boldsymbol{\beta}\omega(t), \tag{6.65}$$

where $\mathbf{G}$ has the special form given by eqn. (6.64), so that, in eqn. (6.65),

$$\boldsymbol{\beta} = \mathbf{B}\mathbf{f} \tag{6.66}$$

and

$$\omega(t) = \mathbf{d}'\mathbf{x}(t). \tag{6.67}$$

The $n \times 1$ mode-controllability matrix, $\boldsymbol{\pi}$, of the system (6.65) is related to the $n \times r$ mode-controllability matrix, $\mathbf{P}$, of the system (6.1) by the equation

$$\boldsymbol{\pi} = \mathbf{P}\mathbf{f}, \tag{6.68}$$

where

$$\mathbf{P} = \mathbf{V}'\mathbf{B}. \tag{6.69}$$

In eqn. (6.69), $\mathbf{V}$ is, of course, the modal matrix of $\mathbf{A}'$. The vector $\mathbf{f}$ is chosen so that only zero elements in $\boldsymbol{\pi}$ correspond to the null rows (if any exist) of $\mathbf{P}$: this ensures that all the controllable modes of the system (6.1) remain controllable in the system (6.65). In order to use the result (5.33), the vector $\mathbf{d}'$ is chosen according to the equation

$$\mathbf{d}' = \sum_{j=1}^{m} k_j \mathbf{v}_j' \quad (1 \leqslant m \leqslant n), \tag{6.70}$$

where the $\mathbf{v}_j$ $(j=1, 2, \ldots, m)$ are the eigenvectors of $\mathbf{A}'$ corresponding to the m controllable modes of system (6.1). Thus, if the k_j in eqn. (6.70) are calculated using the equation

$$k_j = \frac{\prod_{k=1}^{m} (\rho_k - \lambda_j)}{\pi_j \prod_{\substack{k=1 \\ k \neq j}}^{m} (\lambda_k - \lambda_j)} \quad (j = 1, 2, \ldots, m), \tag{6.71}$$

where π_j is the jth element of $\boldsymbol{\pi}$ and the λ_j are eigenvalues of the matrix $\mathbf{A}$ which correspond to the controllable modes of the systems (6.1) and (6.65), then the closed-loop system matrix

$$\mathbf{C} = \mathbf{A} + \mathbf{B}\mathbf{f}\mathbf{d}'$$

will have the eigenvalue spectrum

$$\{\rho_1, \rho_2, \ldots, \rho_m, \lambda_{m+1}, \ldots, \lambda_n\}.$$

It is evident from eqns. (6.63), (6.64), (6.70), and (6.71) that the required control law is

$$\mathbf{z}(t) = \mathbf{f} \sum_{j=1}^{m} \frac{\prod_{k=1}^{m} (\rho_k - \lambda_j)}{\mathbf{v}_j'\mathbf{B}\mathbf{f} \prod_{\substack{k=1 \\ k \neq j}}^{m} (\lambda_k - \lambda_j)} \mathbf{x}(t). \tag{6.72}$$

The design procedure presented in this section can be conveniently illustrated by considering the system

$$\dot{\mathbf{x}}(t)=\begin{bmatrix} 0, & 1 \\ -2, & -3 \end{bmatrix}\mathbf{x}(t)+\begin{bmatrix} 1, & 0 \\ 1, & 2 \end{bmatrix}\mathbf{z}(t). \tag{6.73}$$

In this case, the eigenvalues of the plant matrix are $\lambda_1=-1$ and $\lambda_2=-2$, and it will be assumed that it is desired to generate $\mathbf{z}(t)$ by linear feedback of $\mathbf{x}(t)$ such that the eigenvalues of the closed-loop plant matrix are $\rho_1=-3$ and $\rho_2=-4$. It may be readily verified that

$$\mathbf{V}'=\begin{bmatrix} 2, & 1 \\ -1, & -1 \end{bmatrix}, \tag{6.74}$$

and

$$\mathbf{P}=\begin{bmatrix} 3, & 2 \\ -2, & -2 \end{bmatrix}, \tag{6.75}$$

and that a suitable choice of $\mathbf{f}$ is

$$\mathbf{f}=\begin{bmatrix} 1 \\ 1 \end{bmatrix}. \tag{6.76}$$

It then follows that

$$\boldsymbol{\pi}=\begin{bmatrix} 5 \\ -4 \end{bmatrix} \tag{6.77}$$

and

$$\boldsymbol{\beta}=\begin{bmatrix} 1 \\ 3 \end{bmatrix}. \tag{6.78}$$

The required control law may now be calculated using eqn. (6.72) and is

$$\mathbf{z}(t)=\mathbf{G}\mathbf{x}(t)$$

where

$$\mathbf{G}=\begin{bmatrix} -19/10, & -7/10 \\ -19/10, & -7/10 \end{bmatrix}. \tag{6.79}$$

Although the foregoing numerical example involves only real values of the λ_j and ρ_k, the formula (6.72) is, of course, applicable to situations involving complex eigenvalues, provided that these occur in conjugate pairs. In addition, in this example it so happens that the choice of the vector $\mathbf{f}$ given in eqn. (6.76) leads to a vector $\boldsymbol{\pi}$ which contains no null elements and which therefore indicates that the mode-controllability characteristics of the original multi-input system are unchanged. However, the process of choosing $\mathbf{f}$ such that the mode-controllability characteristics of a multi-input system are not changed can be made systematic by making use of the relationship which exists between $\mathbf{f}$ and $\boldsymbol{\pi}$. Thus, it follows by premultiplying eqn. (6.68) by $\mathbf{P}'$ that

$$\mathbf{P}'\mathbf{P}\mathbf{f}=\mathbf{P}'\boldsymbol{\pi}. \tag{6.80}$$

Now, if $r<n$ and $\mathbf{P}$ has full rank, it is evident that $\mathbf{P}'\mathbf{P}$ in eqn. (6.80) is an $r\times r$ non-singular matrix and therefore that

$$\mathbf{f}=(\mathbf{P}'\mathbf{P})^{-1}\mathbf{P}'\boldsymbol{\pi}, \tag{6.81}$$

provided that $\boldsymbol{\pi}$ satisfies the equation

$$\mathbf{P}(\mathbf{P}'\mathbf{P})^{-1}\mathbf{P}'\boldsymbol{\pi}=\boldsymbol{\pi} \tag{6.82}$$

in view of eqns. (6.68) and (6.81).

The significance of eqn. (6.82) in the design of dyadic modal controllers can be conveniently illustrated by considering a system whose mode-controllability matrix is

$$\mathbf{P}=\begin{bmatrix}1, & 0\\ 1, & 2\\ 0, & 1\end{bmatrix} \tag{6.83}$$

and whose plant matrix has the Jordan canonical form

$$\mathbf{J}=\begin{bmatrix}1, & 0, & 0\\ 0, & 2, & 0\\ 0, & 0, & -1\end{bmatrix}. \tag{6.84}$$

In the case of this system, the matrix eqn. (6.82) reduces to the scalar equation

$$\pi_1-\pi_2+2\pi_3=0, \tag{6.85}$$

where

$$\boldsymbol{\pi}=\begin{bmatrix}\pi_1\\ \pi_2\\ \pi_3\end{bmatrix}. \tag{6.86}$$

It is clear from eqns. (6.85) and (6.86) that the only admissible form for the vector $\boldsymbol{\pi}$ which leads to the existence of a dyadic feedback matrix (6.64) for a system whose mode-controllability matrix is given by equation (6.83) is

$$\boldsymbol{\pi}=\begin{bmatrix}\pi_1\\ \pi_2\\ (\pi_2-\pi_1)/2\end{bmatrix}. \tag{6.87}$$

If $\boldsymbol{\pi}$ is chosen to have the form (6.87), it follows from eqn. (6.81) that, in this example,

$$\mathbf{f}=\begin{bmatrix}\pi_1\\ \\ (\pi_2-\pi_1)/2\end{bmatrix}. \tag{6.88}$$

Since the three eigenvalues of the matrix (6.84) are all distinct, it follows from eqn. (6.87) that the dyadically-controlled system associated with eqns. (6.83) and (6.84) is controllable if and only if $\pi_1 \neq 0$ and $\pi_2 \neq 0$ and $\pi_1 \neq \pi_2$. It is important to note that the choice $\pi_2=0$ in eqn. (6.88) would lead to an equivalent single-input system of the form (6.65) for which the mode associated with the eigenvalue $\lambda_2=2$ is uncontrollable : it is clear, however, from eqns. (6.83) and (6.84) that the original multi-input system is controllable.

6.3. Continuous-time systems : plant matrices with confluent eigenvalues

In the case of systems governed by state equations of the form (6.1) whose plant matrices have a number of sets of confluent eigenvalues, it is possible to generalize the various theories and design procedures presented in §§ 6.2.1 to 6.2.3. However, such generalizations yield highly complicated algorithms which are very unattractive from the computational point of view. In the design of modal controllers for systems of this class, it is therefore invariably better to use sequentially the single-input results presented in § 5.3.1 in the manner of § 6.2.4.

In the case of *dyadic* modal controllers, however, the design procedure presented in § 6.2.5 can be extended by making use of the results derived in § 5.3.1. In effecting this extension, it is important to note that the mode-controllability characteristics of the system (6.1) will be altered by the introduction of dyadic modal control in cases when the Jordan canonical form, $\mathbf{J}$, of the plant matrix $\mathbf{A}$ contains sets of Jordan blocks associated with the same-valued eigenvalues. In fact, the controllability of all but one block of each such set of Jordan blocks will be completely destroyed since a time-invariant single-input linear system is controllable if and only if all the rows of its mode-controllability matrix corresponding to the last rows of the constituent Jordan blocks of $\mathbf{J}$ containing the same-valued eigenvalue are linearly independent.

6.4. Discrete-time systems

The results presented in the foregoing sections of this chapter concerning the modal control of multi-input continuous-time systems may be used (by analogy) in connection with discrete-time systems governed by a state equation of the form

$$\mathbf{x}\{(k+1)T\}=\mathbf{\Psi}(T)\mathbf{x}(kT)+\mathbf{\Delta}(T)\mathbf{z}(kT). \tag{6.89}$$

Indeed, the general theory developed in § 6.2.1 and the various design procedures presented in §§ 6.2.2, 6.2.3, 6.2.4, and 6.2.5 carry over directly from the continuous-time domain to the discrete-time domain simply by using the eigenproperties of the matrices associated with systems governed by state equations of the form (6.89) instead of the corresponding eigenproperties of the matrices associated with systems governed by state equations of the form (6.1).

References

[1] Simon, J. D., and Mitter, S. K., 1968, " A theory of modal control ", *Inf. Control*, **13**, 316.

[2] Gould, L. A., Murphy, A. T., and Berkman, E. F., 1970, " On the Simon–Mitter pole allocation algorithm-explicit gains for repeated eigenvalues ", *I.E.E.E. Trans. autom. Control*, **15**, 259.

[3] Retallack, D. G., and MacFarlane, A. G. J., 1970, " Pole-shifting techniques for multivariable systems ", *Proc. Instn. elect. Engrs*, **117**, 1037.

Modal Control Systems Incorporating State Observers

CHAPTER 7

7.1. Introduction

In a number of design problems concerning the modal control of both continuous-time and discrete-time systems, the state vector of the system concerned is not accessible to direct measurement. However, it is nevertheless possible in many such problems to implement modal control by using *output* rather than *state* feedback if an additional dynamical system known as a *state observer* [1] is introduced into the original system. In this chapter a method of effecting modal control by feedback of the outputs of the original system and of the state observer is presented. The method applies to continuous-time systems modelled by state, output, and control-law equations of the respective forms

$$\dot{\mathbf{x}}(t)=\mathbf{A}\mathbf{x}(t)+\mathbf{B}\mathbf{z}(t), \tag{7.1 a}$$

$$\mathbf{y}(t)=\mathbf{C}'\mathbf{x}(t), \tag{7.1 b}$$

and

$$\mathbf{z}(t)=\mathbf{G}\mathbf{x}(t), \tag{7.1 c}$$

and to discrete-time systems modelled by corresponding state, output, and control-law equations of the respective forms

$$\mathbf{x}\{(k+1)T\}=\mathbf{\Psi}(T)\mathbf{x}(kT)+\mathbf{\Delta}(T)\mathbf{z}(kT), \tag{7.2 a}$$

$$\mathbf{y}(kT)=\mathbf{\Gamma}'(T)\mathbf{x}(kT), \tag{7.2 b}$$

and

$$\mathbf{z}(kT)=\mathbf{\Omega}(T)\mathbf{x}(kT). \tag{7.2 c}$$

In eqns. (7.1 a), (7.1 b), (7.2 a), and (7.2 b), the various vectors and matrices are as defined in eqns. (1.1), (1.2), (1.3), and (1.4) : the gain matrices, $\mathbf{G}$ and $\mathbf{\Omega}(T)$, are $r\times n$ constant matrices which define the feedback laws (7.1 c) and (7.2 c), respectively.

Solutions of the state observer problem for continuous-time systems have, of course, been given by Luenberger [1], [2] and developed by Cumming [3], Newmann [4], [5], and Bongiorno and Youla [6], *inter alia*. However, all these solutions are unsatisfactory in certain respects : thus, for example, the design procedure associated with Luenberger's first solution [1] of this problem is not presented as a closed-form algorithm, whilst the second solution [2] involves the construction of a special-purpose canonical representation of the system as part of the process of designing a state observer of minimal order.† Similarly, the design procedures presented by Cumming [3] and Newmann [4], [5] do not guarantee that the resulting observer has satisfactory dynamical characteristics : furthermore, if, in a given design, it transpires that these characteristics are unsatisfactory, these design procedures give no guidance as to the means of achieving a design with satisfactory characteristics. Finally, the design procedure developed by Bongiorno and Youla [6], whilst permitting the assignment of arbitrary dynamics to the observer, requires the implementation of an algorithm which contains numerous complicated calculations such as the manipulation of matrix polynomials.

In contrast, the modal theory of state observers presented in this chapter is based upon Kalman's [7] classical canonical representation of a multivariable dynamical system : since this representation is simply a description of a system in modal state space, the need to construct a canonical representation of the kind used by Luenberger [2] is thus circumvented. Moreover, it transpires that the modal theory of state observers leads to closed-form formulas for the design of state observers whose dynamical characteristics may be prescribed arbitrarily provided that the following conditions obtain :

(i) the appropriate open-loop plant matrix has n distinct eigenvalues,
(ii) the modes which are to be controlled are controllable by at least one of the input variables and are observable in at least one of the output variables,
(iii) the rank of the output matrix of the system is q.

It will be noted that (ii) implies that the system need not necessarily be controllable and observable.

7.2. Continuous-time state observers

7.2.1. *General theory*

Since all the eigenvalues, λ_j, of $\mathbf{A}$ are distinct, then

$$\mathbf{AU}=\mathbf{U\Lambda}, \tag{7.3}$$

where $\mathbf{U}$ is the non-singular modal matrix of $\mathbf{A}$ and $\mathbf{\Lambda}=\text{diag}\,\{\lambda_1, \lambda_2, \ldots, \lambda_n\}$ is the associated eigenvalue matrix. It will be recalled from Chapter 2 that under the transformation

$$\boldsymbol{\xi}(t)=\mathbf{U}^{-1}\mathbf{x}(t), \tag{7.4}$$

† The minimal order of a Luenberger observer is $n-q$.

eqns. (7.1 *a*) and (7.1 *b*) assume the respective forms

$$\dot{\boldsymbol{\xi}}(t)=\boldsymbol{\Lambda}\boldsymbol{\xi}(t)+\mathbf{P}\mathbf{z}(t), \tag{7.5 a}$$

and

$$\mathbf{y}(t)=\mathbf{R}'\boldsymbol{\xi}(t), \tag{7.5 b}$$

where $\mathbf{P}=\mathbf{U}^{-1}\mathbf{B}$ and $\mathbf{R}=\mathbf{U}'\mathbf{C}$ are respectively the mode-controllability and mode-observability matrices of the system (7.1).

In view of Kalman's canonical decomposition theorem [7], it will in general be possible to partition the state eqn. (7.5 *a*) into the four equations

$$\dot{\boldsymbol{\xi}}_1(t)=\boldsymbol{\Lambda}_1\boldsymbol{\xi}_1(t)+\mathbf{P}_1\mathbf{z}(t), \tag{7.6 a}$$

$$\dot{\boldsymbol{\xi}}_2(t)=\boldsymbol{\Lambda}_2\boldsymbol{\xi}_2(t)+\mathbf{P}_2\mathbf{z}(t), \tag{7.6 b}$$

$$\dot{\boldsymbol{\xi}}_3(t)=\boldsymbol{\Lambda}_3\boldsymbol{\xi}_3(t), \tag{7.6 c}$$

and

$$\dot{\boldsymbol{\xi}}_4(t)=\boldsymbol{\Lambda}_4\boldsymbol{\xi}_4(t), \tag{7.6 d}$$

and to write the output eqn. (7.5 *b*) in the form

$$\mathbf{y}(t)=\mathbf{R}_1'\boldsymbol{\xi}_1(t)+\mathbf{R}_3'\boldsymbol{\xi}_3(t), \tag{7.6 e}$$

where

$$\boldsymbol{\xi}(t)=\begin{bmatrix}\boldsymbol{\xi}_1(t)\\ \boldsymbol{\xi}_2(t)\\ \boldsymbol{\xi}_3(t)\\ \boldsymbol{\xi}_4(t)\end{bmatrix} \tag{7.7}$$

The sub-systems (7.6 *a*, *b*, *c*, *d*) are respectively the n_1th-order controllable and observable, n_2th-order controllable and unobservable, n_3th-order uncontrollable and observable, and n_4th-order uncontrollable and unobservable sub-systems of the system (7.5). It is evident from eqn. (7.6 *e*) that the vectors, $\boldsymbol{\xi}_1(t)$ and $\boldsymbol{\xi}_3(t)$, associated with the observable modes may be recovered by direct computation using a passive state observer in cases when $n_1+n_3\leqslant q$. However, in most cases $n\geqslant n_1+n_3>q$, and a dynamic state observer is therefore required.

Thus, if the system (7.1) is now augmented by the introduction of a state observer governed by state and output equations of the respective forms

$$\dot{\boldsymbol{\eta}}(t)=\boldsymbol{\theta}\boldsymbol{\eta}(t)+\mathbf{M}\mathbf{y}(t)+\mathbf{F}\mathbf{z}(t), \tag{7.8 a}$$

and

$$\boldsymbol{\chi}(t)=\boldsymbol{\eta}(t), \tag{7.8 b}$$

then it may be deduced from eqns. (7.6) and (7.8) that

$$\begin{aligned}\dot{\boldsymbol{\chi}}(t)-\mathbf{T}_1\dot{\boldsymbol{\xi}}_1(t)-\mathbf{T}_3\dot{\boldsymbol{\xi}}_3(t)&=\boldsymbol{\theta}\{\boldsymbol{\chi}(t)-\mathbf{T}_1\boldsymbol{\xi}_1(t)-\mathbf{T}_3\boldsymbol{\xi}_3(t)\}\\ &\quad+(\mathbf{F}-\mathbf{T}_1\mathbf{P}_1)\mathbf{z}(t)+(\boldsymbol{\theta}\mathbf{T}_1-\mathbf{T}_1\boldsymbol{\Lambda}_1+\mathbf{M}\mathbf{R}_1')\boldsymbol{\xi}_1(t)\\ &\quad+(\boldsymbol{\theta}\mathbf{T}_3-\mathbf{T}_3\boldsymbol{\Lambda}_3+\mathbf{M}\mathbf{R}_3')\boldsymbol{\xi}_3(t),\end{aligned} \tag{7.9}$$

where $\boldsymbol{\theta}$, $\mathbf{F}$, $\mathbf{M}$, $\mathbf{T}_1$ and $\mathbf{T}_3$ are constant matrices. If these matrices are chosen such that

$$\mathbf{M}\mathbf{R}_1' = \mathbf{T}_1\boldsymbol{\Lambda}_1 - \boldsymbol{\theta}\mathbf{T}_1, \tag{7.10 a}$$

$$\mathbf{M}\mathbf{R}_3' = \mathbf{T}_3\boldsymbol{\Lambda}_3 - \boldsymbol{\theta}\mathbf{T}_3, \tag{7.10 b}$$

and

$$\mathbf{F} = \mathbf{T}_1\mathbf{P}_1, \tag{7.10 c}$$

then eqn. (7.9) assumes the form

$$\dot{\boldsymbol{\chi}}(t) - \mathbf{T}_1\dot{\boldsymbol{\xi}}_1(t) - \mathbf{T}_3\dot{\boldsymbol{\xi}}_3(t) = \boldsymbol{\theta}\{\boldsymbol{\chi}(t) - \mathbf{T}_1\boldsymbol{\xi}_1(t) - \mathbf{T}_3\boldsymbol{\xi}_3(t)\}. \tag{7.11}$$

The solution of (7.11) is clearly

$$\boldsymbol{\chi}(t) = \mathbf{T}_1\boldsymbol{\xi}_1(t) + \mathbf{T}_3\boldsymbol{\xi}_3(t) + \{\exp(\boldsymbol{\theta}t)\}\{\boldsymbol{\chi}(0) - \mathbf{T}_1\boldsymbol{\xi}_1(0) - \mathbf{T}_3\boldsymbol{\xi}_3(0)\}, \tag{7.12}$$

where $\boldsymbol{\chi}(0)$, $\boldsymbol{\xi}_1(0)$, and $\boldsymbol{\xi}_3(0)$ are the respective initial values of $\boldsymbol{\chi}(t)$, $\boldsymbol{\xi}_1(t)$, and $\boldsymbol{\xi}_3(t)$. Equation (7.12) indicates that the vector $\boldsymbol{\chi}(t)$ is given by an expression of the form

$$\boldsymbol{\chi}(t) = \mathbf{T}_1\boldsymbol{\xi}_1(t) + \mathbf{T}_3\boldsymbol{\xi}_3(t) + \mathbf{f}(t), \tag{7.13}$$

where $\mathbf{f}(t)$ is a vector which will tend to vanish as $t \to \infty$ if $\boldsymbol{\theta}$ is chosen to be a matrix whose eigenvalues all have negative real parts. It may therefore be deduced from eqns. (7.6 *e*) and (7.13) that

$$\begin{bmatrix} \mathbf{y}(t) \\ \boldsymbol{\chi}(t) \end{bmatrix} = \begin{bmatrix} \mathbf{R}_1', & \mathbf{R}_3' \\ \mathbf{T}_1, & \mathbf{T}_3 \end{bmatrix} \begin{bmatrix} \boldsymbol{\xi}_1(t) \\ \boldsymbol{\xi}_3(t) \end{bmatrix} + \begin{bmatrix} \mathbf{0} \\ \mathbf{I} \end{bmatrix} \mathbf{f}(t) \tag{7.14}$$

and consequently that

$$\begin{bmatrix} \boldsymbol{\xi}_1(t) \\ \boldsymbol{\xi}_3(t) \end{bmatrix} = \begin{bmatrix} \mathbf{L}_{11}, & \mathbf{L}_{12} \\ \mathbf{L}_{21}, & \mathbf{L}_{22} \end{bmatrix} \begin{bmatrix} \mathbf{y}(t) \\ \boldsymbol{\chi}(t) \end{bmatrix} - \begin{bmatrix} \mathbf{L}_{12} \\ \mathbf{L}_{22} \end{bmatrix} \mathbf{f}(t) \tag{7.15}$$

provided that the matrix

$$\mathbf{N} = \begin{bmatrix} \mathbf{R}_1', & \mathbf{R}_3' \\ \mathbf{T}_1, & \mathbf{T}_3 \end{bmatrix} \tag{7.16}$$

is of dimension $(n_1+n_3) \times (n_1+n_3)$ and has the inverse

$$\mathbf{N}^{-1} = \begin{bmatrix} \mathbf{L}_{11}, & \mathbf{L}_{12} \\ \mathbf{L}_{21}, & \mathbf{L}_{22} \end{bmatrix}. \tag{7.17}$$

The fact that $[\mathbf{R}_1', \mathbf{R}_3']$ is a $q \times (n_1+n_3)$ matrix implies that $[\mathbf{T}_1, \mathbf{T}_3]$ must be an $(n_1+n_3-q) \times (n_1+n_3)$ matrix and, consequentially, that the respective dimensions of $\mathbf{T}_1$, $\mathbf{T}_3$, $\boldsymbol{\eta}$, $\boldsymbol{\theta}$, $\mathbf{M}$, and $\mathbf{F}$ are $(n_1+n_3-q) \times n_1$, $(n_1+n_3-q) \times n_3$, $(n_1+n_3-q) \times 1$, $(n_1+n_3-q) \times (n_1+n_3-q)$, $(n_1+n_3-q) \times q$, and $(n_1+n_3-q) \times r$. It is thus clear that, in view of the dimension of $\boldsymbol{\eta}$, the state observer (7.8) is of order n_1+n_3-q; since $n_1+n_3-q \leqslant n-q$, it is evident that the order of a state observer designed by the present method is never greater than $n-q$, the minimal order of a Luenberger observer.

Now if $\boldsymbol{\theta}$ is chosen to be diagonal and to have real or complex conjugate eigenvalues which are all distinct from those of $\mathbf{A}$, then the elements of the unique solutions, $\mathbf{T}_1$ and $\mathbf{T}_3$, of eqns. (7.10 *a*) and (7.10 *b*) are

$$t_{ij}^{(l)} = \sum_{k=1}^{q} \frac{m_{ik} r_{jk}^{(l)}}{(\lambda_j^{(l)} - \theta_i)}$$

$$(l=1, 3 \; ; \; i=1, 2, \ldots, n_1+n_3-q \; ; \; j=1, 2, \ldots, n_l), \tag{7.18}$$

where

$$\mathbf{T}_l = [t_{ij}^{(l)}], \tag{7.19 a}$$

$$\boldsymbol{\theta} = \mathrm{diag}\,(\theta_1, \theta_2, \ldots, \theta_{n_1+n_3-q}), \tag{7.19 b}$$

$$\mathbf{M} = [m_{ik}], \tag{7.19 c}$$

$$\mathbf{R}_l = [r_{jk}^{(l)}], \tag{7.19 d}$$

and

$$\boldsymbol{\Lambda}_l = [\lambda_j^{(l)}]. \tag{7.19 e}$$

It is clear that the matrix $\mathbf{F}$ can be found from eqn. (7.10 *c*) by direct matrix multiplication after $\mathbf{T}_1$ has been computed in accordance with eqn. (7.18).

The matrices $\mathbf{T}_1$ and $\mathbf{T}_3$ defined by eqns. (7.18) indicate that the rows of $[\mathbf{T}_1, \mathbf{T}_3]$ and $[\mathbf{R}_1', \mathbf{R}_3']$ will be linearly independent and therefore that the matrix $\mathbf{N}$ defined in eqn. (7.16) will be non-singular in view of the following facts :

(i) $\lambda_j^{(l)} \neq \theta_i$,
(ii) the sub-systems (7.6 *a*) and (7.6 *c*) are observable,
(iii) the rank of the output matrix, $\mathbf{C}'$, of the system (7.1) is q,
(iv) the m_{ik} can always be chosen so that

$$\sum_{k=1}^{q} m_{ik} r_{jk}^{(l)} \neq 0 \quad (l=1, 3 \; ; \; i=1, 2, \ldots, n_1+n_3-q \; ; \; j=1, 2, \ldots, n_l).$$

Equation (7.15), which holds only when $\mathbf{N}$ is non-singular, is therefore valid in these circumstances.

Since eqn. (7.15) indicates that

$$\boldsymbol{\xi}_1(t) = \mathbf{L}_{11}\mathbf{y}(t) + \mathbf{L}_{12}\boldsymbol{\chi}(t) - \mathbf{L}_{12}\mathbf{f}(t) \tag{7.20 a}$$

and

$$\boldsymbol{\xi}_3(t) = \mathbf{L}_{21}\mathbf{y}(t) + \mathbf{L}_{22}\boldsymbol{\chi}(t) - \mathbf{L}_{22}\mathbf{f}(t), \tag{7.20 b}$$

it is apparent that control laws which have the form

$$\mathbf{z}(t) = \mathbf{G}\mathbf{x}(t) = \mathbf{G}\,\mathbf{U}\boldsymbol{\xi}(t) = \mathbf{H}_1\boldsymbol{\xi}_1(t) + \mathbf{H}_2\boldsymbol{\xi}_2(t) + \mathbf{H}_3\boldsymbol{\xi}_3(t) + \mathbf{H}_4\boldsymbol{\xi}_4(t) \tag{7.21}$$

when expressed in terms of the modal state vectors $\boldsymbol{\xi}_1(t)$, $\boldsymbol{\xi}_2(t)$, $\boldsymbol{\xi}_3(t)$, and $\boldsymbol{\xi}_4(t)$, can also be expressed in the form

$$\mathbf{z}(t) = (\mathbf{H}_1\mathbf{L}_{11} + \mathbf{H}_3\mathbf{L}_{21})\mathbf{y}(t) + (\mathbf{H}_1\mathbf{L}_{12} + \mathbf{H}_3\mathbf{L}_{22})\boldsymbol{\chi}(t)$$
$$- (\mathbf{H}_1\mathbf{L}_{12} + \mathbf{H}_3\mathbf{L}_{22})\mathbf{f}(t) + \mathbf{H}_2\boldsymbol{\xi}_2(t) + \mathbf{H}_4\boldsymbol{\xi}_4(t) \tag{7.22}$$

in view of eqns. (7.20). It is apparent that the control law (7.22) cannot be implemented in practice in view of the presence of the vector $\mathbf{f}(t)$ and of the vectors $\boldsymbol{\xi}_2(t)$ and $\boldsymbol{\xi}_4(t)$ associated with the unobservable sub-systems (7.6 *b*) and (7.6 *d*) of the system (7.1). Nevertheless, the fact that $\mathbf{f}(t)\rightarrow\mathbf{0}$ as $\mathbf{t}\rightarrow\infty$ suggests that the control law

$$\mathbf{z}(t)=(\mathbf{H}_1\mathbf{L}_{11}+\mathbf{H}_3\mathbf{L}_{21})\mathbf{y}(t)+(\mathbf{H}_1\mathbf{L}_{12}+\mathbf{H}_3\mathbf{L}_{22})\boldsymbol{\chi}(t) \tag{7.23}$$

constitutes an alternative to (7.22) in cases when

$$\mathbf{H}_2=\mathbf{0} \tag{7.24 a}$$

and

$$\mathbf{H}_4=\mathbf{0}. \tag{7.24 b}$$

The control law (7.23) can be implemented in practice since the system output, $\mathbf{y}(t)$, and the observer output, $\boldsymbol{\chi}(t)$, are both available for control purposes. Equations (7.24) are, of course, identically satisfied if the system (7.1) is observable. However, even if the system is only partially observable, it is still possible to implement the control law (7.23) provided that the gain matrices concerned have the special structure indicated by eqns. (7.21) and (7.24) : such gain matrices arise naturally in the case of control laws of any class derived using modal theory.

The conjecture that the control law (7.23) constitutes a practical alternative to the control law (7.22) can be validated by determining the state equation of the complete, augmented, closed-loop system defined by eqns. (7.6), (7.8), and (7.23). In fact, by making the appropriate substitutions from eqn. (7.23) into eqns. (7.6) and (7.8), the required state equation is found to be

$$\begin{bmatrix}\dot{\boldsymbol{\xi}}_1(t)\\ \dot{\boldsymbol{\xi}}_2(t)\\ \dot{\boldsymbol{\xi}}_3(t)\\ \dot{\boldsymbol{\xi}}_4(t)\\ \dot{\boldsymbol{\eta}}(t)\end{bmatrix}=\begin{bmatrix}\boldsymbol{\Lambda}_1+\mathbf{P}_1\mathbf{Q}_1\mathbf{R}_1', & \mathbf{0}, & \mathbf{P}_1\mathbf{Q}_1\mathbf{R}_3', & \mathbf{0}, & \mathbf{P}_1\mathbf{Q}_2\\ \mathbf{P}_2\mathbf{Q}_1\mathbf{R}_1', & \boldsymbol{\Lambda}_2, & \mathbf{P}_2\mathbf{Q}_1\mathbf{R}_3', & \mathbf{0}, & \mathbf{P}_2\mathbf{Q}_2\\ \mathbf{0}, & \mathbf{0}, & \boldsymbol{\Lambda}_3, & \mathbf{0}, & \mathbf{0}\\ \mathbf{0}, & \mathbf{0}, & \mathbf{0}, & \boldsymbol{\Lambda}_4, & \mathbf{0}\\ \mathbf{M}\mathbf{R}_1'+\mathbf{F}\mathbf{Q}_1\mathbf{R}_1', & \mathbf{0}, & \mathbf{M}\mathbf{R}_3'+\mathbf{F}\mathbf{Q}_1\mathbf{R}_3', & \mathbf{0}, & \boldsymbol{\theta}+\mathbf{F}\mathbf{Q}_2\end{bmatrix}\begin{bmatrix}\boldsymbol{\xi}_1(t)\\ \boldsymbol{\xi}_2(t)\\ \boldsymbol{\xi}_3(t)\\ \boldsymbol{\xi}_4(t)\\ \boldsymbol{\eta}(t)\end{bmatrix}, \tag{7.25}$$

where

$$\mathbf{Q}_1=\mathbf{H}_1\mathbf{L}_{11}+\mathbf{H}_3\mathbf{L}_{21} \tag{7.26 a}$$

and

$$\mathbf{Q}_2=\mathbf{H}_1\mathbf{L}_{12}+\mathbf{H}_3\mathbf{L}_{22}. \tag{7.26 b}$$

The $(n+n_1+n_3-q)$ eigenvalues of the closed-loop plant matrix given in eqn. (7.25) comprise the n_2 eigenvalues of the diagonal matrix $\boldsymbol{\Lambda}_2$, the n_3 eigenvalues of the diagonal matrix $\boldsymbol{\Lambda}_3$, the n_4 eigenvalues of the diagonal matrix $\boldsymbol{\Lambda}_4$, together with the $(2n_1+n_3-q)$ eigenvalues of the matrix

$$\Xi=\begin{bmatrix}\boldsymbol{\Lambda}_1+\mathbf{P}_1\mathbf{Q}_1\mathbf{R}_1', & \mathbf{P}_1\mathbf{Q}_2\\ \mathbf{M}\mathbf{R}_1'+\mathbf{F}\mathbf{Q}_1\mathbf{R}_1', & \boldsymbol{\theta}+\mathbf{F}\mathbf{Q}_2\end{bmatrix}. \tag{7.27}$$

The latter eigenvalues can be determined by first verifying by explicit matrix multiplication that, because of equations (7.10 *a*), (7.10 *c*), (7.16), and (7.26), the matrix Ξ satisfies the equation

$$\begin{bmatrix} \mathbf{I}_{n_1}, & \mathbf{0} \\ \mathbf{T}_1, & \mathbf{I}_{n_1+n_3-q} \end{bmatrix}^{-1} \Xi \begin{bmatrix} \mathbf{I}_{n_1}, & \mathbf{0} \\ \mathbf{T}_1, & \mathbf{I}_{n_1+n_3-q} \end{bmatrix} = \begin{bmatrix} \mathbf{\Lambda}_1 + \mathbf{P}_1 \mathbf{H}_1, & \mathbf{P}_1 \mathbf{Q}_2 \\ \mathbf{0}, & \boldsymbol{\theta} \end{bmatrix}. \quad (7.28)$$

Since eqn. (7.28) expresses a similarity transformation, it follows that the $(2n_1+n_3-q)$ eigenvalues of Ξ are the n_1 eigenvalues of the matrix $(\mathbf{\Lambda}_1+\mathbf{P}_1\mathbf{H}_1)$ together with the (n_1+n_3-q) eigenvalues of the matrix $\boldsymbol{\theta}$. If it were possible to implement the control law

$$\mathbf{z}(t) = \mathbf{H}_1 \boldsymbol{\xi}_1(t) + \mathbf{H}_3 \boldsymbol{\xi}_3(t), \quad (7.29)$$

it can similarly be shown that the eigenvalues of the resulting nth-order closed-loop system would be the n_2 eigenvalues of the diagonal matrix $\mathbf{\Lambda}_2$, the n_3 eigenvalues of the diagonal matrix $\mathbf{\Lambda}_3$, and the n_4 eigenvalues of the diagonal matrix $\mathbf{\Lambda}_4$, together with the n_1 eigenvalues of the matrix $(\mathbf{\Lambda}_1+\mathbf{P}_1\mathbf{H}_1)$. It may therefore now be seen that the presence of the state observer does not affect the closed-loop eigenvalues which would have been obtained by implementing the control law (7.29) if the state vectors $\boldsymbol{\xi}_1(t)$ and $\boldsymbol{\xi}_3(t)$ were available for feedback in the first instance : however, its presence does, of course, introduce the (n_1+n_3-q) additional arbitrarily-assignable eigenvalues of the diagonal matrix $\boldsymbol{\theta}$.

If it is desired to assign complex conjugate values to certain of the eigenvalues of $\boldsymbol{\theta}$, eqn. (7.8 *a*) will contain complex matrices and will therefore not be in a form suitable for immediate physical realization. However, it is of course always possible in such circumstances to introduce a new state vector into eqn. (7.8 *a*) which is related to $\boldsymbol{\eta}(t)$ by a non-singular linear transformation and which is such that all the matrices in the resulting transformed equation are real : the required state observer can then be physically realised by implementing this transformed version of eqn. (7.8 *a*).

7.2.2. *Illustrative example*

The foregoing design procedure can be conveniently illustrated by considering a system of the class (7.1) with

$$\mathbf{A} = \begin{bmatrix} -2, & -1, & 1 \\ 1, & 0, & 1 \\ -1, & 0, & 1 \end{bmatrix}, \quad (7.30\,a)$$

$$\mathbf{B} = \begin{bmatrix} 1 \\ 1 \\ 1 \end{bmatrix}, \quad (7.30\,b)$$

and

$$\mathbf{C}' = \begin{bmatrix} 1, & 0, & 0 \\ 0, & 1, & 0 \end{bmatrix}, \quad (7.30\,c)$$

and with the associated eigenstructure

$$\mathbf{\Lambda} = \text{diag}\,\{1, \quad -1+i, \quad -1-i\}, \tag{7.31 a}$$

$$\mathbf{U} = \begin{bmatrix} 0, & 5, & 5 \\ 1, & -3-4i, & -3+4i \\ 1, & 2+i, & 2-i \end{bmatrix}, \tag{7.31 b}$$

$$\mathbf{V} = \tfrac{1}{10} \begin{bmatrix} -2, & 1+i, & 1-i \\ 2, & i, & -i \\ 8, & -i, & i \end{bmatrix}, \tag{7.31 c}$$

$$\mathbf{P} = \tfrac{1}{10} \begin{bmatrix} 8 \\ 1+i \\ 1-i \end{bmatrix}, \tag{7.31 d}$$

and

$$\mathbf{R} = \begin{bmatrix} 0, & 1 \\ 5, & -3-4i \\ 5, & -3+4i \end{bmatrix}, \tag{7.31 e}$$

where $\mathbf{V} = (\mathbf{U}^{-1})'$. Neither the mode-controllability matrix, $\mathbf{P}$, nor the mode-observability matrix, $\mathbf{R}$, of this system has null rows and it therefore follows that

$$n_1 = 3$$

and

$$n_2 = n_3 = n_4 = 0.$$

Since the system has two independent outputs ($q=2$), it is evident that the required state observer will be a first-order dynamical system governed by a state equation of the form

$$\dot{\eta}(t) = \theta_1 \eta(t) + [m_{11}, m_{12}]\mathbf{y}(t) + f_1 z(t). \tag{7.32}$$

In eqn. (7.32), θ_1, m_{11}, and m_{12} can be chosen arbitrarily subject to the constraints discussed in § 7.2.1 : one such choice is

$$\theta_1 = -2, \tag{7.33 a}$$

$$m_{11} = m_{12} = 1. \tag{7.33 b}$$

The transformation matrix, $\mathbf{T}_1$, can readily be calculated using eqn. (7.18) and is found to be

$$\mathbf{T}_1 = \tfrac{1}{3}\,[1, \quad -3-9i, \quad -3+9i]. \tag{7.34}$$

It should be noted that $\mathbf{R}_3'$ and $\mathbf{T}_3$ do not exist in this example since $n_3 = 0$. Thus, the matrix $\mathbf{N}$ and the scalar f_1 are given by the respective equations

$$\mathbf{N} = \begin{bmatrix} \mathbf{R}_1' \\ \\ \mathbf{T}_1 \end{bmatrix} = \begin{bmatrix} 0, & 5, & 5 \\ 1, & -3-4i, & -3+4i \\ 1/3, & -1-3i, & -1+3i \end{bmatrix} \tag{7.35 a}$$

and

$$f_1 = \mathbf{T}_1 \mathbf{P}_1 = 2/3, \tag{7.35 b}$$

where $\mathbf{P}_1=\mathbf{P}$ and $\mathbf{R}_1=\mathbf{R}$ since $n_1=n$ in this example. Furthermore, it follows from (7.35 *a*) that

$$\mathbf{N}^{-1}=\begin{bmatrix}\mathbf{L}_{11}, & \mathbf{L}_{12}\\ \\ \mathbf{L}_{21}, & \mathbf{L}_{22}\end{bmatrix}=\tfrac{1}{10}\begin{bmatrix}6, & 18, & -24\\ 1, & -i, & 3i\\ 1, & i, & -3i\end{bmatrix} \tag{7.36 a}$$

so that

$$\mathbf{L}_{11}=\tfrac{1}{10}\,[6, \quad 18] \tag{7.36 b}$$

and

$$\mathbf{L}_{12}=\tfrac{1}{10}\,[-24]. \tag{7.36 c}$$

It can be deduced from eqns. (7.8), (7.33), and (7.35 *b*) that the required state observer for the system (7.29) is described by the state and output equations

$$\dot{\eta}_1(t)=-2\eta_1(t)+y_1(t)+y_2(t)+\tfrac{2}{3}\,z(t), \tag{7.37 a}$$

and

$$\chi_1(t)=\eta_1(t). \tag{7.37 b}$$

Now consider that it is desired to control the system defined by eqns. (7.30) in such a way that the unstable eigenvalue $\lambda_1=1$ is changed to the stable eigenvalue $\rho_1=-3$, whilst leaving the remaining eigenvalues in the matrix (7.31 *a*) unchanged. If $\mathbf{x}(t)$ were available for direct feedback, the formula (5.34) indicates that this eigenvalue shift could be obtained by generating the input variable to the system (7.30) in accordance with the control law

$$z(t)=K_1\mathbf{v}_1{}'\mathbf{x}(t)=K_1\xi_1(t), \tag{7.38}$$

where

$$K_1=\frac{\rho_1-\lambda_1}{p_1}, \tag{7.39}$$

p_1 is the first element of the vector (7.31 *d*), $\mathbf{v}_1$ is the first column of the matrix (7.31 *c*), and $\xi_1(t)$ is the state variable associated with the unstable but controllable and observable mode of the system (7.30). In the present numerical example it is evident from eqns. (7.31 *d*) and (7.39) that

$$K_1=-5 \tag{7.40}$$

and therefore that the control law (7.38) is

$$z(t)=-5\xi_1(t), \tag{7.41}$$

so that, in the terminology of eqn. (7.21),

$$\mathbf{H}_1=[-5, \quad 0, \quad 0]. \tag{7.42}$$

It can be deduced from eqns. (7.36) and (7.42) that the control law (7.23) is

$$z(t)=[-3, \quad -9]\mathbf{y}(t)+12\chi_1(t) \tag{7.43}$$

in this example. Since eqn. (7.30 *c*) indicates that

$$\mathbf{y}(t)=\begin{bmatrix}1, & 0, & 0\\ \\ 0, & 1, & 0\end{bmatrix}\mathbf{x}(t), \tag{7.44}$$

it follows from eqns. (7.37 *b*) and (7.44) that eqn. (7.43) has the explicit form

$$z(t)=-3x_1(t)-9x_2(t)+12\eta_1(t) \tag{7.45}$$

when expressed in terms of the state variables of the system (7.30) and of the observer (7.37). Finally, it follows from eqns. (7.30), (7.37), and (7.45) that the state equation of the complete, closed-loop system incorporating the observer is

$$\begin{bmatrix}\dot{x}_1(t)\\ \dot{x}_2(t)\\ \dot{x}_3(t)\\ \dot{\eta}_1(t)\end{bmatrix}=\begin{bmatrix}-5, & -10, & 1, & 12\\ -2, & -9, & 1, & 12\\ -4, & -9, & 1, & 12\\ -1, & -5, & 0, & 6\end{bmatrix}\begin{bmatrix}x_1(t)\\ x_2(t)\\ x_3(t)\\ \eta_1(t)\end{bmatrix}. \tag{7.46}$$

It may readily be verified that the eigenvalues of the plant matrix in eqn. (7.46) are the members of the set $\{-3, -1+i, -1-i, -2\}$, as required. Thus, the inclusion of the state observer and the implementation of the control law (7.43) results in the same closed-loop eigenvalues as would be obtained by the implementation of the control law (7.38) if the state vector $\mathbf{x}(t)$ were available for feedback in the first instance.

7.3. Discrete-time state observers

The results presented in the foregoing sections of this chapter concerning the design of state observers for continuous-time systems modelled by governing equations of the form (7.1) may be used (by analogy) in connection with the design of state observers for discrete-time systems modelled by governing equations of the form (7.2).

Thus, consider that the discrete-time system (7.2) is augmented by the introduction of a discrete-time state observer governed by state and output equations of the respective forms

$$\boldsymbol{\eta}\{(k+1)T\}=\boldsymbol{\theta}(T)\boldsymbol{\eta}(kT)+\mathbf{M}(T)\mathbf{y}(kT)+\mathbf{F}(T)\mathbf{z}(kT) \tag{7.47 a}$$

and

$$\boldsymbol{\chi}(kT)=\boldsymbol{\eta}(kT). \tag{7.47 b}$$

Since the system output, $\mathbf{y}(kT)$, and the observer output, $\boldsymbol{\chi}(kT)$, are both available for control purposes, it is possible to implement a control law of the form

$$\mathbf{z}(kT)=\{\mathbf{H}_1(T)\mathbf{L}_{11}(T)+\mathbf{H}_3(T)\mathbf{L}_{21}(T)\}\mathbf{y}(kT) \\ +\{\mathbf{H}_1(T)\mathbf{L}_{12}(T)+\mathbf{H}_3(T)\mathbf{L}_{22}(T)\}\boldsymbol{\chi}(kT) \tag{7.48}$$

if the matrices $\boldsymbol{\theta}(T)$, $\mathbf{M}(T)$, and $\mathbf{F}(T)$ are chosen in the manner of § 7.2.1 : in eqn. (7.48), the matrices $\mathbf{L}_{11}(T)$, $\mathbf{L}_{12}(T)$, $\mathbf{L}_{21}(T)$, and $\mathbf{L}_{22}(t)$ are determined by inverting a matrix $\mathbf{N}(T)$ which is analagous to the matrix $\mathbf{N}$ given in eqn.

(7.16). In this way, it is possible to assign arbitrary values to the eigenvalues of the matrix $\boldsymbol{\theta}(T)$, whilst at the same time ensuring that the presence of the state observer (7.47) does not affect the closed-loop eigenvalues which would have been obtained by implementing the control law

$$\mathbf{z}(kT) = \mathbf{H}_1(T)\boldsymbol{\xi}_1(kT) + \mathbf{H}_3(T)\boldsymbol{\xi}_3(kT) \tag{7.49}$$

if the state vectors $\boldsymbol{\xi}_1(kT)$ and $\boldsymbol{\xi}_3(kT)$ associated with the observable modes of the system (7.2) were available for feedback in the first instance.

References

[1] Luenberger, D. G., 1964, " Observing the state of a linear system ", *I.E.E.E. Trans. milit. Electron.*, **8,** 74.

[2] Luenberger, D. G., 1966, " Observers for multivariable systems ", *I.E.E.E. Trans. autom. Control*, **11,** 190.

[3] Cumming, S. D. G., 1969, " Design of observers of reduced dynamics ", *Electron. Lett.*, **5,** 213.

[4] Newmann, M. M., 1969, " Optimal and sub-optimal control using an observer when some of the state variables are not measurable ", *Int. J. Control*, **9,** 281.

[5] Newmann, M. M., 1969, " Design algorithms for minimal-order Luenberger observers ", *Electron. Lett.*, **5,** 390.

[6] Bongiorno, J. J., and Youla, P. C., 1968, " On observers in multivariable control systems ", *Int. J. Control*, **8,** 221.

[7] Kalman, R. E., 1961, " On the general theory of control systems ", *Proc. 1st I.F.A.C. Congress*, Moscow, 1960 (London : Butterworths), **1,** 481.

Modal control systems incorporating integral feedback

CHAPTER 8

8.1. Introduction

In order to design control systems which possess good steady-state as well as good transient response characteristics, it is of course frequently necessary to introduce integral as well as proportional feedback of some or all of the state variables of the system under control. However, it has been shown by Porter and Power [1], [2] that the introduction of integral feedback may destroy the controllability of otherwise controllable systems. Accordingly, in this chapter, some necessary and sufficient conditions are derived which must be satisfied if a system is to remain controllable in the presence of feedback of the integrals of state variables. These conditions are obtained by examining the structure of the mode-controllability matrices of systems of this class : the properties of these matrices may then be used in the manner of §§ 6.2.4 and 6.2.5 to design appropriate feedback loops for systems incorporating integral control action.

8.2. Continuous-time systems incorporating integral feedback

8.2.1. *Mode-controllability matrices*

Thus, consider a system governed by a state equation of the form

$$\dot{\mathbf{x}}(t) = \mathbf{A}\mathbf{x}(t) + \mathbf{B}\mathbf{z}(t), \tag{8.1}$$

where $\mathbf{A}$ and $\mathbf{B}$ are respectively $n \times n$ and $n \times r$ constant real matrices, $\mathbf{x}(t)$ is an $n \times 1$ state vector, and $\mathbf{z}(t)$ is an $n \times r$ input vector : in eqn. (8.1), $\mathbf{A}$ is assumed initially to be non-singular. If the integrals of $l(\leqslant n)$ of the elements of $\mathbf{x}(t)$ are to be fed back, the state vector of the resulting system will consist of $\mathbf{x}(t)$ augmented by the $l \times 1$ vector, $\boldsymbol{\chi}(t)$, where

$$\dot{\boldsymbol{\chi}}(t) = \mathbf{E}_l \mathbf{x}(t) \tag{8.2}$$

and $\mathbf{E}_l$ is an $l \times n$ matrix which consists of an appropriate selection of l rows from the $n \times n$ unit matrix $\mathbf{I}_n$. The state equation of the augmented system is then clearly

$$\begin{bmatrix} \dot{\mathbf{x}}(t) \\ \dot{\boldsymbol{\chi}}(t) \end{bmatrix} = \begin{bmatrix} \mathbf{A}, & \mathbf{0} \\ \mathbf{E}_l, & \mathbf{0} \end{bmatrix} \begin{bmatrix} \mathbf{x}(t) \\ \boldsymbol{\chi}(t) \end{bmatrix} + \begin{bmatrix} \mathbf{B} \\ \mathbf{0} \end{bmatrix} \mathbf{z}(t), \tag{8.3}$$

where it is convenient to introduce the following abbreviations :

$$\hat{\mathbf{A}} = \begin{bmatrix} \mathbf{A}, & \mathbf{0} \\ \mathbf{E}_l, & \mathbf{0} \end{bmatrix} \tag{8.4}$$

and

$$\hat{\mathbf{B}} = \begin{bmatrix} \mathbf{B} \\ \mathbf{0} \end{bmatrix}. \tag{8.5}$$

Now if the Jordan canonical form of $\mathbf{A}$ is the non-singular matrix, $\mathbf{J}$, the generalized modal matrix of $\mathbf{A}$ is $\mathbf{U}$, and the generalized modal matrix of $\mathbf{A}'$ is $\mathbf{V}$, then

$$\mathbf{A}\mathbf{U} = \mathbf{U}\mathbf{J}, \tag{8.6}$$

where the generalized modal matrices can be normalized so that

$$\mathbf{V}'\mathbf{U} = \mathbf{I}_n \tag{8.7}$$

and the primes denote transposition. By the same token, if the Jordan canonical form of $\hat{\mathbf{A}}$ is $\hat{\mathbf{J}}$, the generalized modal matrix of $\hat{\mathbf{A}}$ is $\hat{\mathbf{U}}$, and the generalized modal matrix of $\hat{\mathbf{A}}'$ is $\hat{\mathbf{V}}$, then

$$\hat{\mathbf{A}}\hat{\mathbf{U}} = \hat{\mathbf{U}}\hat{\mathbf{J}} \tag{8.8}$$

and

$$\hat{\mathbf{V}}'\hat{\mathbf{U}} = \mathbf{I}_{n+l}. \tag{8.9}$$

Since $\hat{\mathbf{A}}$ is given by eqn. (8.4), and since $\mathbf{A}$ is assumed to be non-singular, it may be inferred that

$$\hat{\mathbf{J}} = \begin{bmatrix} \mathbf{J}, & \mathbf{0} \\ \mathbf{0}, & \mathbf{0} \end{bmatrix}. \tag{8.10}$$

It can therefore be deduced from eqns. (8.6), (8.7), (8.8), (8.9), and (8.10) that

$$\hat{\mathbf{U}} = \begin{bmatrix} \mathbf{U}, & \mathbf{0} \\ \mathbf{E}_l\mathbf{U}\mathbf{J}^{-1}, & \mathbf{I}_l \end{bmatrix} \tag{8.11 a}$$

and

$$\hat{\mathbf{V}}' = \begin{bmatrix} \mathbf{V}' & \mathbf{0} \\ -\mathbf{E}_l\mathbf{U}\mathbf{J}^{-1}\mathbf{U}^{-1}, & \mathbf{I}_l \end{bmatrix}. \tag{8. 11 b}$$

The mode-controllability matrix, $\hat{\mathbf{P}}$, of the system governed by eqn. (8.3) can now be readily determined, since

$$\hat{\mathbf{P}} = \hat{\mathbf{V}}'\hat{\mathbf{B}}. \tag{8.12}$$

In fact, it follows immediately from eqns. (8.5) and (8.11 *b*) that

$$\hat{\mathbf{P}} = \begin{bmatrix} \mathbf{V}'\mathbf{B} \\ \\ -\mathbf{E}_l\mathbf{U}\mathbf{J}^{-1}\mathbf{U}^{-1}\mathbf{B} \end{bmatrix}$$

which, in view of eqn. (8.6) can be written in the simpler form

$$\hat{\mathbf{P}} = \begin{bmatrix} \mathbf{V}'\mathbf{B} \\ \\ -\mathbf{E}_l\mathbf{A}^{-1}\mathbf{B} \end{bmatrix}. \tag{8.13}$$

Equation (8.13) gives the mode-controllability matrix of the system (8.3) in the required explicit form.

Now a system is controllable if and only if all the rows of its mode-controllability matrix corresponding to the last rows of Jordan blocks containing the same-valued eigenvalue are linearly independent. Since $\mathbf{A}$ is assumed to be non-singular, it follows that $\mathbf{A}$ has no zero-valued eigenvalues. It is therefore evident that the rows of the matrix $\hat{\mathbf{P}}$ given by eqn. (8.13) fall into the two following disjoint classes :

(i) the n rows of $\mathbf{V}'\mathbf{B}$ *none* of which corresponds to a Jordan block associated with a zero-valued eigenvalue ;

(ii) the l rows of $\mathbf{E}_l\mathbf{A}^{-1}\mathbf{B}$, *each* of which corresponds to a 1×1 Jordan block associated with a same-valued (zero) eigenvalue.

It may therefore be concluded that the system (8.3) will be controllable if and only if the system (8.1) (whose mode-controllability matrix is of course $\mathbf{V}'\mathbf{B}$) is controllable and the l rows of $\mathbf{E}_l\mathbf{A}^{-1}\mathbf{B}$ are linearly independent. These results clearly indicate that the introduction of integral action may destroy the controllability of a system. In particular, the condition involving $\mathbf{E}_l\mathbf{A}^{-1}\mathbf{B}$ is equivalent to the condition that

$$\text{rank } (\mathbf{E}_l\mathbf{A}^{-1}\mathbf{B}) = l. \tag{8.14}$$

It is important to recall that the foregoing analysis applies only to the special case when $\mathbf{A}$ is non-singular. However, if $(\mathbf{A}, \mathbf{B})$ is a controllable pair, Wonham's theorem [3] guarantees the existence of an $r \times n$ matrix $\mathbf{K}$ which is such that the matrix $(\mathbf{A} + \mathbf{BK})$ possesses n non-zero eigenvalues. The analysis culminating in (8.14) may therefore be readily generalized to embrace the case when $\mathbf{A}$ is singular by regarding the input vector, $\mathbf{z}(t)$, in eqn. (8.3) as being given by the equation

$$\mathbf{z}(t) = \mathbf{Kx}(t) + \mathbf{s}(t), \tag{8.15}$$

where $\mathbf{K}$ is an appropriate matrix and $\mathbf{s}(t)$ is an $r \times 1$ input vector. Equation (8.3) may then be replaced by the equation

$$\begin{bmatrix} \dot{\mathbf{x}}(t) \\ \dot{\boldsymbol{\chi}}(t) \end{bmatrix} = \begin{bmatrix} (\mathbf{A}+\mathbf{BK}), & \mathbf{0} \\ \mathbf{E}_l, & \mathbf{0} \end{bmatrix} \begin{bmatrix} \mathbf{x}(t) \\ \boldsymbol{\chi}(t) \end{bmatrix} + \begin{bmatrix} \mathbf{B} \\ \mathbf{0} \end{bmatrix} \mathbf{s}(t) \tag{8.16}$$

and it is thus evident from (8.14) and (8.16) that the mode-controllability matrix in this more general case is given by

$$\hat{\mathbf{P}} = \begin{bmatrix} \bar{\mathbf{V}}'\mathbf{B} \\ -\mathbf{E}_l(\mathbf{A}+\mathbf{BK})^{-1}\mathbf{B} \end{bmatrix}, \tag{8.17}$$

where $\bar{\mathbf{V}}$ is the modal matrix of $(\mathbf{A}+\mathbf{BK})'$. It therefore follows from (8.17) that in the general case when $\mathbf{A}$ is an $n \times n$ matrix, the system (8.3) incorporating integral feedback will be controllable if and only if the system (8.1) is controllable and there exists a matrix, $\mathbf{K}$, such that $(\mathbf{A}+\mathbf{BK})$ is non-singular and the l rows of the matrix $\mathbf{E}_l(\mathbf{A}+\mathbf{BK})^{-1}\mathbf{B}$ are linearly independent.

8.2.2. *Dyadic modal control*

In § 6.2.5, simple closed-form formulae were derived for the feedback gains of dyadic modal control systems. It will be recalled from Chapter 6 that dyadic systems possess the very important property that the determination of the feedback matrix

$$\mathbf{G} = \mathbf{fd}' \tag{8.18}$$

does not involve the solution of simultaneous non-linear algebraic equations which arise in general in the case of non-dyadic systems.

It is the purpose of the present section to extend the results of § 6.2.5 by presenting results which make it possible to design dyadic modal controllers which incorporate *integral* as well as *proportional* action. However, it will be shown that, in view of the results given in § 8.2.1 concerning the controllability of time-invariant linear systems incorporating integral feedback, such an extension is severely limited in regard to the number of integrated variables which may be fed back : this limitation is the price that has to be paid for the linearity (and consequential computational tractability) which results from the use of a dyadic feedback matrix of the form (8.18).

Thus, consider the class of closed-loop systems defined by the equations

$$\dot{\mathbf{x}}(t) = \mathbf{Ax}(t) + \mathbf{Bz}(t), \tag{8.19}$$

$$\dot{\boldsymbol{\chi}}(t) = \mathbf{E}_l\mathbf{x}(t), \tag{8.20}$$

and

$$\mathbf{z}(t) = \mathbf{G} \begin{bmatrix} \mathbf{x}(t) \\ \boldsymbol{\chi}(t) \end{bmatrix}, \tag{8.21}$$

where $\mathbf{x}(t)$ is an $n \times 1$ state vector, $\boldsymbol{\chi}(t)$ is an $l \times 1$ state vector whose elements are the integrals of l $(l \leqslant n)$ of the elements of $\mathbf{x}(t)$, $\mathbf{z}(t)$ is the $r \times 1$ input vector, $\mathbf{A}$ is a real non-singular $n \times n$ matrix with n distinct eigenvalues, $\mathbf{B}$ is a real $n \times r$ matrix, $\mathbf{E}_l$ is an $l \times n$ matrix consisting of an appropriate selection of l rows from the $n \times n$ unit matrix, and $\mathbf{G}$ is an $r \times (n+l)$ dyadic feedback matrix of the form defined in eqn. (8.18), where $\mathbf{f}$ and $\mathbf{d}$ are respectively $r \times 1$ and $(n+l) \times 1$ vectors.

Now, following the method of § 6.2.5, it may readily be shown that the system governed by the state eqns. (8.19) and (8.20) may be transformed to the equivalent single-input form

$$\dot{\boldsymbol{\zeta}}(t) = \boldsymbol{\Gamma}\boldsymbol{\zeta}(t) + \boldsymbol{\beta}\omega(t) \tag{8.22}$$

when $\mathbf{G}$ has the dyadic form given by eqn. (8.18). In eqn. (8.22),

$$\boldsymbol{\zeta}(t) = \begin{bmatrix} \mathbf{x}(t) \\ \boldsymbol{\chi}(t) \end{bmatrix}, \tag{8.23}$$

$$\boldsymbol{\Gamma} = \begin{bmatrix} \mathbf{A}, & \mathbf{0} \\ \mathbf{E}_l, & \mathbf{0} \end{bmatrix}, \tag{8.24}$$

$$\boldsymbol{\beta} = \begin{bmatrix} \mathbf{B} \\ \mathbf{0} \end{bmatrix} \mathbf{f}, \tag{8.25}$$

and

$$\omega(t) = \mathbf{d}'\boldsymbol{\zeta}(t). \tag{8.26}$$

In order to proceed with the determination of the dyadic feedback matrix which defines the control law (8.21), it is necessary to assume that the system defined by eqn. (8.19) is such that $(\mathbf{A}, \mathbf{B})$ constitutes a controllable pair : it is shown in § 8.2.1 that this property of the initial system (8.19) is a necessary condition for the controllability of the augmented system (8.19), (8.20), (8.21) which is created by the introduction of integral action. In addition, it is necessary to take into account the fact that the results presented in § 8.2.1 indicate that the mode-controllability matrix of the system (8.22) is given by the equation

$$\hat{\mathbf{p}} = \begin{bmatrix} \mathbf{V}'\mathbf{B}\mathbf{f} \\ -\mathbf{E}_l\mathbf{A}^{-1}\mathbf{B}\mathbf{f} \end{bmatrix}. \tag{8.27}$$

It is evident that the elements of $\hat{\mathbf{p}}$ given by eqn. (8.27) fall into the following two disjoint classes :

(i) the elements of $\mathbf{V}'\mathbf{B}\mathbf{f}$ none of which corresponds to a Jordan block associated with a zero-valued eigenvalue ;

(ii) the l elements of $\mathbf{E}_l\mathbf{A}^{-1}\mathbf{B}\mathbf{f}$, each of which correspond to a 1×1 Jordan block associated with the same-valued (zero) eigenvalue.

Since $\hat{\mathbf{p}}$ is a vector, it may therefore be concluded that the pair $(\boldsymbol{\Gamma}, \boldsymbol{\beta})$ associated with the dyadically-controlled system (8.22) will be controllable if and only if the pair $(\mathbf{A}, \mathbf{B}\mathbf{f})$ is controllable and

$$\text{rank } (\mathbf{E}_l\mathbf{A}^{-1}\mathbf{B}\mathbf{f}) = l. \tag{8.28}$$

Now

$$\text{rank } (\mathbf{E}_l\mathbf{A}^{-1}\mathbf{B}\mathbf{f}) \leqslant 1 \tag{8.29}$$

so that (8.28) and (8.29) are clearly equivalent to the inequality

$$l \leqslant 1$$

which has only the two possible solutions $l=0$ or $l=1$. If the former case obtains, then it is impossible to implement dyadic modal control action incorporating integral feedback : only the case $l=1$ (corresponding to feeding back the integral of only *one* state variable) therefore remains. These limitations on the possible forms of dyadic modal controllers incorporating integral action are those to which reference is made at the beginning of this section. Since most schemes [4], [5], [6] that have been proposed for eigenvalue assignment in the case of multi-input systems of the form (8.19) incorporate feedback matrices having (implicitly or explicitly) the dyadic structure (8.18), it is most important that these properties of systems of this class should be well known to the designer.

However, now that these limitations of the dyadic controller have been established, simple closed-form formulae can be obtained for the feedback gains in the case when (8.29) is satisfied by $l=1$. In this case, the plant matrix (8.24) is of dimension $(n+1)\times(n+1)$ and has the form

$$\mathbf{\Gamma}=\begin{bmatrix}\mathbf{A}, & \mathbf{0}\\ \mathbf{e}', & 0\end{bmatrix}, \tag{8.30}$$

where $\mathbf{e}'$ is a $1\times n$ vector : if $\mathbf{x}'(t)=[x_1(t), x_2(t), \ldots, x_n(t)]$ and it is desired to feed back the integral of $x_k(t)$, then all the elements of $\mathbf{e}'$ are zero except the kth which is unity.

Now, if the eigenvalues of $\mathbf{\Gamma}$ are $\hat{\lambda}_j$ $(j=1, 2, \ldots, n+1)$ and the eigenvalues of $\mathbf{A}$ are λ_j $(j=1, 2, \ldots, n)$, then it is evident from (8.24) that

$$\{\hat{\lambda}_1, \hat{\lambda}_2, \ldots, \hat{\lambda}_n, \hat{\lambda}_{n+1}\}=\{\lambda_1, \lambda_2, \ldots, \lambda_n, 0\}, \tag{8.31}$$

where the sets of the $\hat{\lambda}_j$ and λ_j in (8.31) are ordered. In addition, it can be readily verified that the $(n+1)\times(n+1)$ modal matrix, $\hat{\mathbf{V}}$, of $\mathbf{\Gamma}'$ is given by the equation

$$\hat{\mathbf{V}}'=\begin{bmatrix}\mathbf{V}', & \mathbf{0}\\ -\mathbf{e}'\mathbf{A}^{-1}, & 1\end{bmatrix}, \tag{8.32}$$

where $\mathbf{V}$ is the $n\times n$ modal matrix of $\mathbf{A}'$. It is now evident that the $(n+1)\times 1$ mode-controllability matrix, $\boldsymbol{\pi}$, of the system (8.22) is

$$\boldsymbol{\pi}=\hat{\mathbf{V}}'\boldsymbol{\beta}=\begin{bmatrix}\mathbf{V}', & \mathbf{0}\\ -\mathbf{e}'\mathbf{A}^{-1}, & 1\end{bmatrix}\begin{bmatrix}\mathbf{B}\\ 0\end{bmatrix}\mathbf{f}$$

which has the explict form

$$\boldsymbol{\pi}=\begin{bmatrix}\mathbf{v}_1'\mathbf{Bf}\\ \mathbf{v}_2'\mathbf{Bf}\\ \vdots\\ \mathbf{v}_n'\mathbf{Bf}\\ -\mathbf{e}'\mathbf{A}^{-1}\mathbf{Bf}\end{bmatrix}. \tag{8.33}$$

The first n elements of $\boldsymbol{\pi}$ correspond to the eigenvalues $\lambda_1, \lambda_2, \ldots, \lambda_n$ of $\boldsymbol{\Gamma}$, and the last element to the zero-valued eigenvalue of $\boldsymbol{\Gamma}$: in eqn. (8.33) the $\mathbf{v}_j$ are, of course, the eigenvectors of $\mathbf{A}'$ and $\mathbf{f}$ is chosen so that $\boldsymbol{\pi}$ contains no zero elements.

The required closed-form formula for $\mathbf{G}$ can be conveniently obtained by generalizing the method of § 6.2.5 and choosing the vector $\mathbf{d}'$ according to the equation

$$\mathbf{d}' = \sum_{j=1}^{n+1} K_j \hat{\mathbf{v}}_j', \tag{8.34}$$

where the K_j are the proportional-controller gains and the $\hat{\mathbf{v}}_j$ are the eigenvectors of $\boldsymbol{\Gamma}'$. Now it is clear from eqn. (8.32) that

$$\hat{\mathbf{v}}_j' = [\mathbf{v}_j', 0] \quad (j = 1, 2, \ldots, n), \tag{8.35 a}$$

where the $\mathbf{v}_j$ are the eigenvectors of $\mathbf{A}'$, and that

$$\hat{\mathbf{v}}_{n+1}' = [-\mathbf{e}'\mathbf{A}^{-1}, 1]. \tag{8. 35b}$$

These expressions, when substituted into equation (8.34), indicate that the vector $\mathbf{d}'$ has the form

$$\mathbf{d}' = \sum_{j=1}^{n} K_j \hat{\mathbf{v}}_j' + K_{n+1}[-\mathbf{e}'\mathbf{A}^{-1}, 1].$$

Since equation (8.20) implies that

$$\chi(t) = \int_{t_0}^{t} x_k(t)\, dt,$$

it follows that the dyadic control law

$$\mathbf{z}(t) = \mathbf{G}\boldsymbol{\zeta}(t) = \mathbf{f}\mathbf{d}' \begin{bmatrix} \mathbf{x}(t) \\ \\ \chi(t) \end{bmatrix}$$

defined in eqns. (8.18), (8.21), and (8.23) is given by the equation

$$\mathbf{z}(t) = \mathbf{f}\left[\left\{\left(\sum_{j=1}^{n} K_j \mathbf{v}_j'\right) - K_{n+1}\mathbf{e}'\mathbf{A}^{-1}\right\} \mathbf{x}(t) + K_{n+1} \int_{t_0}^{t} x_k(t)\, dt\right]. \tag{8.36}$$

This control law is clearly of the proportional-plus-integral type.

The formulae derived in § 5.2.2 may now be invoked to determine the K_j $(j = 1, 2, \ldots, n+1)$ in eqn. (8.36) : in fact these formulae immediately yield the results

$$K_j = -\frac{\prod_{k=1}^{n+1} (\rho_k - \lambda_j)}{\pi_j \lambda_j \prod_{\substack{k=1 \\ k \neq j}}^{n} (\lambda_k - \lambda_j)} \quad (j = 1, 2, \ldots, n), \tag{8.37 a}$$

and

$$K_{n+1} = \frac{\prod_{k=1}^{n+1} \rho_k}{\pi_{n+1} \prod_{k=1}^{n} \lambda_k}. \tag{8.37 b}$$

In eqn. (8.37), the π_j ($j=1, 2, \ldots, n+1$) are of course the elements of the vector $\boldsymbol{\pi}$ defined in eqn. (8.33), and the ρ_k ($k=1, 2, \ldots, n+1$) are any arbitrary desired set of real or conjugate-complex eigenvalues of the closed-loop plant matrix.

It is now clear that, given the eigenstructure of the system (8.19) and the desired eigenvalue spectrum of the closed-loop dyadic modal control system, the foregoing results make it possible to calculate the required feedback gains by simple arithmetic. The choice of which state variable, $x_k(t)$, is to be subjected to integral as well as to proportional feedback action will, of course, be made on the basis of practical considerations concerning priorities regarding the avoidance of steady-state offset amongst the state variables of a given system when subjected to given disturbances.

The theoretical results presented in this section can be conveniently illustrated by considering a controllable but unstable system of the class (8.1) governed by the state equation

$$\begin{bmatrix} \dot{x}_1(t) \\ \dot{x}_2(t) \end{bmatrix} = \begin{bmatrix} 0, & 1 \\ -2, & 3 \end{bmatrix} \begin{bmatrix} x_1(t) \\ x_2(t) \end{bmatrix} + \begin{bmatrix} 1, & 1 \\ 0, & 2 \end{bmatrix} \begin{bmatrix} z_1(t) \\ z_2(t) \end{bmatrix} + \begin{bmatrix} 1 \\ 0 \end{bmatrix}, \tag{8.38}$$

where the last vector in the right-hand member of eqn. (8.38) represents an input disturbance. It is desired to design a controller which will not only stabilize the system but will also be such that $x_1 = 0$ in the steady state.

In view of the latter requirement it is appropriate to feed back the integral of $x_1(t)$, so that $\mathbf{e}' = [1, 0]$ in this case. The augmented state vector, $\boldsymbol{\zeta}(t)$, and the augmented plant matrix, $\boldsymbol{\Gamma}$, defined in eqns. (8.23) and (8.24) are accordingly

$$\boldsymbol{\zeta}(t) = \begin{bmatrix} x_1(t) \\ x_2(t) \\ \chi(t) \end{bmatrix} \tag{8.39 a}$$

and

$$\boldsymbol{\Gamma} = \begin{bmatrix} 0, & 1, & 0 \\ -2, & 3, & 0 \\ 1, & 0, & 0 \end{bmatrix}. \tag{8.39 b}$$

If $\mathbf{f}$ is chosen to have the value

$$\mathbf{f} = \begin{bmatrix} -1 \\ 1 \end{bmatrix} \tag{8.40}$$

it follows that the vector $\boldsymbol{\pi}$ (see eqn. (8.33)) is given by

$$\boldsymbol{\pi} = \begin{bmatrix} -2 \\ 2 \\ 1 \end{bmatrix}, \tag{8.41}$$

because

$$\boldsymbol{\beta} = \begin{bmatrix} 1, & 1 \\ 0, & 2 \\ 0, & 0 \end{bmatrix} \mathbf{f} \tag{8.42}$$

and

$$\hat{\mathbf{V}} = \begin{bmatrix} 2, & -1, & 0 \\ -1, & 1, & 0 \\ -3/2, & 1/2, & 1 \end{bmatrix} \tag{8.43}$$

in this example : the value of the vector $\mathbf{f}$ as given in eqn. (8.40) is clearly a suitable choice since none of the elements of the resulting vector $\boldsymbol{\pi}$ is zero.

If it is desired to design the controller such that $\rho_1 = -1$, $\rho_2 = -2$, and $\rho_3 = -3$, then the formulae (8.37) indicate that

$$K_1 = -12, \quad K_2 = -15, \quad K_1 = -3, \tag{8.44}$$

since $\lambda_1 = 1$, $\lambda_2 = 2$, and the π_j $(j = 1, 2, 3)$ are given by eqn. (8.41). In addition, eqns. (8.43) and (8.44) imply that the required vector $\mathbf{d}'$ (see eqn. (8.34)) in this case is

$$\mathbf{d}' = [-9/2, \ -9/2, \ -3]. \tag{8.45}$$

It now follows from eqns. (8.40) and (8.45) that the dyadic feedback matrix is

$$\mathbf{G} = \mathbf{f}\mathbf{d}' = \begin{bmatrix} 9/2, & 9/2, & 3 \\ & & \\ -9/2, & -9/2, & -3 \end{bmatrix}. \tag{8.46}$$

The validity of this result can be checked by using eqns. (8.38), (8.39), and (8.46) to form the augmented version of eqn. (8.38). Indeed, the augmented equation is readily seen to be

$$\begin{bmatrix} \dot{x}_1(t) \\ \dot{x}_2(t) \\ \dot{\chi}(t) \end{bmatrix} = \begin{bmatrix} 0, & 1, & 0 \\ -11, & -6, & -6 \\ 1, & 0, & 0 \end{bmatrix} \begin{bmatrix} x_1(t) \\ x_2(t) \\ \chi(t) \end{bmatrix} + \begin{bmatrix} 1 \\ 0 \\ 0 \end{bmatrix}, \tag{8.47}$$

where it will be recalled that

$$\chi(t) = \int_{t_0}^{t} x_1(t)\, dt. \tag{8.48}$$

The eigenvalues of the closed-loop plant matrix in eqn. (8.47) are $\rho_1 = -1$, $\rho_2 = -2$, $\rho_3 = -3$, as required, and the steady-state solution of this equation is found by simple calculation to be given by

$$\begin{bmatrix} x_1 \\ x_2 \\ \chi \end{bmatrix} = \begin{bmatrix} 0 \\ -1 \\ 1 \end{bmatrix}. \tag{8.49}$$

It will be noted that $x_1 = 0$ in the steady state, as desired.

8.2.3 *Multi-input modal control*

The central result of § 8.2.2 is that, if the integral of a single state variable only is fed back, a desired eigenvalue spectrum may be realized by means of a dyadic feedback matrix of the form (8.18). It was pointed out, however, that this simple form cannot be applied if it is required to feed back the integrals of $l(>1)$ state variables. The purpose of the present section is to outline an algorithm for eigenvalue assignment in the general case $1<l\leqslant r$.

Thus, using the notation of § 8.2.1, it is required to determine the gain matrix, $\mathbf{G}$, in a non-dyadic feedback law of the form

$$\mathbf{z}(t)=\mathbf{G}\begin{bmatrix}\mathbf{x}(t)\\ \boldsymbol{\chi}(t)\end{bmatrix} \tag{8.50}$$

such that the controllable modes of the system governed by the state eqn. (8.3) may be assigned arbitrary eigenvalues. The plant matrix of this system (see eqn. (8.4)) is

$$\hat{\mathbf{A}}=\begin{bmatrix}\mathbf{A}, & \mathbf{0}\\ \mathbf{E}_l, & \mathbf{0}\end{bmatrix} \tag{8.51}$$

and those modes of the system (8.3) which are controllable may be identified by examining the structure of the appropriate mode-controllability matrix (8.13) or (8.17). The matrix $\mathbf{G}$ can then be determined in the manner of § 6.2.4 by sequential application of the single-input modal control algorithm.

8.3. Discrete-time systems incorporating integral feedback

It will be recalled from Chapter 2 that, in the case of continuous-time systems governed by the state eqn. (8.1), the states $\mathbf{x}\{(k+1)T\}$ and $\mathbf{x}(kT)$ are related by the difference equation

$$\mathbf{x}\{(k+1)T\}=\boldsymbol{\Psi}(T)\mathbf{x}(kT)+\boldsymbol{\Delta}(T)\mathbf{z}(kT) \tag{8.52}$$

where the input vector, $\mathbf{z}(t)$, is a piecewise-constant function defined by the equation

$$\mathbf{z}(t)=\mathbf{z}(kT) \quad (kT\leqslant t<(k+1)T\ ;\ k=0,\ 1,\ 2,\ \ldots). \tag{8.53}$$

If the discrete-time integrals of l $(l\leqslant n)$ of the elements of $\mathbf{x}(kT)$ are to be fed back, the resulting system will consist of $\mathbf{x}(kT)$ augmented by the $l\times 1$ vector $\boldsymbol{\chi}(kT)$ where

$$\boldsymbol{\chi}\{(k+1)T\}=\boldsymbol{\chi}(kT)+T\mathbf{E}_l\mathbf{x}(kT) \tag{8.54}$$

and $\mathbf{E}_l$ is an $l\times n$ matrix which consists of an appropriate selection of l rows from the $n\times n$ unit matrix $\mathbf{I}_n$. The state equation of the augmented discrete-time system is then clearly

$$\begin{bmatrix}\mathbf{x}\{(k+1)T\}\\ \boldsymbol{\chi}\{(k+1)T\}\end{bmatrix}=\begin{bmatrix}\boldsymbol{\Psi}(T), & \mathbf{0}\\ T\mathbf{E}_l, & \mathbf{I}_l\end{bmatrix}\begin{bmatrix}\mathbf{x}(kT)\\ \boldsymbol{\chi}(kT)\end{bmatrix}+\begin{bmatrix}\boldsymbol{\Delta}(T)\\ \mathbf{0}\end{bmatrix}\mathbf{z}(kT), \tag{8.55}$$

and it may readily be shown in the manner of § 8.2.1 that the mode-controllability matrix of the system (8.55) is

$$\hat{\mathbf{P}}(T)=\begin{bmatrix}\mathbf{V}'\boldsymbol{\Delta}(T)\\ -T\mathbf{E}_l\{\boldsymbol{\Psi}(T)-\mathbf{I}_n\}^{-1}\boldsymbol{\Delta}(T)\end{bmatrix}, \tag{8.56}$$

where $\mathbf{V}'$ is the generalized modal matrix of $\boldsymbol{\Psi}'(T)$: the formula (8.56) is, of course, only valid when $\boldsymbol{\Psi}(T)$ has no unit-valued eigenvalues. The mode-controllability matrix (8.56) is the discrete-time analogue of the matrix (8.13).

Now, it will be recalled that a system is controllable if and only if all the rows of its mode-controllability matrix corresponding to the last rows of the Jordan blocks containing the same-valued eigenvalues are linearly independent. Therefore, since it is assumed that $\boldsymbol{\Psi}(T)$ has no unit-valued eigenvalues, it may be inferred from the structure of the matrix $\hat{\mathbf{P}}(T)$ given by (8.56) that the augmented discrete-time system (8.55) will be controllable if and only if the system (8.54) is controllable and the l rows of $T\mathbf{E}_l\{\boldsymbol{\Psi}(T)-\mathbf{I}_n\}^{-1}\boldsymbol{\Delta}(T)$ are linearly independent.

The results for the discrete-time system when $\boldsymbol{\Psi}(T)$ has unit-valued eigenvalues are precisely analogous to those which apply to the continuous-time system discussed in § 8.2.1 when $\mathbf{A}$ has zero-valued eigenvalues.

References

[1] Porter, B., and Power, H. M., 1970, " Controllability of multivariable systems incorporating integral feedback ", *Electron. Lett.*, **6,** 689.

[2] Power, H. M., and Porter, B., 1970, " Necessary and sufficient conditions for controllability of multivariable systems incorporating integral feedback ", *Electron. Lett.*, **6,** 815.

[3] Wonham, W. M., 1967, " On pole-assignment in multi-input controllable systems ", *I.E.E.E. Trans. autom. Control*, **12,** 660.

[4] Gould, L. A., Murphy, A. T., and Berkman, E. F., 1970, " On the Simon–Mitter pole allocation algorithm—explicit gains for repeated eigenvalues ", *I.E.E.E. Trans. autom. Control*, **15,** 259.

[5] Retallack, D. G., and Macfarlane, A. G. J., 1970, " Pole-shifting techniques for multivariable systems ", *Proc. Instn elect. Engrs*, **117,** 1037.

[6] Davison, E. J., 1968, " On pole assignment in multivariable systems ", *I.E.E.E. Trans. autom. Control*, **13,** 747.

Sensitivity Characteristics of Modal Control Systems

CHAPTER 9

9.1. Introduction

In this chapter, simple formulae are derived which enable the principal sensitivity characteristics of modal control systems to be calculated. These sensitivity characteristics may be categorized as follows :

(i) the sensitivity of the elements of the feedback matrix to changes in the values of the required closed-loop eigenvalues, to changes in the elements of the plant matrix, and to changes in the elements of the input matrix ;

(ii) the sensitivity of all the closed-loop eigenvalues to changes in the elements of the feedback matrix, to changes in the elements of the plant matrix, and to changes in the elements of the input matrix.

9.2. Continuous-time systems

9.2.1. *Introduction*

In this section, simple and explicit derivations are given for the following first-order and second-order sensitivity vectors and coefficients associated with *single*-input systems governed by state and control law equations of the respective forms (5.1 a) and (5.34) :

(i) $\partial \mathbf{g}/\partial a_{kl}$, $\partial^2 \mathbf{g}/\partial a_{kl}\partial a_{st}$;

(ii) $\partial \mathbf{g}/b_j$, $\partial^2 \mathbf{g}/\partial b_j \partial b_k$;

(iii) $\partial \mathbf{g}/\partial \rho_j$, $\partial^2 \mathbf{g}/\partial \rho_j \partial \rho_k$;

(iv) $\partial \rho_i/\partial g_j$, $\partial^2 \rho_i/\partial g_j \partial g_k$;

(v) $\partial \rho_i/\partial a_{kl}$, $\partial^2 \rho_i/\partial a_{kl}\partial a_{st}$;

(vi) $\partial \rho_i/\partial b_j$, $\partial^2 \rho_i/\partial b_j \partial b_k$.

The results derived in this section can be applied to *multi*-input systems designed sequentially in the manner of § 6.2.4.

9.2.2. *First-order sensitivity vector*, $\partial \mathbf{g}/\partial a_{kl}$

It will be recalled from Chapter 5 that the effect on a system governed by a state equation of the form

$$\dot{\mathbf{x}}(t) = \mathbf{A}\mathbf{x}(t) + \mathbf{b}z(t) \tag{9.1}$$

of generating the input variable, $z(t)$, according to the control law

$$z(t) = \sum_{j=1}^{m} \left[\frac{\prod_{i=1}^{m} (\rho_i - \lambda_j)}{p_j \prod_{\substack{i=1 \\ i \neq j}}^{m} (\lambda_i - \lambda_j)} \mathbf{v}_j' \right] \mathbf{x}(t) \tag{9.2}$$

is to alter the eigenvalues, λ_j ($j = 1, 2, \ldots, m$), of the uncontrolled plant matrix, $\mathbf{A}$, to some new values ρ_j ($j = 1, 2, \ldots, m$). The ensuing sensitivity theories are restricted to systems whose plant matrices have distinct eigenvalues.

An alternative form of eqn. (9.2), suitable for the development of the following sensitivity theory, is

$$z(t) = \mathbf{g}'\mathbf{x}(t) \tag{9.3}$$

where $\mathbf{g}$ is the $n \times 1$ feedback vector defined by

$$\mathbf{g} = \mathbf{V}_m \boldsymbol{\varkappa}. \tag{9.4 a}$$

In eqn. (9.4 *a*)

$$\mathbf{V}_m = [\mathbf{v}_1, \mathbf{v}_2, \ldots, \mathbf{v}_m], \tag{9.4 b}$$

$$\boldsymbol{\varkappa} = [K_1, K_2, \ldots, K_m]', \tag{9.4 c}$$

and

$$K_j = \prod_{i=1}^{m} (\rho_i - \lambda_j) / p_j \left[\prod_{\substack{i=1 \\ i \neq j}}^{m} (\lambda_i - \lambda_j) \right] \quad (j = 1, 2, \ldots, m), \tag{9.4 d}$$

where

$$p_j = \mathbf{v}_j'\mathbf{b} \neq 0 \quad (j = 1, 2, \ldots, m). \tag{9.4 e}$$

In eqn. (9.4 *b*), $\mathbf{V}_m$ is the $n \times m$ matrix formed by selecting an appropriate set, $\{\mathbf{v}_1, \mathbf{v}_2, \ldots, \mathbf{v}_m\}$, of m eigenvectors of $\mathbf{A}'$. It is the object of this section to determine the first derivative of the elements of the gain vector $\mathbf{g}$ with respect to changes in a_{kl}, the generic element of the plant matrix $\mathbf{A}$.

The sensitivity vector, $\partial \mathbf{g}/\partial a_{kl}$, may be determined by differentiating eqn. (9.4 *a*) with respect to a_{kl} : the resulting equation is

$$\frac{\partial \mathbf{g}}{\partial a_{kl}} = \mathbf{V}_m \frac{\partial \boldsymbol{\varkappa}}{\partial a_{kl}} + \frac{\partial \mathbf{V}_m}{\partial a_{kl}} \boldsymbol{\varkappa} \quad (k, l = 1, 2, \ldots, n). \tag{9.5}$$

In order to determine an expression for $\partial \boldsymbol{\varkappa}/\partial a_{kl}$ in eqn. (9.5), it is convenient to write eqn. (9.4 *d*) in the equivalent form

$$\ln K_j = -\ln p_j + \sum_{i=1}^{m} \ln (\rho_i - \lambda_j) - \sum_{\substack{i=1 \\ i \neq j}}^{m} \ln (\lambda_i - \lambda_j). \tag{9.6}$$

If eqn. (9.6) is now differentiated with respect to a_{kl} it follows that

$$\frac{1}{K_j}\frac{\partial K_j}{\partial a_{kl}} = -\frac{1}{p_j}\frac{\partial p_j}{\partial a_{kl}} - \frac{\partial \lambda_j}{\partial a_{kl}} \sum_{i=1}^{m} \frac{1}{(\rho_i - \lambda_j)} - \sum_{\substack{i=1 \\ i \neq j}}^{m} \frac{1}{(\lambda_i - \lambda_j)} \left(\frac{\partial \lambda_i}{\partial a_{kl}} - \frac{\partial \lambda_j}{\partial a_{kl}} \right) \quad (j = 1, 2, \ldots, m), \tag{9.7 a}$$

where it follows from eqn. (9.4 *e*) that

$$\frac{\partial p_j}{\partial a_{kl}} = \frac{\partial \mathbf{v}_j'}{\partial a_{kl}} \mathbf{b}. \tag{9.7 b}$$

Expressions for the first-order eigenvalue and eigenvector sensitivities, $\partial \lambda_i / \partial a_{kl}$ and $\partial \mathbf{v}_j / \partial a_{kl}$, are given in eqns. (3.7) and (3.25 *b*) respectively, where the arbitrary coefficient ζ_j^{kl} may be chosen to be zero. It is evident that an expression for $\partial \mathbf{g} / \partial a_{kl}$ can be obtained in terms of the open- and closed-loop modal characteristics of the system by using eqns. (3.7), (3.25 *b*), (9.2), (9.3), and (9.4). However, the resulting closed-form expression is unwieldy and therefore, for the purposes of computation, the simplest approach is, in general, to compute $\partial \mathbf{v}'_j / \partial a_{kl}$, $\partial p_j / \partial a_{kl}$, and $\partial \lambda_j / \partial a_{kl}$ $(j = 1, 2, \ldots, m)$ using eqns. (3.25 *b*), (9.7 *b*), and (3.7) respectively, and then to use these results to compute $\partial K_j / \partial a_{kl}$ $(j = 1, 2, \ldots, m)$ using eqn. (9.7 *a*). The required first-order gain sensitivity can then be computed using the expression given in eqn. (9.5).

9.2.3. *Second-order sensitivity vector*, $\partial^2 \mathbf{g} / \partial a_{kl} \partial a_{st}$.

In this section an expression for the first derivative of the first-order gain vector sensitivity $\partial \mathbf{g} / \partial a_{kl}$ with respect to changes in the element a_{st} of the plant matrix $\mathbf{A}$ is determined. This second-order sensitivity vector may be determined by differentiating eqn. (9.5) with respect to a_{st}: the resulting equation is

$$\frac{\partial^2 \mathbf{g}}{\partial a_{kl} \partial a_{st}} = \mathbf{V}_m \frac{\partial^2 \boldsymbol{\varkappa}}{\partial a_{kl} \partial a_{st}} + \frac{\partial \mathbf{V}_m}{\partial a_{st}} \frac{\partial \boldsymbol{\varkappa}}{\partial a_{kl}} + \frac{\partial \mathbf{V}_m}{\partial a_{kl}} \frac{\partial \boldsymbol{\varkappa}}{\partial a_{st}} + \frac{\partial^2 \mathbf{V}_m}{\partial a_{kl} \partial a_{st}} \boldsymbol{\varkappa} \quad (k, l, s, t = 1, 2, \ldots, n). \tag{9.8}$$

In order to evaluate this expression for $\partial^2 \mathbf{g} / \partial a_{kl} \partial a_{st}$, it is first necessary to determine the value of the second-order sensitivity coefficient, $\partial^2 K_j / \partial a_{kl} \partial a_{st}$, by differentiating eqn. (9.7 *a*) with respect to a_{st}. The resulting expression is given by the equation

$$\begin{aligned} \frac{1}{K_j} \frac{\partial^2 K_j}{\partial a_{kl} \partial a_{st}} &= \frac{1}{K_j^2} \frac{\partial K_j}{\partial a_{kl}} \frac{\partial K_j}{\partial a_{st}} - \frac{1}{p_j} \frac{\partial^2 p_j}{\partial a_{kl} \partial a_{st}} + \frac{1}{p_j^2} \frac{\partial p_j}{\partial a_{kl}} \frac{\partial p_j}{\partial a_{st}} \\ &\quad - \frac{\partial \lambda_j}{\partial a_{kl}} \frac{\partial \lambda_j}{\partial a_{st}} \sum_{i=1}^{m} \frac{1}{(\rho_i - \lambda_j)^2} - \frac{\partial^2 \lambda_j}{\partial a_{kl} \partial a_{st}} \sum_{i=1}^{m} \frac{1}{(\rho_i - \lambda_j)} \\ &\quad - \sum_{\substack{i=1 \\ i \neq j}}^{m} \frac{1}{(\lambda_i - \lambda_j)} \left(\frac{\partial^2 \lambda_i}{\partial a_{kl} \partial a_{st}} - \frac{\partial^2 \lambda_j}{\partial a_{kl} \partial a_{st}} \right) \\ &\quad + \sum_{\substack{i=1 \\ i \neq j}}^{m} \frac{1}{(\lambda_i - \lambda_j)^2} \left(\frac{\partial \lambda_i}{\partial a_{kl}} - \frac{\partial \lambda_j}{\partial a_{kl}} \right) \left(\frac{\partial \lambda_i}{\partial a_{st}} - \frac{\partial \lambda_j}{\partial a_{st}} \right) \quad (j = 1, 2, \ldots, m), \end{aligned} \tag{9.9 a}$$

where it follows from eqn. (9.7 *b*) that

$$\frac{\partial^2 p_j}{\partial a_{kl}\partial a_{st}}=\frac{\partial^2 \mathbf{v}_j'}{\partial a_{kl}\partial a_{st}}\,\mathbf{b}. \tag{9.9 b}$$

In addition to the first-order sensitivity characteristics used (or derived) in § 9.2.2, the computation of $\partial^2 K_j/\partial a_{kl}\partial a_{st}$ requires the expressions for the second-order eigenvalue and eigenvector sensitivities, $\partial^2\lambda_i/\partial a_{kl}\partial a_{st}$ and $\partial^2\mathbf{v}_j'/\partial a_{kl}\partial a_{st}$, which were given in eqns. (3.39) and (3.44 *b*) respectively.

Although a closed-form expression for the required sensitivity vector $\partial^2\mathbf{g}/\partial a_{kl}\partial a_{st}$ can be obtained using eqns. (9.5) to (9.9), such an expression would certainly be cumbersome and would not lead to a clearer insight into the physical significance of the numerical magnitude of the sensitivity vector. The simplest approach for computation is to calculate $\partial^2\mathbf{v}_j'/\partial a_{kl}\partial a_{st}$, $\partial^2 p_j/\partial a_{kl}\partial a_{st}$, and $\partial^2\lambda_j/\partial a_{kl}\partial a_{st}$ $(j=1, 2, \ldots, m)$ using eqns. (3.44 *b*), (9.9 *b*), and (3.39), respectively, and then to use these results, together with results obtained in § 9.2.2, to compute $\partial^2 K_j/\partial a_{kl}\partial a_{st}$ $(j=1, 2, \ldots, m)$ using eqn. (9.9 *a*). The required second-order gain sensitivity can then be computed using the expression given in eqn. (9.8).

9.2.4. *First-order sensitivity vector,* $\partial\mathbf{g}/\partial b_j$.

In this section an expression for the first derivative of the elements of the gain vector **g** with respect to changes in b_j, the jth element of the system input matrix **b**, is determined. This sensitivity vector, $\partial\mathbf{g}/\partial b_j$, may be conveniently obtained by writing eqn. (9.4 *a*) in the form

$$\mathbf{g}=\sum_{l=1}^{m} K_l\mathbf{v}_l. \tag{9.10}$$

Thus, if eqn. (9.10) is differentiated with respect to b_j, it follows that

$$\frac{\partial\mathbf{g}}{\partial b_j}=\sum_{l=1}^{m}\frac{\partial K_l}{\partial p_l}\frac{\partial p_l}{\partial b_j}\mathbf{v}_l \quad (j=1, 2, \ldots, n). \tag{9.11}$$

In order to obtain a more explicit expression for $\partial\mathbf{g}/\partial b_j$, it is convenient to write eqn. (9.4 *e*) in the form

$$p_l=\sum_{j=1}^{n} v_l^j b_j \quad (l=1, 2, \ldots, m), \tag{9.12}$$

where v_l^j is the jth element of $\mathbf{v}_l$. Differentiation of eqns. (9.4 *d*) and (9.12) with respect to b_j then yields the expressions

$$\frac{\partial K_l}{\partial p_l}=-\frac{K_l}{p_l} \tag{9.13 a}$$

and

$$\frac{\partial p_l}{\partial b_j}=v_l^j. \tag{9.13 b}$$

The required sensitivity vector, $\partial\mathbf{g}/\partial b_j$, can be obtained by using the results given in eqns. (9.11) and (9.13): the former equation then assumes the simplified form

$$\frac{\partial \mathbf{g}}{\partial b_j} = -\sum_{l=1}^{m} \frac{K_l}{p_l} v_l^j \mathbf{v}_l \quad (j=1, 2, \ldots, n). \tag{9.14}$$

9.2.5. *Second-order sensitivity vector*, $\partial^2 \mathbf{g}/\partial b_j \partial b_k$.

Differentiation of eqn. (9.11) with respect to b_k, the kth element of the input matrix $\mathbf{b}$, indicates that

$$\frac{\partial^2 \mathbf{g}}{\partial b_j \partial b_k} = \sum_{l=1}^{m} \frac{\partial^2 K_l}{\partial p_l^2} \frac{\partial p_l}{\partial b_k} \frac{\partial p_l}{\partial b_j} \mathbf{v}_l, \tag{9.15}$$

where $\partial^2 K_l/\partial p_l^2$ follows directly by differentiating eqn. (9.13 *a*) with respect to p_l: thus, it is found that

$$\frac{\partial^2 K_l}{\partial p_l^2} = \frac{2K_l}{p_l^2}. \tag{9.16}$$

The required second-order sensitivity vector can then be obtained by using the results given in eqns. (9.13 *b*), (9.15) and (9.16): the resulting expression for this vector is

$$\frac{\partial^2 \mathbf{g}}{\partial b_j \partial b_k} = 2\sum_{l=1}^{m} \frac{K_l}{p_l^2} v_l^j v_l^k \mathbf{v}_l \quad (j, k=1, 2, \ldots, n). \tag{9.17}$$

9.2.6. *First-order sensitivity vector*, $\partial \mathbf{g}/\partial \rho_j$.

In this section an expression for $\partial \mathbf{g}/\partial \rho_j$, the first derivative of the elements of the gain vector $\mathbf{g}$ with respect to changes in ρ_j, the jth closed-loop eigenvalue, is determined. This sensitivity vector follows directly by differentiating eqns. (9.10) and (9.4 *d*) with respect to ρ_j: the resulting expressions are found to be

$$\frac{\partial \mathbf{g}}{\partial \rho_j} = \sum_{l=1}^{m} \frac{\partial K_l}{\partial \rho_j} \mathbf{v}_l \quad (j=1, 2, \ldots, m) \tag{9.18}$$

and

$$\frac{\partial K_l}{\partial \rho_j} = \frac{K_l}{(\rho_j - \lambda_l)} \quad (j, l=1, 2, \ldots, m). \tag{9.19}$$

Equations (9.18) and (9.19) indicate that the required gain-vector sensitivity is given by the expression

$$\frac{\partial \mathbf{g}}{\partial \rho_j} = \sum_{l=1}^{m} \frac{K_l}{(\rho_j - \lambda_l)} \mathbf{v}_l \quad (l=1, 2, \ldots, m). \tag{9.20}$$

9.2.7. *Second-order sensitivity vector*, $\partial^2 \mathbf{g}/\partial \rho_j \partial \rho_k$

Differentiation of eqn. (9.18) with respect to ρ_k, the kth closed-loop eigenvalue, indicates that

$$\frac{\partial^2 \mathbf{g}}{\partial \rho_j \partial \rho_k} = \frac{\partial}{\partial \rho_k}\left(\sum_{l=1}^{m} \frac{K_l}{(\rho_j - \lambda_l)} \mathbf{v}_l\right). \tag{9.21}$$

The required sensitivity vector is obtained by using the results given in eqns. (9.19) and (9.21) : thus, it is found that

$$\frac{\partial^2 \mathbf{g}}{\partial \rho_j \partial \rho_k} = \begin{cases} \sum_{l=1}^{m} \frac{K_l}{(\rho_j - \lambda_l)(\rho_k - \lambda_l)} \mathbf{v}_l & (j \neq k, j, k = 1, 2, \ldots, m), \quad (9.22\ a) \\ 0 & (j = k = 1, 2, \ldots, m). \quad (9.22\ b) \end{cases}$$

9.2.8. *First-order sensitivity coefficient,* $\partial \rho_i / \partial g_j$

The sensitivity coefficient $\partial \rho_i / \partial g_j$ is a measure of the sensitivity of the ith closed-loop eigenvalue to manufacturing or installation errors in the jth feedback gain. It follows from eqns. (9.1) and (9.3) that the resulting closed-loop plant matrix, $\mathbf{C}$, is given by the expression

$$\mathbf{C} = \mathbf{A} + \mathbf{b}\mathbf{g}'. \qquad (9.23\ a)$$

Equation (9.23 *a*) may clearly be expressed in the alternative form

$$c_{kj} = a_{kj} + b_k g_j \quad (k, j = 1, 2, \ldots, n) \qquad (9.23\ b)$$

in terms of the elements of the appropriate matrices. The required sensitivity coefficients, $\partial \rho_i / \partial g_j$, can be written in the form

$$\frac{\partial \rho_i}{\partial g_i} = \sum_{k=1}^{n} \sum_{l=1}^{n} \frac{\partial \rho_i}{\partial c_{kl}} \frac{\partial c_{kl}}{\partial g_j} \quad (i, j = 1, 2, \ldots, n), \qquad (9.24)$$

where it follows from eqn. (9.23 *b*) that

$$\frac{\partial c_{kl}}{\partial g_j} = \begin{cases} b_k & (l = j\ ;\ k = 1, 2, \ldots, n), \\ 0 & (l \neq j\ ;\ l, k, = 1, 2, \ldots, n). \end{cases} \qquad (9.25)$$

If the modal matrix of $\mathbf{C}$ is expressed in the form

$$\boldsymbol{\Upsilon} = [\boldsymbol{\upsilon}_1, \boldsymbol{\upsilon}_2, \ldots, \boldsymbol{\upsilon}_n], \qquad (9.26\ a)$$

and the corresponding modal matrix of $\mathbf{C}'$ is similarly expressed in the form

$$\boldsymbol{\Xi} = [\boldsymbol{\xi}_1, \boldsymbol{\xi}_2, \ldots, \boldsymbol{\xi}_n], \qquad (9.26\ b)$$

then it may be deduced from eqn. (3.7) that the first-order closed-loop eigenvalue sensitivity to changes in c_{kl} is given by the expression

$$\frac{\partial \rho_i}{\partial c_{kl}} = \xi_i^k v_i^l \quad (i, k, l = 1, 2, \ldots, n), \qquad (9.27)$$

where ξ_i^k is the kth element of ξ_i and v_i^l is the lth element of $\boldsymbol{\upsilon}_i$. It then follows from eqns. (9.24), (9.25) and (9.26) that the required sensitivity coefficient is

$$\begin{aligned} \frac{\partial \rho_i}{\partial g_j} &= v_i^j \sum_{k=1}^{n} \xi_i^k b_k \\ &= v_i^j \boldsymbol{\xi}_i' \mathbf{b} \quad (i, j = 1, 2, \ldots, n). \end{aligned} \qquad (9.28)$$

9.2.9. *Second-order sensitivity coefficient,* $\partial^2\rho_i/\partial g_j\partial g_k$

In order to determine an expression for the sensitivity coefficient, $\partial^2\rho_i/\partial g_j\partial g_k$, it is convenient to combine eqns. (9.24) and (9.25) in the form

$$\frac{\partial\rho_i}{\partial g_j}=\sum_{l=1}^{n}\frac{\partial\rho_i}{\partial c_{lj}}b_l \tag{9.29}$$

and then to differentiate eqn. (9.29) with respect to g_k whilst making use of eqn. (9.25) again : the resulting equation is

$$\begin{aligned}\frac{\partial^2\rho_i}{\partial g_j\partial g_k}&=\sum_{l=1}^{n}\sum_{m=1}^{n}\frac{\partial}{\partial c_{mk}}\left(\frac{\partial\rho_i}{\partial c_{lj}}\right)\frac{\partial c_{mk}}{\partial g_k}b_l\\&=\sum_{l=1}^{n}\sum_{m=1}^{n}\frac{\partial^2\rho_i}{\partial c_{mk}\partial c_{lj}}b_lb_m \quad (i,j,k=1,2,\ldots,n).\end{aligned} \tag{9.30}$$

The second-order closed-loop eigenvalue sensitivity coefficient, $\partial^2\rho_i/\partial c_{mk}\partial c_{lj}$, can be determined by using the appropriate form of eqn. (3.35).

9.2.10. *First-order sensitivity coefficient,* $\partial\rho_i/\partial a_{kl}$

In this case, the required sensitivity coefficient is $\partial\rho_i/\partial a_{kl}$, and since a change in the element a_{kl} produces an equal change in the element c_{kl} it follows directly from eqn. (9.27) that

$$\frac{\partial\rho_i}{\partial a_{kl}}=\frac{\partial\rho_i}{\partial c_{kl}}=\xi_i^k v_i^{\,l} \quad (i,k,l=1,2,\ldots,n). \tag{9.31}$$

9.2.11. *Second-order sensitivity coefficient,* $\partial^2\rho_i/\partial a_{kl}\partial a_{st}$

The second-order sensitivity coefficient, $\partial^2\rho_i/\partial a_{kl}\partial a_{st}$, is equal to $\partial^2\rho_i/\partial c_{kl}\partial c_{st}$ and can thus be determined directly using eqn. (3.35).

9.2.12. *First-order sensitivity coefficient,* $\partial\rho_i/\partial b_j$

The required sensitivity coefficient, $\partial\rho_i/\partial b_j$, may be expressed in the form

$$\frac{\partial\rho_i}{\partial b_j}=\sum_{k=1}^{n}\sum_{l=1}^{n}\frac{\partial\rho_i}{\partial c_{kl}}\frac{\partial c_{kl}}{\partial b_j} \quad (i,j=1,2,\ldots,n), \tag{9.32}$$

where it follows from eqn. (9.23 *b*) that

$$\frac{\partial c_{kl}}{\partial b_j}=\begin{cases}g_l \ (j=k;\ j,k,l=1,2,\ldots,n),\\ 0 \ (j\neq k;\ j,k,l=1,2,\ldots,n).\end{cases} \tag{9.33}$$

It then follows from eqns. (9.27), (9.32) and (9.33), that the required sensitivity coefficient is

$$\begin{aligned}\frac{\partial\rho_i}{\partial b_j}&=\xi_i^j\sum_{l=1}^{n}v_i^{\,l}g_l\\&=\xi_i^j\mathbf{v}_i'\mathbf{g} \quad (i,j=1,2,\ldots,n).\end{aligned} \tag{9.34 a}$$

It can be deduced from eqns. (9.4 *b*, *c*) and (9.34 *a*) that

$$\frac{\partial \rho_i}{\partial b_j} = 0 \quad (i > m), \tag{9.34 b}$$

since

$$\boldsymbol{\upsilon}_i = \mathbf{u}_i \quad (i = m+1, m+2, \ldots, n)$$

and

$$\mathbf{u}_j' \mathbf{v}_i = \delta_{ij}.$$

This result indicates that, to a first-order of approximation, the closed-loop eigenvalues corresponding to modes which are not subject to modal control are insensitive to input matrix changes. This is, in fact, an exact result which may be verified using the results developed in Chapter 5.

9.2.13. *Second-order sensitivity coefficient,* $\partial^2 \rho_i / \partial b_j \partial b_k$

In order to determine an expression for $\partial^2 \rho_i / \partial b_j \partial b_k$, it is convenient to combine eqns. (9.32) and (9.33) in the form

$$\frac{\partial \rho_i}{\partial b_j} = \sum_{l=1}^{n} \frac{\partial \rho_i}{\partial c_{jl}} \frac{\partial c_{jl}}{\partial b_j} \tag{9.35}$$

and then to differentiate eqn. (9.35) with respect to b_k whilst making use of eqn. (9.33) again : the resulting equation is

$$\begin{aligned} \frac{\partial^2 \rho_i}{\partial b_j \partial b_k} &= \sum_{l=1}^{n} \sum_{m=1}^{n} \frac{\partial}{\partial c_{km}} \left\{ \frac{\partial \rho_i}{\partial c_{jl}} \frac{\partial c_{km}}{\partial b_k} g_l \right\} \\ &= \sum_{l=1}^{n} \sum_{m=1}^{n} \frac{\partial^2 \rho_i}{\partial c_{jl} \partial c_{km}} g_l g_m \quad (i, j, k = 1, 2, \ldots, n). \end{aligned} \tag{9.36}$$

The second-order closed-loop eigenvalue sensitivity coefficient, $\partial^2 \rho_i / \partial c_{jl} \partial c_{km}$, can be determined by using the appropriate form of eqn. (3.35).

9.2.14. *Calculation of estimated feedback gain vectors and closed-loop eigenvalues*

In order to calculate the first- and second-order estimates of the gain vector, **g**, due to changes in the elements of the plant matrix **A**, due to changes in the elements of the vector **b**, and due to changes in the closed-loop eigenvalues ρ_j ; and also to calculate the first- and second-order estimates of the closed-loop eigenvalues, ρ_i, due to changes in the elements of the feedback gain vector **g**, due to changes in the elements of the plant matrix **A**, and due to changes in the elements of the vector **b**, it is necessary to substitute the sensitivity coefficients into appropriate Taylor series expansions.

These estimates are obtained by evaluating the expression

$$\hat{\mathbf{g}} = \mathbf{g} + \left[\frac{\partial \mathbf{g}}{\partial e_j} \delta e_j + \frac{\partial \mathbf{g}}{\partial e_k} \delta e_k \right] + \tfrac{1}{2} \left[\frac{\partial^2 \mathbf{g}}{\partial e_j^2} \delta e_j^2 + \frac{2 \partial^2 \mathbf{g}}{\partial e_j \partial e_k} \delta e_j \delta e_k + \frac{\partial^2 \mathbf{g}}{\partial e_k^2} \delta e_k^2 \right], \tag{9.37}$$

where, as appropriate,

$$e_j = a_{kl} \text{ and } e_k = a_{st},$$

or

$$e_j = b_j \text{ and } e_k = b_k,$$

or

$$e_j = \rho_j \text{ and } e_k = \rho_k\ ;$$

and also by evaluating the expression

$$\hat{\ddot{\rho}}_i = \rho_i + \left[\frac{\partial \rho_i}{\partial f_j}\,\delta f_j + \frac{\partial \rho_i}{\partial f_k}\,\delta f_k\right]$$

$$+\tfrac{1}{2}\left[\frac{\partial^2 \rho_i}{\partial f_j^{\,2}}\,\delta f_j^{\,2} + 2\,\frac{\partial^2 \rho_i}{\partial f_j \partial f_k}\,\delta f_j \delta f_k + \frac{\partial^2 \rho_i}{\partial f_k^{\,2}}\,\delta f_k^{\,2}\right], \tag{9.38}$$

where, as appropriate,

$$f_j = g_j \quad \text{and} \quad f_k = g_k,$$

or

$$f_j = a_{kl} \quad \text{and} \quad f_k = a_{st},$$

or

$$f_j = b_j \quad \text{and} \quad f_k = b_k.$$

9.2.15. *Illustrative example*

The use of the foregoing results can be illustrated by investigating the sensitivity characteristics of a third-order system governed by the respective state and control-law equations

$$\dot{\mathbf{x}}(t) = \begin{bmatrix} -2, & -1, & 1 \\ 1, & 0, & 1 \\ -1, & 0, & 1 \end{bmatrix} \mathbf{x}(t) + \begin{bmatrix} 1 \\ 1 \\ 1 \end{bmatrix} z(t) \tag{9.39 a}$$

and

$$z(t) = [2{\cdot}5,\ -1{\cdot}5,\ -5]\mathbf{x}(t). \tag{9.39 b}$$

This illustration may be conveniently effected by considering the following selection of sensitivity vectors and sensitivity coefficients relating to the system governed by eqn. (9.39) :

(i) $\partial \mathbf{g}/\partial a_{12}$, $\partial^2 \mathbf{g}/\partial a_{12}^{\,2}$;

(ii) $\partial \mathbf{g}/\partial b_3$, $\partial^2 \mathbf{g}/\partial b_3^{\,2}$;

(iii) $\partial \mathbf{g}/\partial \rho_2$, $\partial \mathbf{g}/\partial \rho_3$, $\partial^2 \mathbf{g}/\partial \rho_2^{\,2}$, $\partial^2 \mathbf{g}/\partial \rho_3^{\,2}$, $\partial^2 \mathbf{g}/\partial \rho_2 \partial \rho_3$;

(iv) $\partial \rho_i/\partial g_1$, $\partial^2 \rho_i/\partial g_1^{\,2}$ $(i = 1, 2, 3)$;

(v) $\partial \rho_i/\partial a_{12}$, $\partial^2 \rho_i/\partial a_{12}^{\,2}$ $(i = 1, 2, 3)$;

(vi) $\partial \rho_i/\partial b_2$, $\partial^2 \rho_i/\partial b_2^{\,2}$ $(i = 1, 2, 3)$.

The eigenstructure of the system (9.39 *a*) is given by eqns. (5.36) and (5.37), and the effect of implementing the control law (9.39 *b*) is to create a closed-loop system whose state equation has the form

$$\dot{\mathbf{x}}(t) = \mathbf{C}\mathbf{x}(t), \tag{9.40}$$

where

$$\mathbf{C} = \begin{bmatrix} 0{\cdot}5, & -2{\cdot}5, & -4 \\ 3{\cdot}5, & -1{\cdot}5, & -4 \\ 1{\cdot}5, & -1{\cdot}5, & -4 \end{bmatrix}. \tag{9.41}$$

It may be verified that the eigenvalue matrix of the closed-loop plant matrix (9.41) is

$$\mathbf{R} = \begin{bmatrix} -1, & 0, & 0 \\ 0, & -2+2i, & 0 \\ 0, & 0, & -2-2i \end{bmatrix}, \tag{9.42 a}$$

and that the corresponding modal matrices as defined in eqns. (9.26) are respectively

$$\boldsymbol{\Psi} = \begin{bmatrix} 1, & 5+i, & 5-i \\ -1, & 1-3i, & 1+3i \\ 1, & 3, & 3 \end{bmatrix} \tag{9.42 b}$$

and

$$\boldsymbol{\Xi} = \begin{bmatrix} -0{\cdot}90, & 0{\cdot}15-0{\cdot}20i, & 0{\cdot}15+0{\cdot}20i \\ -0{\cdot}30, & 0{\cdot}05+0{\cdot}10i, & 0{\cdot}05-0{\cdot}10i \\ 1{\cdot}60, & -0{\cdot}10+0{\cdot}30i, & -0{\cdot}10-0{\cdot}30i \end{bmatrix}. \tag{9.42 c}$$

It may also be verified that

$$\left.\begin{aligned} \partial\mathbf{g}/\partial a_{12} &= \begin{bmatrix} 2{\cdot}875 \\ 1{\cdot}125 \\ -4{\cdot}0 \end{bmatrix}, \\ \partial^2\mathbf{g}/\partial {a_{12}}^2 &= \begin{bmatrix} 5{\cdot}4375 \\ 2{\cdot}5625 \\ -8{\cdot}0 \end{bmatrix}, \end{aligned}\right\} \tag{9.43}$$

$$\left.\begin{aligned} \partial\mathbf{g}/\partial b_3 &= \begin{bmatrix} -1{\cdot}5 \\ 0{\cdot}5 \\ 6{\cdot}0 \end{bmatrix}, \\ \partial^2\mathbf{g}/\partial {b_3}^2 &= \begin{bmatrix} 1{\cdot}0 \\ -4{\cdot}0 \\ -9{\cdot}0 \end{bmatrix}, \end{aligned}\right\} \tag{9.44}$$

$$\left.\begin{aligned}
\partial \mathbf{g}/\partial \rho_2 &= \begin{bmatrix} -0{\cdot}5-1{\cdot}0i \\ 0{\cdot}5 \\ 1{\cdot}0+1{\cdot}0i \end{bmatrix}, \\
\partial \mathbf{g}/\partial \rho_3 &= \begin{bmatrix} -0{\cdot}5+1{\cdot}0i \\ 0{\cdot}5 \\ 1{\cdot}0-1{\cdot}0i \end{bmatrix}, \\
\partial^2 \mathbf{g}/\partial \rho_2{}^2 &= \begin{bmatrix} -0{\cdot}36154+0{\cdot}49231i \\ -0{\cdot}33846-0{\cdot}092308i \\ 0{\cdot}14615-0{\cdot}36923i \end{bmatrix}, \\
\partial^2 \mathbf{g}/\partial \rho_3{}^2 &= \begin{bmatrix} -0{\cdot}36154-0{\cdot}49231i \\ -0{\cdot}33846+0{\cdot}092308i \\ 0{\cdot}14615+0{\cdot}36923i \end{bmatrix}, \\
\partial^2 \mathbf{g}/\partial \rho_2 \partial \rho_3 &= \begin{bmatrix} 0{\cdot}5 \\ 0 \\ -0{\cdot}5 \end{bmatrix},
\end{aligned}\right\} \tag{9.45}$$

$$\left.\begin{aligned}
\partial \rho_1/\partial g_1 &= 0{\cdot}4, \\
\partial^2 \rho_1/\partial g_1{}^2 &= -0{\cdot}608, \\
\partial \rho_2/\partial g_1 &= 0{\cdot}3+1{\cdot}1i, \\
\partial^2 \rho_2/\partial g_1{}^2 &= 0{\cdot}304-0{\cdot}922i, \\
\partial \rho_3/\partial g_1 &= 0{\cdot}3-1{\cdot}1i, \\
\partial^2 \rho_3/\partial g_1{}^2 &= 0{\cdot}304+0{\cdot}922i,
\end{aligned}\right\} \tag{9.46}$$

$$\left.\begin{aligned}
\partial \rho_1/\partial a_{12} &= 0{\cdot}9, \\
\partial^2 \rho_1/\partial a_{12}{}^2 &= 0{\cdot}612, \\
\partial \rho_2/\partial a_{12} &= -0{\cdot}45-0{\cdot}65i, \\
\partial^2 \rho_2/\partial a_{12}{}^2 &= -0{\cdot}306+0{\cdot}2455i, \\
\partial \rho_3/\partial a_{12} &= -0{\cdot}45+0{\cdot}65i, \\
\partial^2 \rho_3/\partial a_{12}{}^2 &= -0{\cdot}306-0{\cdot}2455i,
\end{aligned}\right\} \tag{9.47}$$

$$\left.\begin{aligned}
\partial \rho_1/\partial b_2 &= 0{\cdot}3, \\
\partial^2 \rho_1/\partial b_2{}^2 &= -0{\cdot}192, \\
\partial \rho_2/\partial b_2 &= -0{\cdot}9-0{\cdot}05i, \\
\partial^2 \rho_2/\partial b_2{}^2 &= 0{\cdot}096-0{\cdot}18425i, \\
\partial \rho_3/\partial b_2 &= -0{\cdot}9+0{\cdot}05i, \\
\partial^2 \rho_3/\partial b_2{}^2 &= 0{\cdot}096+0{\cdot}18425i.
\end{aligned}\right\} \tag{9.48}$$

The expression given in eqn. (9.37) and the numerical values given in eqns. (9.43) to (9.45) can be used to obtain first- and second-order estimates of $\mathbf{g}$ resulting from changes in the system parameters a_{12}, b_3, ρ_2, and ρ_3. Similarly, the expression given in eqn. (9.38) and the numerical values given in eqns. (9.46) to (9.48) can be used to obtain first- and second-order estimates of ρ_1, ρ_2, and ρ_3 resulting from changes in the system parameters g_1, a_{12}, and b_2. These first- and second-order estimates, together with their directly-calculated values, are shown in figs. 9-1 to 9-6. It is evident from these figures that the second-order estimates are, in general, significantly better than the corresponding first-order estimates, and are never worse.

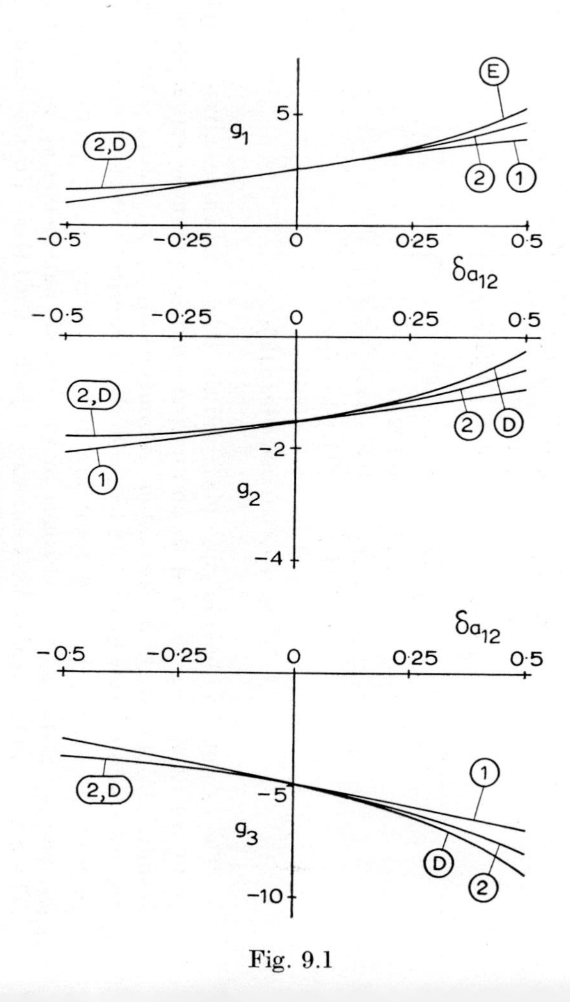

g1
g2
g3
δa12
5
−2
−4
−5
−10
−0·5
−0·25
0
0·25
0·5
2,D
1
2
E
D

Fig. 9.1

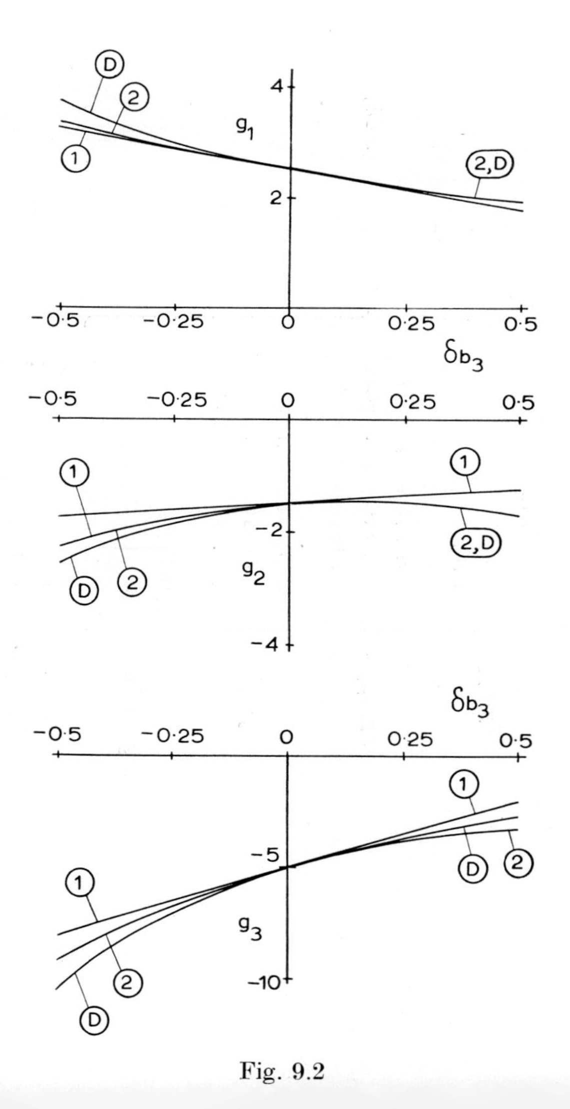

g1
g2
g3
δb3
4
2
−2
−4
−5
−10
−0·5
−0·25
0
0·25
0·5
2,D
1
2
D

Fig. 9.2

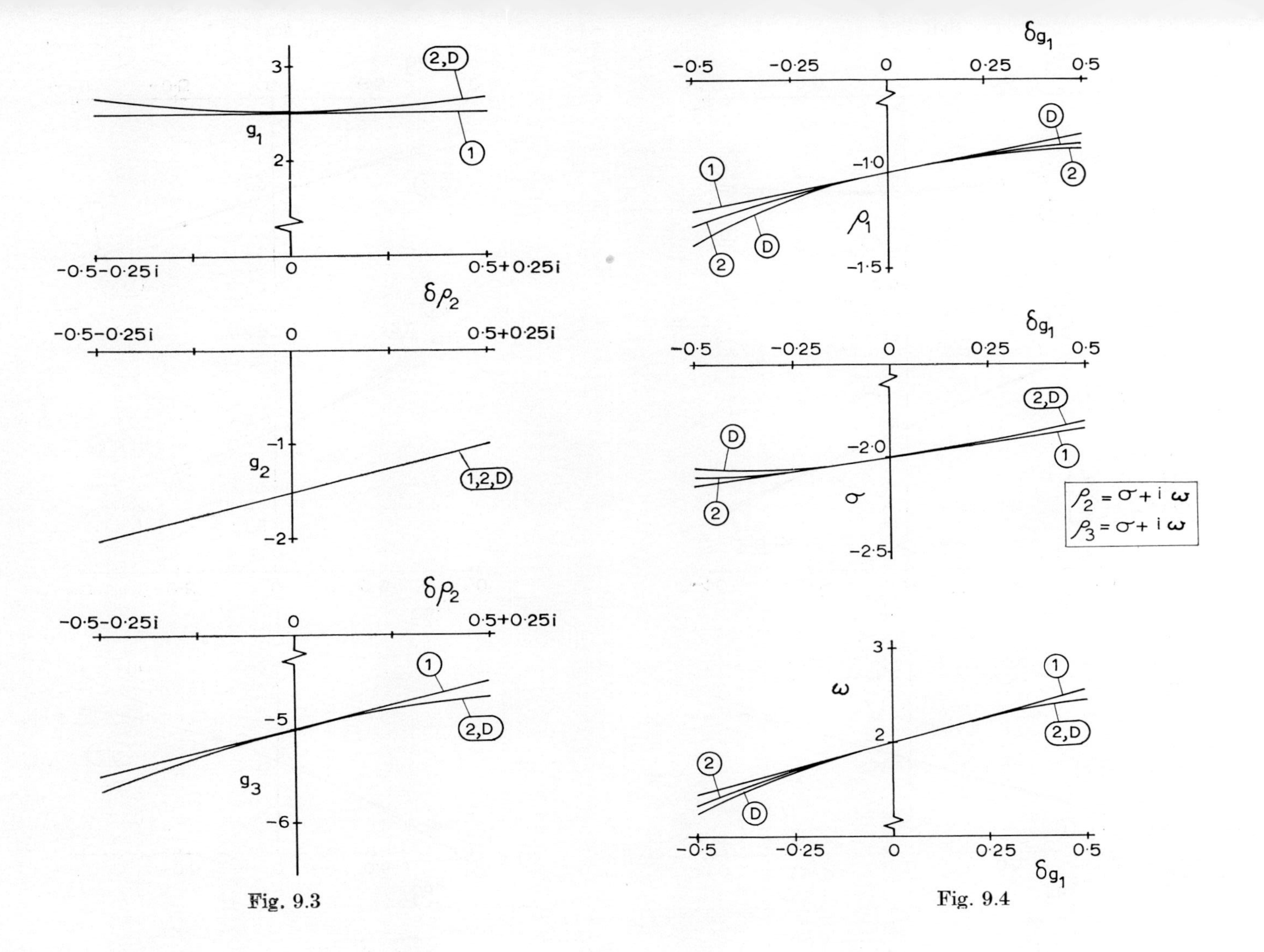

g_1
g_2
g_3
$\delta\rho_2$
-0·5-0·25i
0·5+0·25i
δg_1
ρ_1
σ
ω
$\rho_2 = \sigma + i\,\omega$
$\rho_3 = \sigma + i\,\omega$

Fig. 9.3

Fig. 9.4

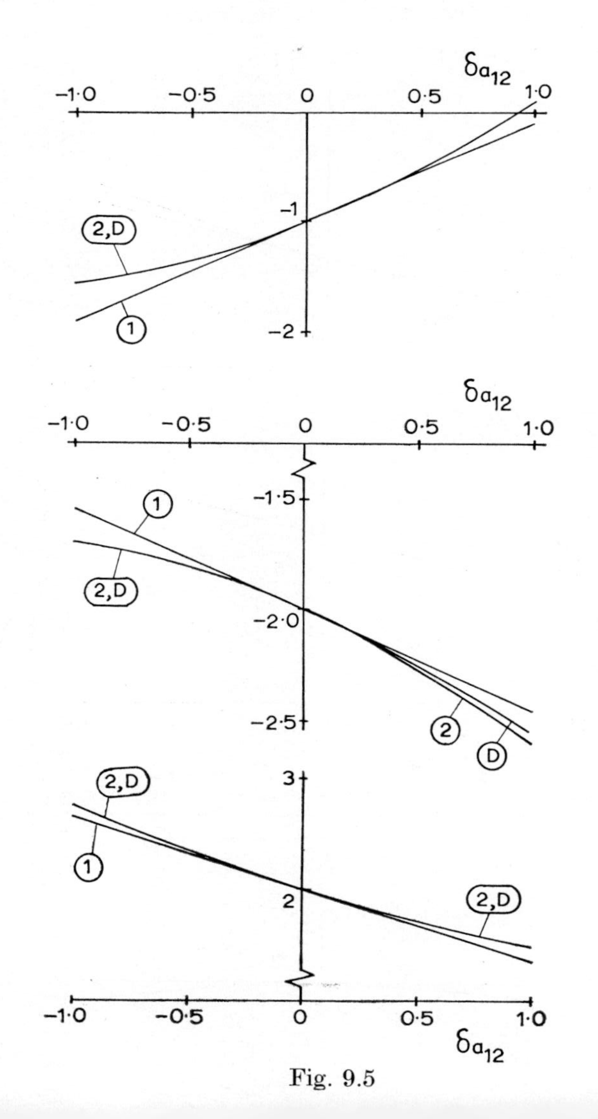

$\delta_{a_{12}}$
-1·0
-0·5
0
0·5
1·0
-1
-2
2,D
1
-1·5
-2·0
-2·5
2
D
3
2

Fig. 9.5

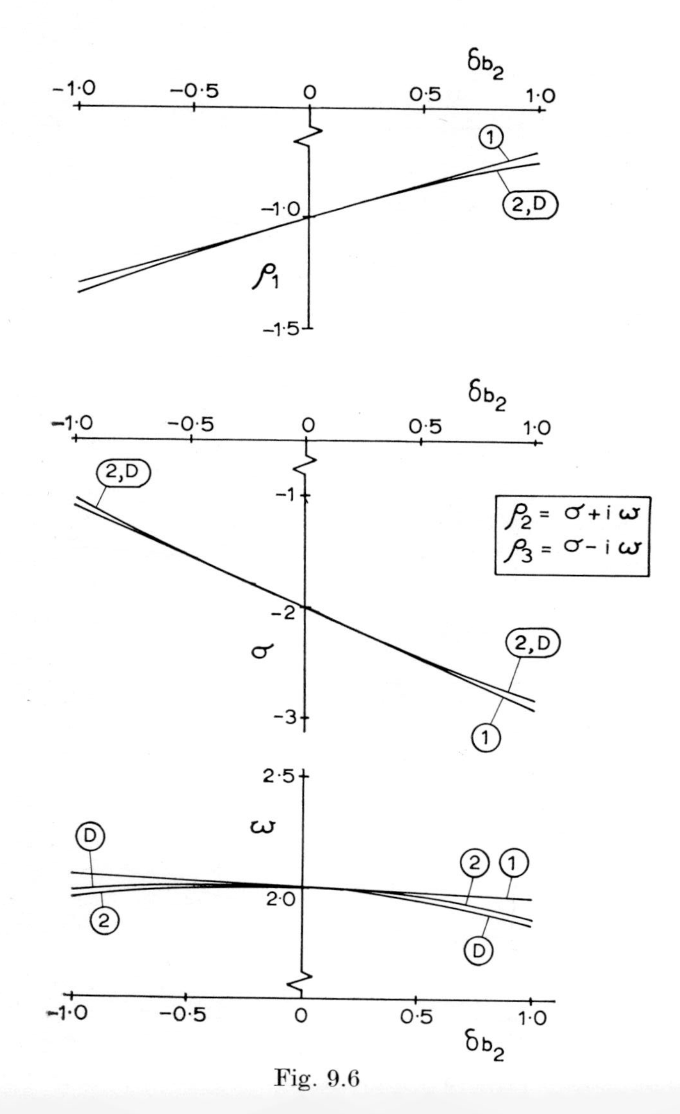

δ_{b_2}
-1·0
-0·5
0
0·5
1·0
-1·0
-1·5
ρ_1
1
2,D
-1
-2
-3
σ
$\rho_2 = \sigma + i\,\omega$
$\rho_3 = \sigma - i\,\omega$
2·5
2·0
ω
2
D

Fig. 9.6

9.3. Discrete-time systems

The results presented in the foregoing sections of this chapter concerning the sensitivity properties of continuous-time systems may be used (by analogy) in connection with discrete-time systems governed by state and control-law equations of the respective forms

$$\mathbf{x}\{(k+1)T\} = \mathbf{\Psi}(T)\mathbf{x}(kT) + \mathbf{\Delta}(T)\mathbf{x}(kT) \tag{9.49}$$

and

$$\mathbf{z}(kT) = \mathbf{\Omega}(T)\mathbf{x}(kT). \tag{9.50}$$

Indeed, the results presented in this chapter for continuous-time systems may be used immediately to calculate the discrete-time analogues of the various sensitivity coefficients given in §§ 9.2.2 to 9.2.13.

PART 2 APPLICATIONS

Synthesis of aircraft lateral autostabilization systems

CHAPTER 10

10.1 Introduction

THIS chapter is concerned with the design of a lateral autostabilizer for the control of the bank-angle mode and the roll-rate mode of an aircraft. In order to keep the presentation as simple as possible, the Dutch-roll of the aircraft is suppressed by neglecting side-slipping, yawing, and rolling interactions. However, a simple representation of a servomotor is included in the uncontrolled system. Typical aerodynamic data for an aircraft flying at $M = 0{\cdot}9$ and at altitudes from sea level to 40 000 ft are used in the calculations.

10.2 Mathematical model of uncontrolled system

In this chapter, the results of § 5.2.1 are used to synthesize appropriate control laws for the uncontrolled system shown in fig. 10-1. This system consists of a single-degree-of-freedom representation of the rolling dynamics of an aircraft, together with a first-order representation of the aileron servomotor. The open-loop response of bank-angle, $\phi(t)$, to aileron demand angle, $\xi_D(t)$, is governed by the equations [1]

$$p(t) = \frac{d\phi(t)}{dt}, \tag{10.1}$$

$$I_x \frac{dp(t)}{dt} = L_\xi \xi_A(t) + L_p p(t), \tag{10.2}$$

and

$$T \frac{d\xi_A(t)}{dt} = \xi_D(t) - \xi_A(t). \tag{10.3}$$

Equation (10.1) is a kinematic relation which defines the roll rate $p(t)$, eqn. (10.2) is the equation of motion of an aircraft in its simple rolling mode, and eqn. (10.3) is the governing equation of the servomotor.

Fig. 10-1

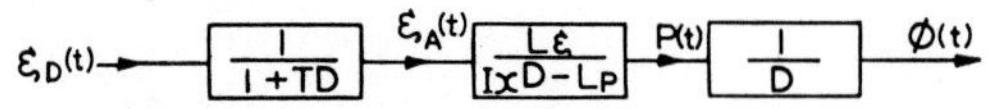

Block diagram of uncontrolled system.

If state variables $x_1(t)$, $x_2(t)$, $x_3(t)$, an input variable $z(t)$, and system parameters δ, μ, ν are introduced in accordance with the definitions

$$x_1(t) = \phi(t), \tag{10.4 a}$$

$$x_2(t) = p(t), \tag{10.4 b}$$

$$x_3(t) = \xi_A(t), \tag{10.4 c}$$

$$z(t) = \xi_D(t), \tag{10.5}$$

and

$$\delta = -L_p/I_x, \tag{10.6 a}$$

$$\mu = -L_\xi/I_x, \tag{10.6 b}$$

$$\nu = 1/T, \tag{10.6 c}$$

then eqns. (10.1), (10.2), and (10.3) can be written as the single vector-matrix differential equation

$$\begin{bmatrix} \dot{x}_1(t) \\ \dot{x}_2(t) \\ \dot{x}_3(t) \end{bmatrix} = \begin{bmatrix} 0, & 1, & 0 \\ 0, & -\delta, & -\mu \\ 0, & 0, & -\nu \end{bmatrix} \begin{bmatrix} x_1(t) \\ x_2(t) \\ x_3(t) \end{bmatrix} + \begin{bmatrix} 0 \\ 0 \\ \nu \end{bmatrix} z(t). \tag{10.7}$$

Equation (10.7) has the form

$$\dot{\mathbf{x}}(t) = \mathbf{A}\mathbf{x}(t) + \mathbf{b}z(t), \tag{10.8}$$

where the plant matrix of the uncontrolled system is

$$\mathbf{A} = \begin{bmatrix} 0, & 1, & 0 \\ 0, & -\delta, & -\mu \\ 0, & 0, & -\nu \end{bmatrix}, \tag{10.9}$$

the input matrix is

$$\mathbf{b} = \begin{bmatrix} 0 \\ 0 \\ \nu \end{bmatrix}, \tag{10.10}$$

and the state vector is

$$\mathbf{x}(t) = \begin{bmatrix} x_1(t) \\ x_2(t) \\ x_3(t) \end{bmatrix}. \tag{10.11}$$

The transient response characteristics of the uncontrolled aircraft are determined by the eigenvalues and eigenvectors of the matrix **A**. The eigenvalues are given by the roots of the equation

$$|\lambda \mathbf{I} - \mathbf{A}| = \begin{vmatrix} \lambda, & -1, & 0 \\ 0, & \lambda+\delta, & \mu \\ 0, & 0, & \lambda+\nu \end{vmatrix} = 0, \tag{10.12}$$

so that

$$\left.\begin{aligned} \lambda_1 &= 0, \\ \lambda_2 &= -\delta, \\ \lambda_3 &= -\nu, \end{aligned}\right\} \tag{10.13}$$

whilst the corresponding eigenvectors are

$$\mathbf{u}_1 = \begin{bmatrix} 1 \\ 0 \\ 0 \end{bmatrix}, \quad \mathbf{u}_2 = \begin{bmatrix} -1/\delta \\ 1 \\ 0 \end{bmatrix}, \quad \mathbf{u}_3 = \begin{bmatrix} -\mu/\nu(\nu-\delta) \\ \mu/(\nu-\delta) \\ 1 \end{bmatrix}. \tag{10.14}$$

The quantities λ_1, λ_2, and λ_3 are also the eigenvalues of the transposed matrix $\mathbf{A}'$, but the corresponding eigenvectors of $\mathbf{A}'$ are

$$\mathbf{v}_1 = \begin{bmatrix} 1 \\ 1/\delta \\ -\mu/\nu\delta \end{bmatrix}, \quad \mathbf{v}_2 = \begin{bmatrix} 0 \\ 1 \\ \mu/(\delta-\nu) \end{bmatrix}, \quad \mathbf{v}_3 = \begin{bmatrix} 0 \\ 0 \\ 1 \end{bmatrix}. \tag{10.15}$$

The eigenvectors defined in (10.14) and (10.15) satisfy the required orthogonality and normalization conditions.

The vector **b** defined in (10.10) can be expressed in the form

$$\mathbf{b} = p_1\mathbf{u}_1 + p_2\mathbf{u}_2 + p_3\mathbf{u}_3 \tag{10.16}$$

where

$$p_j = \mathbf{v}_j'\mathbf{b} \quad (j=1, 2, 3). \tag{10.17}$$

It thus follows from (10.10), (10.15), and (10.17) that

$$\left.\begin{aligned} p_1 &= -\mu/\delta, \\ p_2 &= \mu\nu/(\delta-\nu), \\ p_3 &= \nu, \end{aligned}\right\} \tag{10.18}$$

in the case of the system governed by eqn. (10.7). The general theory given in Chapter 5 indicates the manner in which these values of p_1, p_2, and p_3 are used in the procedure for synthesizing modal control laws. In the following sections of this chapter, this procedure is applied to the design of feedback laws for the control of a number of different aircraft flight modes. It is assumed that idealized transducers are available for the measurement of bank-angle, roll rate, and aileron angle.

10.3. Design of modal controllers

10.3.1. *Bank-angle mode control*

If it is desired to apply control loops to the uncontrolled system so that the plant matrix of the resulting controlled system has eigenvalues ρ_1, λ_2, and λ_3, then eqn. (5.2) indicates that the required input variable is given by

$$z(t)=K_1\mathbf{v}_1'\mathbf{x}(t), \tag{10.19}$$

where, according to eqn. (5.13),

$$K_1=\frac{\rho_1-\lambda_1}{p_1}=\frac{-\rho_1\delta}{\mu}. \tag{10.20}$$

It follows by substituting the expressions for $\mathbf{v}_1'$ and K_1 given in (10.15) and (10.20) into eqn. (10.19) that the eigenvalue associated with the bank-angle mode will be altered from λ_1 to ρ_1 (leaving λ_2 and λ_3 unchanged) if a control law defined by

$$z(t)=-\frac{\rho_1\delta}{\mu}x_1(t)-\frac{\rho_1}{\mu}x_2(t)+\frac{\rho_1}{\nu}x_3(t) \tag{10.21}$$

is implemented.

10.3.2. *Roll-rate mode control*

Similarly, if it is desired to apply control loops so that the plant matrix of the controlled system has eigenvalues λ_1, ρ_2, and λ_3, then eqn. (5.2) indicates that the required input variable is given by

$$z(t)=K_2\mathbf{v}_2'\mathbf{x}(t), \tag{10.22}$$

where, according to eqn. (5.13),

$$K_2=\frac{\rho_2-\lambda_2}{p_2}=\frac{(\rho_2+\delta)(\delta-\nu)}{\mu\nu}. \tag{10.23}$$

It follows by substituting the expressions for $\mathbf{v}_2'$ and K_2 given in (10.15) and (10.23) into eqn. (10.22) that the eigenvalue associated with the roll-rate mode will be altered from λ_2 to ρ_2 (leaving λ_1 and λ_3 unchanged) if a control law defined by

$$z(t)=\frac{(\rho_2+\delta)(\delta-\nu)}{\mu\nu}x_2(t)+\frac{(\rho_2+\delta)}{\nu}x_3(t) \tag{10.24}$$

is implemented.

10.3.3. *Bank-angle and roll-rate mode control*

In each of the foregoing examples, only one of the eigenvalues of the plant matrix of the uncontrolled system was altered. However, it is possible to apply feedback loops to the uncontrolled system so that two or more of the eigenvalues are changed by designing the loops sequentially in the manner described in Chapter 6. Thus, if it is desired to synthezise a controlled system having a plant matrix with eigenvalues ρ_1, ρ_2, and λ_3, then this can be achieved by resolving the input variable, $z(t)$, appearing in eqn. (10.7) into two components, $z_1(t)$ and $z_2(t)$, so that

$$z(t)=z_1(t)+z_2(t). \tag{10.25}$$

If the input variable defined in (10.21) is chosen as $z_1(t)$, then eqn. (10.8) assumes the form

$$\dot{\mathbf{x}}(t) = \mathbf{C}\mathbf{x}(t) + \mathbf{b}z_2(t) \tag{10.26}$$

where the matrix

$$\mathbf{C} = \begin{bmatrix} 0, & 1, & 0 \\ 0, & -\delta, & -\mu \\ -\rho_1\delta\nu/\mu, & -\rho_1\nu/\mu, & \rho_1 - \nu \end{bmatrix} \tag{10.27}$$

has eigenvalues ρ_1, λ_2, and λ_3. It now remains to generate $z_2(t)$ so that the matrix of the controlled system finally has the eigenvalues ρ_1, ρ_2, and λ_3.

The vector $\mathbf{b}$ can be expressed in the form

$$\mathbf{b} = q_1\mathbf{l}_1 + q_2\mathbf{l}_2 + q_3\mathbf{l}_3, \tag{10.28}$$

where $\mathbf{l}_1$, $\mathbf{l}_2$, and $\mathbf{l}_3$ are the eigenvectors of $\mathbf{C}$ and, by analogy with eqn. (10.17),

$$q_j = \mathbf{m}_j{}'\mathbf{b} \quad (j = 1, 2, 3), \tag{10.29}$$

where $\mathbf{m}_1$, $\mathbf{m}_2$, and $\mathbf{m}_3$ are the normalized eigenvectors of $\mathbf{C}'$. It may now be inferred from eqn. (5.2) that the required input-variable component, $z_2(t)$, must have the form

$$z_2(t) = K_2\mathbf{m}_2{}'\mathbf{x}(t), \tag{10.30}$$

where

$$\mathbf{m}_2 = \begin{bmatrix} \rho_1\nu/(\rho_1 - \nu + \delta) \\ 1 \\ \mu/(\rho_1 - \nu + \delta) \end{bmatrix} \tag{10.31}$$

and, according to eqn. (5.13),

$$K_2 = \frac{\rho_2 - \lambda_2}{q_2} = \frac{(\rho_2 + \delta)(\rho_1 - \nu + \delta)}{\mu\nu}. \tag{10.32}$$

It can therefore be deduced from (10.30), (10.31), and (10.32) that

$$z_2(t) = \frac{\rho_1(\rho_2 + \delta)}{\mu}x_1(t) + \frac{(\rho_2 + \delta)(\rho_1 - \nu + \delta)}{\mu\nu}x_2(t) + \frac{(\rho_2 + \delta)}{\nu}x_3(t) \tag{10.33}$$

and then from (10.21), (10.25), and (10.33) that

$$z(t) = z_1(t) + z_2(t) = \frac{\rho_1\rho_2}{\mu}x_1(t) + \frac{(\rho_2 + \delta)(\rho_1 - \nu + \delta) - \rho_1\nu}{\mu\nu}x_2(t) + \frac{(\rho_1 + \rho_2 + \delta)}{\nu}x_3(t). \tag{10.34}$$

If this control law (10.34) is implemented, then the eigenvalues associated with the bank-angle mode and the roll-rate mode will be altered from λ_1 to ρ_1 and from λ_2 to ρ_2, respectively, leaving the eigenvalue λ_3 unchanged. The same sequential synthesis procedure could of course be used to determine a third input-variable component, $z_3(t)$, which would alter λ_3 to some new value ρ_3, leaving ρ_1 and ρ_2 unchanged.

Table 10.1

Altitude		Eigenvalues of uncontrolled system			Eigenvalues of controlled system			g_1	g_2	g_3	Comments
Feet	μ	$\lambda_1 = 0$	$\lambda_2 = -\delta$	$\lambda_3 = -\nu$	ρ_1	ρ_2	ρ_3	deg/deg	deg/deg/sec	deg/deg	
0	102·06	0	−4·104	−50	−5	−4·104	−50	0·201	0·049	0·100	Control of bank angle mode ($\rho_1 \neq \lambda_1$, $\rho_2 = \lambda_2$, $\rho_3 = \lambda_3$)
20,000	47·31	0	−2·028	−50	−5	−2·028	−50	0·214	0·106	0·100	
40,000	19·06	0	−0·875	−50	−5	0·875	−50	0·230	0·262	0·100	
0	102·06	0	−4·104	−50	0	−4·104	−50	0	0	0	Control of roll rate mode ($\rho_2 \neq \lambda_2$, $\rho_1 = \lambda_1$, $\rho_3 = \lambda_3$)
20,000	47·31	0	−2·028	−50	0	−4·104	−50	0	0·042	0·041	
40,000	19·06	0	−0·875	−50	0	−4·104	−50	0	0·166	0·065	
0	102·06	0	−4·104	−50	−5	−4·104	−50	0·201	0·049	0·100	Control of bank angle and roll rate modes ($\rho_1 \neq \lambda_1$, $\rho_2 \neq \lambda_2$, $\rho_3 = \lambda_3$)
20,000	47·31	0	−2·028	−50	−5	−4·104	−50	0·438	0·152	0·141	
40,000	19·06	0	−0·875	−50	−5	−4·104	−50	1·077	0·446	0·165	

10.4. Numerical example

In each of the types of modal control considered in the previous section, the required control law (see eqns. (10.21), (10.24), and (10.34) has the form

$$z(t)=g_1x_1(t)+g_2x_2(t)-g_3x_3(t),$$

that is,

$$\xi_{\rm D}(t)=g_1\phi(t)+g_2p(t)-g_3\xi_{\rm A}(t),$$

so that, in each case, the controlled system has a block diagram of the form shown in fig. 10-2.

Fig. 10-2

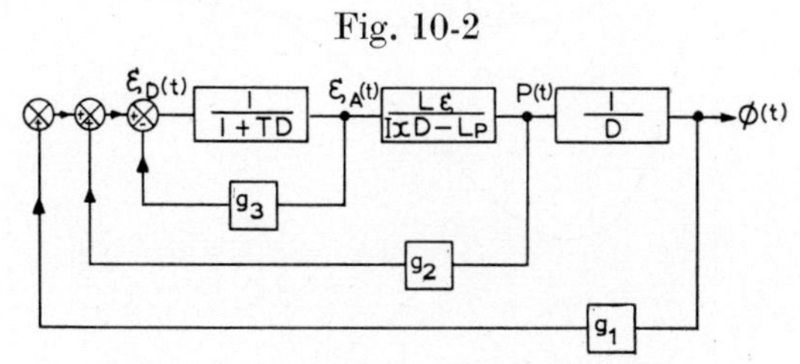

Block diagram of controlled system.

Fig. 10-3

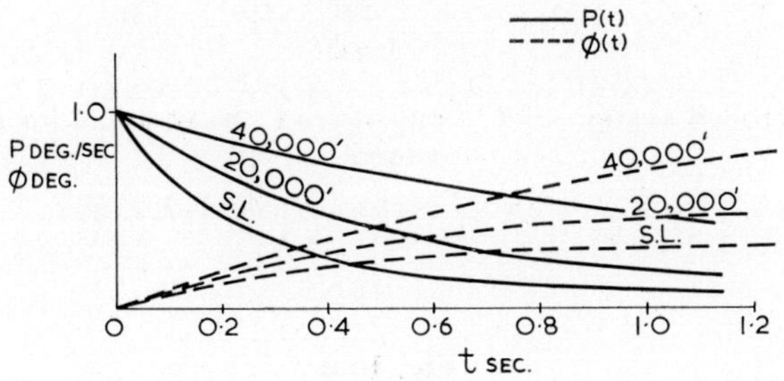

Response of uncontrolled system to 1°/sec ‘ out-of-trim ’ roll rate.

Fig. 10-4

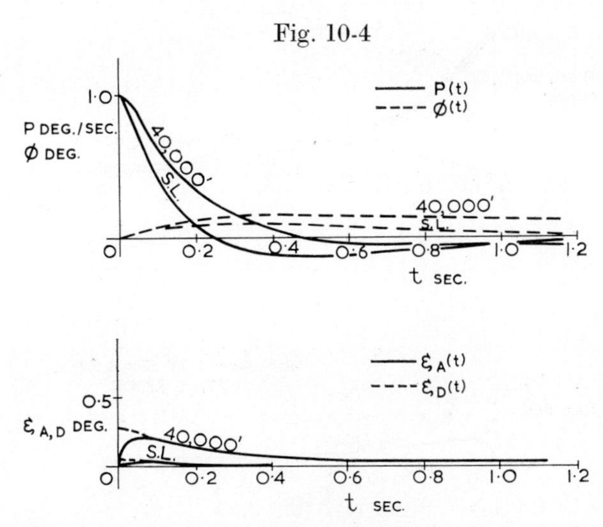

Response of controlled system to 1°/sec ‘ out-of-trim ’ roll rate for bank-angle mode control.

Fig. 10-5

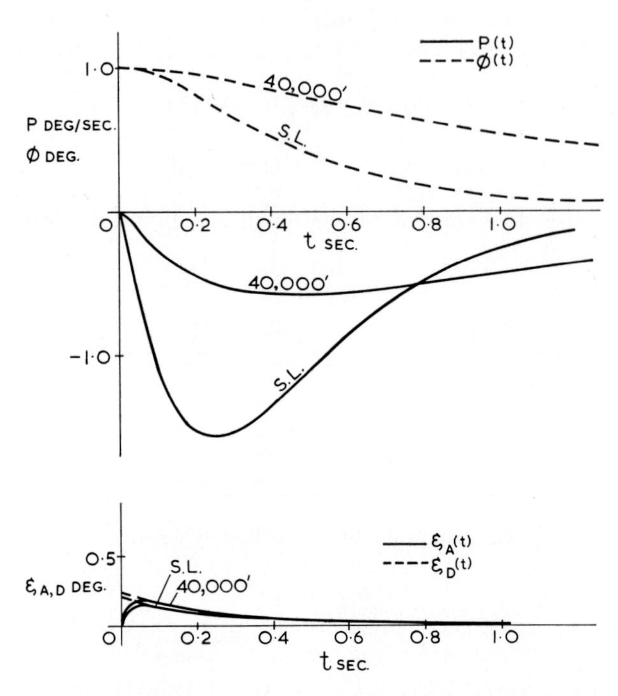

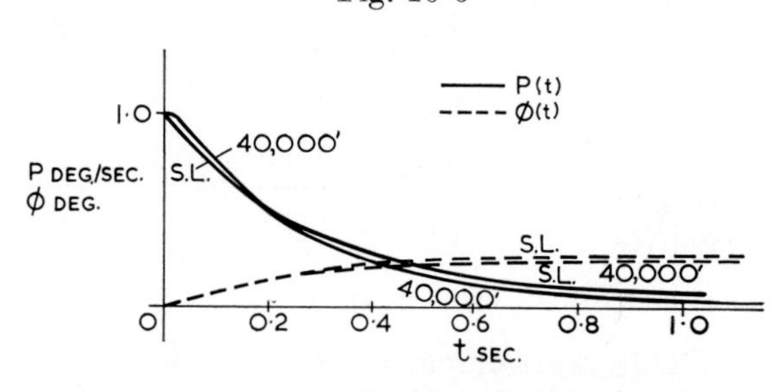

Response of controlled system to 1° ‘ out-of-trim ’ bank angle for bank-angle mode control.

Fig. 10-6

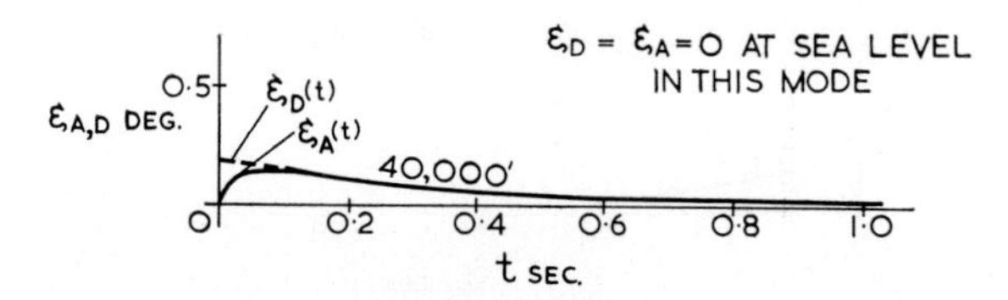

Response of controlled system to 1°/sec ‘ out-of-trim ’ roll rate for roll-rate mode control.

Fig. 10-7

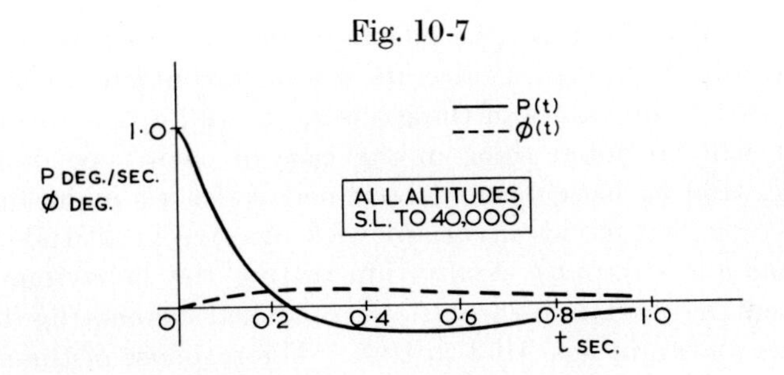

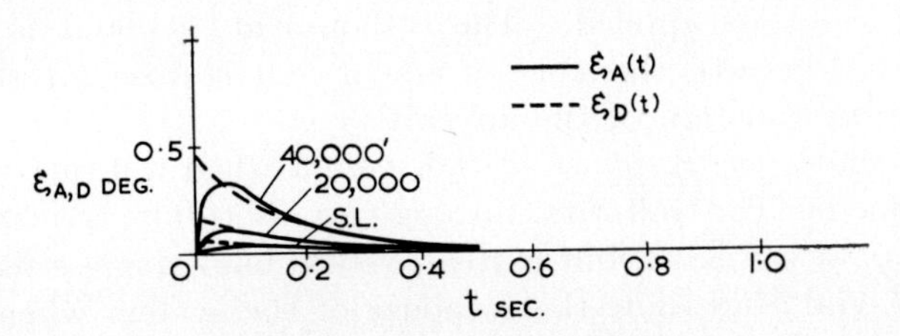

Response of controlled system to 1°/sec 'out-of-trim' roll rate for bank-angle and roll-rate mode control.

Fig. 10-8

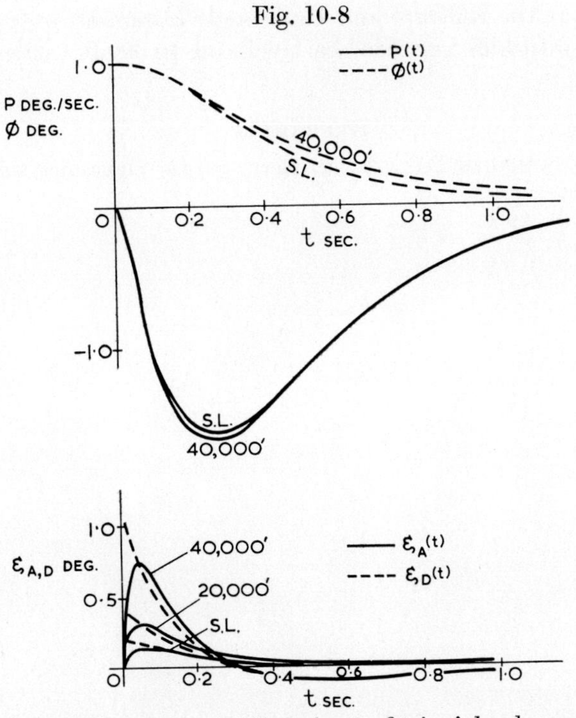

Response of controlled system to 1°/sec 'out-of-trim' bank angle for bank-angle and roll-rate mode control.

It is convenient to illustrate the effects of the various types of control by applying the results to a typical aircraft whose uncontrolled eigenvalues are given in table 10-1 : the values of the gains g_1, g_2, and g_3 are also presented in this table. It will be noted that, in the case of each type of control, the values of g_1, g_2, and g_3 have been determined so as to maintain the eigenvalues of the controlled modes invariant with changes in altitude.

Typical analogue computer results illustrating the behaviour of the uncontrolled system (fig. 10-1) and the various controlled systems (fig. 10-2) defined in table 10-1 are shown in figs. 10-3 to 10-8. The response of the uncontrolled aircraft to an out-of-trim roll-rate disturbance is shown in fig. 10-3 for these altitudes. The increase in the roll-rate mode time constant with altitude is apparent (corresponding to the decrease in the absolute magnitude of the eigenvalue λ_2 shown in table 10-1), as is the increase in steady-state bank-angle with altitude.

Figures 10-4 and 10-5 show the response of the system when bank-angle mode control ($\lambda_1 \rightarrow \rho_1$) is applied. The response at 40 000 ft is still inferior to that at sea level because this type of modal control does not alter the basic roll-rate mode time constant of the aircraft.

Figure 10-6 shows the response of the system when roll-rate mode control ($\lambda_2 \rightarrow \rho_2$) is applied. The roll-rate response at 40 000 ft is now practically the same as that at sea level, but a steady-state bank-angle still results.

Figures 10-7 and 10-8 show the response of the system when both bank-angle mode control ($\lambda_1 \rightarrow \rho_1$) and roll-rate mode control ($\lambda_2 \rightarrow \rho_2$) are applied simultaneously. These figures show the response of the aircraft to an out-of-trim roll-rate and an out-of-trim bank-angle disturbance respectively : it will be noted that the roll-rate and bank-angle responses are now practically identical for all altitudes between sea level and 40 000 ft.

Reference

[1] Engineering Sciences Data Unit, Aero. Series, Dynamics sub-series, 67003, 1967.

Synthesis of aircraft longitudinal autostabilization systems

CHAPTER 11

11.1. Introduction

THIS chapter is concerned with the design of a longitudinal autostabilizer for the control of the short-period oscillatory mode and the pitch-angle mode of an aircraft. The appropriate aerodynamic data for an aircraft flying at $M=0\cdot9$ and at altitudes from sea level to 40 000 ft are used in the computations as in Chapter 10.

11.2. Mathematical model of uncontrolled system

In this chapter, the results of § 5.2.2 are used to synthesize appropriate control laws for a system modelled as a third-order representation of the pitching dynamics of an aircraft. The open-loop response of pitch angle, $\theta(t)$, incidence angle, $\alpha(t)$, and pitch rate, $q(t)$, to elevator angle, $\eta(t)$, is governed by the equations [1]

$$\dot{\theta}(t)=q(t), \tag{1.11}$$

$$mV(\dot{\alpha}(t)-q(t))=Z_{\alpha}\alpha(t)+Z_{\eta}\eta(t), \tag{11.2}$$

and

$$I_{y}\dot{q}(t)=M_{\dot{\alpha}}\dot{\alpha}(t)+M_{\alpha}\alpha(t)+M_{\eta}\eta(t)+M_{q}q(t). \tag{11.3}$$

Equation (11.1) is a kinematic relation which defines the pitch rate, $q(t)$, eqn. (11.2) is the normal-force equation, and eqn. (11.3) is the pitching equation of motion.

If state variables $x_1(t)$, $x_2(t)$, $x_3(t)$, and an input variable, $z(t)$, are introduced in accordance with the definitions

$$x_1(t)=\theta(t), \tag{11.4 a}$$

$$x_2(t)=\alpha(t), \tag{114. b}$$

$$x_3(t) = q(t), \tag{11.4 c}$$

and

$$z(t) = \eta(t), \tag{11.5}$$

then eqns. (11.1), (11.2), and (11.3) can be written as the single vector-matrix differential equation

$$\begin{bmatrix} 1, & 0, & 0 \\ 0, & 1, & 0 \\ 0, & -M_{\dot{\alpha}}/I_y, & 1 \end{bmatrix} \begin{bmatrix} \dot{x}_1(t) \\ \dot{x}_2(t) \\ \dot{x}_3(t) \end{bmatrix} = \begin{bmatrix} 0, & 0, & 1 \\ 0, & Z_\alpha/mV, & 1 \\ 0, & M_\alpha/I_y, & M_q/I_y \end{bmatrix} \begin{bmatrix} x_1(t) \\ x_2(t) \\ x_3(t) \end{bmatrix} + \begin{bmatrix} 0 \\ Z_\eta/mV \\ M_\eta/I_y \end{bmatrix} z(t). \tag{11.6}$$

Equation (11.6) can then be pre-multiplied by the matrix

$$\begin{bmatrix} 1, & 0, & 0 \\ 0, & 1, & 0 \\ 0, & M_{\dot{\alpha}}/I_y, & 1 \end{bmatrix}$$

to give

$$\dot{\mathbf{x}}(t) = \begin{bmatrix} 0, & 0, & 1 \\ 0, & Z_\alpha/mV, & 1 \\ 0, & (M_\alpha + M_{\dot{\alpha}} Z_\alpha/mV)/I_y, & (M_q + M_{\dot{\alpha}})/I_y \end{bmatrix} \mathbf{x}(t)$$

$$+ \begin{bmatrix} 0 \\ Z_\eta/mV \\ (M_\eta + M_{\dot{\alpha}} Z_\eta/mV)/I_y \end{bmatrix} z(t) \tag{11.7}$$

which has the form

$$\dot{\mathbf{x}}(t) = \mathbf{A}\mathbf{x}(t) + \mathbf{b}z(t).$$

11.3. Design of modal controllers

Two forms of modal control system are synthesized. The first example involves the modal control of the longitudinal short-period oscillatory mode associated with the two conjugate complex eigenvalues of the plant matrix, $\mathbf{A}$; the second involves the modal control of both the longitudinal short-period oscillatory mode and the pitch-angle mode. In the first case, the two state variables are incidence angle and pitch rate, whilst in the second case the pitch angle is an additional state variable. Typical aerodynamic data for an aircraft flying at $M = 0{\cdot}9$ at altitudes from sea level to 40 000 ft are used in the calculations and determine the appropriate values of the elements of $\mathbf{A}$ and $\mathbf{b}$. The required eigenvalues characterizing the longitudinal short-period oscillation of the closed-loop system are chosen to correspond to a damping ratio of $\zeta = 0{\cdot}7$ and an undamped natural frequency of $\omega_n = 3$ rad/sec. In the second example, the zero-valued eigenvalue is required to become $-0{\cdot}2$ for the pitch-angle mode corresponding to a time-to-half amplitude of 3·5 sec.

The data for the uncontrolled system and the other relevant parameters required in the synthesis procedure are presented in tables 11-1 and 11-2. The necessary calculations are straightforward and involve the computation of the following quantities :

(i) the eigenvalues and eigenvectors of $\mathbf{A}$ and $\mathbf{A}'$ [2],

(ii) the mode-controllability indices p_j (eqn. (5.21)),

(iii) the proportional-controller gains K_j (eqn. (5.33)),

(iv) the feedback gain vector $\mathbf{g}$ (eqn. (5.34)).

Table 11-1. Modal control of the longitudinal short-period oscillatory mode

Altitude		Sea level	20 000 ft	40 000 ft
$\mathbf{A}$		$\begin{bmatrix} -0{\cdot}602, & +1 \\ -15{\cdot}988, & -0{\cdot}5 \end{bmatrix}$	$\begin{bmatrix} -0{\cdot}298, & +1 \\ -7{\cdot}363, & -0{\cdot}247 \end{bmatrix}$	$\begin{bmatrix} -0{\cdot}128, & +1 \\ -2{\cdot}970, & -0{\cdot}107 \end{bmatrix}$
$\mathbf{b}$		$\begin{bmatrix} -0{\cdot}0848 \\ -9{\cdot}439 \end{bmatrix}$	$\begin{bmatrix} -0{\cdot}0420 \\ -4{\cdot}340 \end{bmatrix}$	$\begin{bmatrix} -0{\cdot}0181 \\ -1{\cdot}749 \end{bmatrix}$
Eigenvalues of $\mathbf{A}$	λ_1	$-0{\cdot}551+i3{\cdot}998$	$-0{\cdot}273+i2{\cdot}713$	$-0{\cdot}118+i1{\cdot}723$
	λ_2	$-0{\cdot}551-i3{\cdot}998$	$-0{\cdot}273+i2{\cdot}713$	$-0{\cdot}118+i1{\cdot}723$
Required eigen-values of controlled system	ρ_1	$-2{\cdot}1+i2{\cdot}142$	$-2{\cdot}1+i2{\cdot}142$	$-2{\cdot}1+i2{\cdot}142$
	ρ_2	$-2{\cdot}1-i2{\cdot}142$	$-2{\cdot}1-i2{\cdot}142$	$-2{\cdot}1-i2{\cdot}142$
Eigenvectors of $\mathbf{A}$	$\mathbf{u}_1$	$\begin{bmatrix} +0{\cdot}00319-i0{\cdot}2501 \\ +1 \end{bmatrix}$	$\begin{bmatrix} +0{\cdot}00343-i0{\cdot}3685 \\ +1 \end{bmatrix}$	$\begin{bmatrix} +0{\cdot}00367-i0{\cdot}5803 \\ +1 \end{bmatrix}$
	$\mathbf{u}_2$	$\begin{bmatrix} +0{\cdot}00319+i0{\cdot}2501 \\ +1 \end{bmatrix}$	$\begin{bmatrix} +0{\cdot}00344+i0{\cdot}3685 \\ +1 \end{bmatrix}$	$\begin{bmatrix} +0{\cdot}00367+i0{\cdot}5803 \\ +1 \end{bmatrix}$
Eigenvectors of $\mathbf{A}'$	$\mathbf{v}_1$	$\begin{bmatrix} +i1{\cdot}9994 \\ +0{\cdot}5-i0{\cdot}00638 \end{bmatrix}$	$\begin{bmatrix} +i1{\cdot}3567 \\ +0{\cdot}5-i0{\cdot}00465 \end{bmatrix}$	$\begin{bmatrix} +i0{\cdot}8617 \\ +0{\cdot}5-i0{\cdot}00316 \end{bmatrix}$
	$\mathbf{v}_2$	$\begin{bmatrix} -i1{\cdot}9994 \\ +0{\cdot}5+i0{\cdot}00638 \end{bmatrix}$	$\begin{bmatrix} -i1{\cdot}3567 \\ +0{\cdot}5+i0{\cdot}00465 \end{bmatrix}$	$\begin{bmatrix} -i0{\cdot}8617 \\ +0{\cdot}5+i0{\cdot}00316 \end{bmatrix}$
Mode-controllability matrix	$\mathbf{p}_1$ $\mathbf{p}_2$	$\begin{bmatrix} -4{\cdot}7195-i0{\cdot}1094 \\ -4{\cdot}7195+i0{\cdot}1094 \end{bmatrix}$	$\begin{bmatrix} -2{\cdot}1698-i0{\cdot}0368 \\ -2{\cdot}1698+i0{\cdot}0368 \end{bmatrix}$	$\begin{bmatrix} -0{\cdot}8744-i0{\cdot}0101 \\ -0{\cdot}8744+i0{\cdot}0101 \end{bmatrix}$
Proportional-controller gains	K_1	$+0{\cdot}3336+i0{\cdot}2306$	$+0{\cdot}8411-i0{\cdot}0624$	$+2{\cdot}2458-i1{\cdot}8677$
	K_2	$+0{\cdot}3336-i0{\cdot}2306$	$+0{\cdot}8411+i0{\cdot}0624$	$+2{\cdot}2458+i1{\cdot}8677$
Feedback gains	g_1	$-0{\cdot}922$	$+0{\cdot}169$	$+3{\cdot}219$
	g_2	$+0{\cdot}336$	$+0{\cdot}840$	$+2{\cdot}234$

Tables 11-2. Modal control of the longitudinal short-period oscillatory mode and the pitch-angle mode

Altitude		Sea level	20 000 ft	40 000 ft
A		$\begin{bmatrix} 0, & 0, & +1 \\ 0, & -0{\cdot}602, & +1 \\ 0, & -15{\cdot}988, & -0{\cdot}5 \end{bmatrix}$	$\begin{bmatrix} 0, & 0, & +1 \\ 0, & -0{\cdot}298, & +1 \\ 0, & -7{\cdot}363, & -0{\cdot}247 \end{bmatrix}$	$\begin{bmatrix} 0, & 0, & +1 \\ 0, & -0{\cdot}128, & +1 \\ 0, & -2{\cdot}970, & -0{\cdot}107 \end{bmatrix}$
b		$\begin{bmatrix} 0 \\ -0{\cdot}0848 \\ -9{\cdot}439 \end{bmatrix}$	$\begin{bmatrix} 0 \\ -0{\cdot}0420 \\ -4{\cdot}340 \end{bmatrix}$	$\begin{bmatrix} 0 \\ -0{\cdot}0181 \\ -1{\cdot}749 \end{bmatrix}$
Eigenvalues of **A**	λ_1 λ_2 λ_3	$-0{\cdot}551+i3{\cdot}998$ $-0{\cdot}551-i3{\cdot}998$ 0	$-0{\cdot}273+i2{\cdot}713$ $-0{\cdot}273-i2{\cdot}713$ 0	$-0{\cdot}118+i1{\cdot}723$ $-0{\cdot}118+i1{\cdot}723$ 0
Required eigen-values of controlled system	ρ_1 ρ_2 ρ_3	$-2{\cdot}1+i2{\cdot}142$ $-2{\cdot}1-i2{\cdot}142$ $-0{\cdot}2$	$-2{\cdot}1+i2{\cdot}142$ $-2{\cdot}1-i2{\cdot}142$ $-0{\cdot}2$	$-2{\cdot}1+i2{\cdot}142$ $-2{\cdot}1-i2{\cdot}142$ $-0{\cdot}2$
Eigenvectors of **A**	$\mathbf{u}_1$	$\begin{bmatrix} -0{\cdot}0338 -i0{\cdot}2454 \\ +0{\cdot}00319-i0{\cdot}2501 \\ +1 \end{bmatrix}$	$\begin{bmatrix} -0{\cdot}0367 -i0{\cdot}3649 \\ +0{\cdot}00343-i0{\cdot}3685 \\ +1 \end{bmatrix}$	$\begin{bmatrix} -0{\cdot}0394 -i0{\cdot}5776 \\ -0{\cdot}0394 -i0{\cdot}5803 \\ +1 \end{bmatrix}$
	$\mathbf{u}_2$	$\begin{bmatrix} -0{\cdot}0338 +i0{\cdot}2454 \\ +0{\cdot}00319+i0{\cdot}2501 \\ 1 \end{bmatrix}$	$\begin{bmatrix} -0{\cdot}0367 +i0{\cdot}3649 \\ +0{\cdot}00343+i0{\cdot}3685 \\ +1 \end{bmatrix}$	$\begin{bmatrix} -0{\cdot}0394 +i0{\cdot}5776 \\ +0{\cdot}00367+i0{\cdot}5803 \\ +1 \end{bmatrix}$
	$\mathbf{u}_3$	$\begin{bmatrix} +1 \\ 0 \\ 0 \end{bmatrix}$	$\begin{bmatrix} +1 \\ 0 \\ 0 \end{bmatrix}$	$\begin{bmatrix} +1 \\ 0 \\ 0 \end{bmatrix}$
Eigenvectors of **A′**	$\mathbf{v}_1$	$\begin{bmatrix} 0 \\ +i1{\cdot}9994 \\ +0{\cdot}5-i0{\cdot}00638 \end{bmatrix}$	$\begin{bmatrix} 0 \\ +i1{\cdot}3567 \\ +0{\cdot}5-i0{\cdot}00465 \end{bmatrix}$	$\begin{bmatrix} 0 \\ +i0{\cdot}8617 \\ +0{\cdot}5-i0{\cdot}00316 \end{bmatrix}$
	$\mathbf{v}_2$	$\begin{bmatrix} 0 \\ -i1{\cdot}9994 \\ +0{\cdot}5+i0{\cdot}00638 \end{bmatrix}$	$\begin{bmatrix} 0 \\ -i1{\cdot}3567 \\ +0{\cdot}5+i0{\cdot}00465 \end{bmatrix}$	$\begin{bmatrix} 0 \\ -i0{\cdot}8617 \\ +0{\cdot}5+i0{\cdot}00316 \end{bmatrix}$
	$\mathbf{v}_3$	$\begin{bmatrix} +1 \\ -0{\cdot}9815 \\ +0{\cdot}0370 \end{bmatrix}$	$\begin{bmatrix} +1 \\ -0{\cdot}9901 \\ +0{\cdot}0401 \end{bmatrix}$	$\begin{bmatrix} +1 \\ -0{\cdot}9954 \\ +0{\cdot}0431 \end{bmatrix}$
Mode-control-lability matrix	$\mathbf{p}_1$ $\mathbf{p}_2$ $\mathbf{p}_3$	$\begin{bmatrix} -4{\cdot}7195-i0{\cdot}1094 \\ -4{\cdot}7195+i0{\cdot}1094 \\ -0{\cdot}2656 \end{bmatrix}$	$\begin{bmatrix} -2{\cdot}1698-i0{\cdot}0368 \\ -2{\cdot}1698+i0{\cdot}0368 \\ -0{\cdot}1323 \end{bmatrix}$	$\begin{bmatrix} -0{\cdot}8744-i0{\cdot}0101 \\ -0{\cdot}8744+i0{\cdot}0101 \\ -0{\cdot}0573 \end{bmatrix}$
Proportional-controller gains	K_1 K_2 K_3	$+0{\cdot}3426+i0{\cdot}2127$ $+0{\cdot}3426-i0{\cdot}2127$ $+0{\cdot}4161$	$+0{\cdot}8304-i0{\cdot}1234$ $+0{\cdot}8304+i0{\cdot}1234$ $+1{\cdot}8302$	$+2{\cdot}0123-i2{\cdot}1124$ $+2{\cdot}0123+i2{\cdot}1124$ $+10{\cdot}5290$
Feedback gains	g_1 g_2 g_3	$+0{\cdot}416$ $-1{\cdot}259$ $+0{\cdot}361$	$+1{\cdot}830$ $-1{\cdot}477$ $+0{\cdot}902$	$+10{\cdot}529$ $-6{\cdot}841$ $+2{\cdot}453$

Typical analogue computer results for the altitude 20 000 ft are presented in fig. 11-1. Figures 11-1 (*a*) and 11-1 (*d*) show clearly the lightly-damped short-period oscillation for initial 'out-of-trim' incidence angle and pitch-rate, respectively. Modal control of the short-period oscillation is demonstrated in figs. 11-1 (*b*) and 11-1 (*e*) where the effects of the increased damping ratio and natural frequency are demonstrated: a steady-state pitch angle results in each of these cases since there is no pitch-angle feedback with this type of control. Figures 11-1 (*c*), 11-1 (*f*), and 11-1 (*g*) illustrate the effect of control of both the short-period oscillation and the pitch-angle mode: the initial conditions are for 'out-of-trim' incidence angle, pitch rate, and pitch angle, respectively.

Fig. 11-1

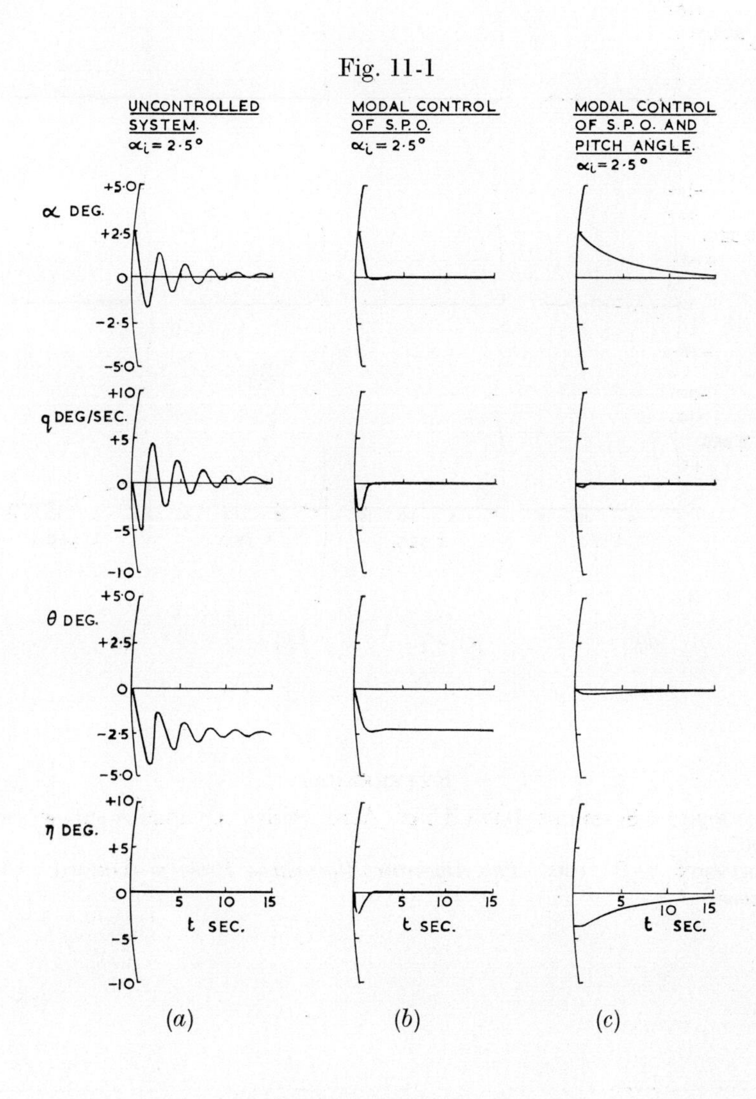

(*a*) (*b*) (*c*)

Fig. 11·1 (*continued*)

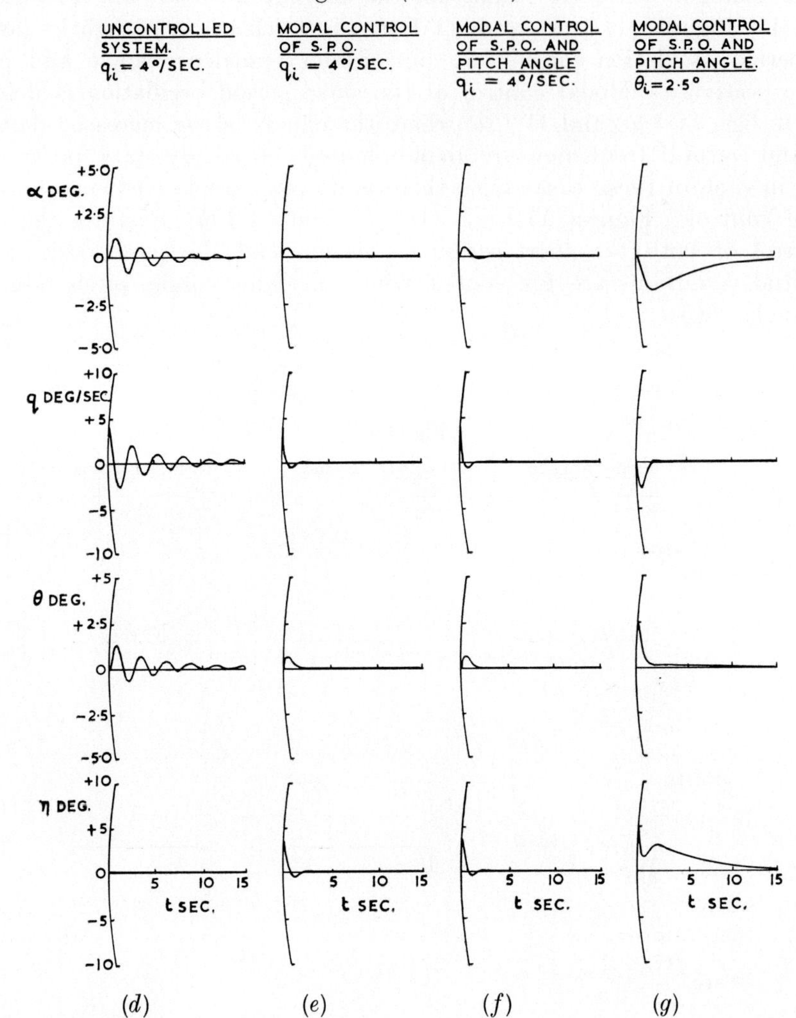

(*d*) (*e*) (*f*) (*g*)

References

[1] Engineering Sciences Data Unit, Aero. Series, Dynamics sub-series, 67003, 1967.

[2] Wilkinson, J. H., 1965, *The Algebraic Eigenvalue Problem* (Oxford : Clarendon Press).

Synthesis of helicopter stabilization systems

CHAPTER 12

12.1. Introduction

In this chapter, appropriate control laws for the stabilization of a helicopter in hovering flight are synthesized using the modal control design procedures developed in Chapters 5 and 6. Although the synthesis techniques employed in this chapter are applicable to cases when longitudinal-to-lateral and lateral-to-longitudinal coupling effects are present, these effects are neglected in order to keep the presentation as simple as possible.

12.2. Mathematical model of uncontrolled system

The mathematical model of the hover dynamics of a helicopter has been described in detail by Murphy and Narendra [1]: the state equation of the uncontrolled system has the form

$$\begin{bmatrix} \dot{\mathbf{x}}_1(t) \\ \dot{\mathbf{x}}_2(t) \end{bmatrix} = \begin{bmatrix} \mathbf{A}_{11}, & \mathbf{A}_{12} \\ \mathbf{A}_{21}, & \mathbf{A}_{22} \end{bmatrix} \begin{bmatrix} \mathbf{x}_1(t) \\ \mathbf{x}_2(t) \end{bmatrix} + \begin{bmatrix} \mathbf{B}_{11}, & \mathbf{B}_{12} \\ \mathbf{B}_{21}, & \mathbf{B}_{22} \end{bmatrix} \begin{bmatrix} \mathbf{z}_1(t) \\ \mathbf{z}_2(t) \end{bmatrix}. \tag{12.1}$$

In eqn. (12.1), the longitudinal state vector, $\mathbf{x}_1(t)$, and the lateral state vector, $\mathbf{x}_2(t)$, are respectively given by

$$\mathbf{x}_1(t) = \begin{bmatrix} x_1(t) \\ x_2(t) \\ x_3(t) \\ x_4(t) \end{bmatrix} = \begin{bmatrix} u(t)/V_R \\ w(t)/V_R \\ q(t) \\ \theta(t) \end{bmatrix}, \tag{12.2 a}$$

and

$$\mathbf{x}_2(t) = \begin{bmatrix} x_5(t) \\ x_6(t) \\ x_7(t) \\ x_8(t) \\ x_9(t) \end{bmatrix} = \begin{bmatrix} v(t)/V_R \\ p(t) \\ r(t) \\ \phi(t) \\ \psi(t) \end{bmatrix}. \tag{12.2 b}$$

In addition, the longitudinal input vector, $\mathbf{z}_1(t)$, and the lateral input vector, $\mathbf{z}_2(t)$, are respectively given by

$$\mathbf{z}_1(t)=\begin{bmatrix} U_{\mathrm{P}}(t) \\ U_{\mathrm{C}}(t) \end{bmatrix}, \tag{12.3 a}$$

and

$$\mathbf{z}_2(t)=\begin{bmatrix} U_{\mathrm{R}}(t) \\ U_{\mathrm{T}}(t) \end{bmatrix}. \tag{12.3 b}$$

In eqns. (12.2), $u(t)$ is the longitudinal velocity, $w(t)$ is the vertical velocity, $q(t)$ is the pitch rate, $\theta(t)$ is the pitch angle, V_{R} is the rotor-tip velocity, $v(t)$ is the lateral velocity, $p(t)$ is the roll rate, $r(t)$ is the yaw rate, $\phi(t)$ is the roll angle, and $\psi(t)$ is the yaw angle : in eqns. (12.3), $U_{\mathrm{P}}(t)$ is the longitudinal cyclic pitch-control angle, $U_{\mathrm{C}}(t)$ is the main-rotor collective pitch-control angle, $U_{\mathrm{R}}(t)$ is the lateral cyclic pitch-control angle, and $U_{\mathrm{T}}(t)$ is the tail-rotor collective pitch-control angle.

In order to decouple the longitudinal and lateral modes it is necessary to assume that the coupling matrices $\mathbf{A}_{12}$, $\mathbf{A}_{21}$, $\mathbf{B}_{12}$, and $\mathbf{B}_{21}$ in eqn. (12.1) are all null. The matrices $\mathbf{A}_{11}$, $\mathbf{A}_{22}$, $\mathbf{B}_{11}$, and $\mathbf{B}_{22}$ (as given by Murphy and Narendra [1]) are

$$\mathbf{A}_{11}=\begin{bmatrix} -0{\cdot}016, & 0, & +0{\cdot}0025, & -0{\cdot}05 \\ 0, & -0{\cdot}3242, & +0{\cdot}0002, & 0 \\ +1{\cdot}97, & +1, & -0{\cdot}542, & 0 \\ 0, & 0, & +1, & 0 \end{bmatrix}, \tag{12.4 a}$$

$$\mathbf{A}_{22}=\begin{bmatrix} -0{\cdot}033, & -0{\cdot}0025, & +0{\cdot}0009, & +0{\cdot}05, & 0 \\ -7{\cdot}25, & -1{\cdot}96, & +0{\cdot}01, & 0, & 0 \\ +5{\cdot}59, & -0{\cdot}0043, & -0{\cdot}303, & 0, & 0 \\ 0, & +1, & 0, & 0, & 0 \\ 0, & 0, & +1, & 0, & 0 \end{bmatrix}, \tag{12.4 b}$$

$$\mathbf{B}_{11}=\begin{bmatrix} +0{\cdot}005, & +0{\cdot}05 \\ -0{\cdot}424, & 0 \\ +0{\cdot}69, & -6{\cdot}15 \\ 0, & 0 \end{bmatrix}, \tag{12.4 c}$$

and

$$\mathbf{B}_{22}=\begin{bmatrix} +0{\cdot}05, & +0{\cdot}022 \\ +21{\cdot}81, & +0{\cdot}3475 \\ +0{\cdot}174, & -7{\cdot}48 \\ 0, & 0 \\ 0, & 0 \end{bmatrix}. \tag{12.4 d}$$

The eigenvalues of $\mathbf{A}_{11}$ are

$$\left.\begin{aligned} \lambda_1 &= +0{\cdot}0887+0{\cdot}3552i, \\ \lambda_2 &= +0{\cdot}0887-0{\cdot}3552i, \\ \lambda_3 &= -0{\cdot}7354, \\ \lambda_4 &= -0{\cdot}3240, \end{aligned}\right\} \tag{12.5}$$

and the corresponding eigenvectors of $\mathbf{A}_{11}'$ are

$$\mathbf{v}_1 = \begin{bmatrix} +1\cdot0000 \\ +0\cdot2898+0\cdot1873i \\ +0\cdot0531+0\cdot1803i \\ -0\cdot0331+0\cdot1325i \end{bmatrix}, \tag{12.6 a}$$

$$\mathbf{v}_2 = \begin{bmatrix} +1\cdot0000 \\ +0\cdot2898-0\cdot1873i \\ +0\cdot0531-0\cdot1803i \\ -0\cdot0331-0\cdot1325i \end{bmatrix}, \tag{12.6 b}$$

$$\mathbf{v}_3 = \begin{bmatrix} +0\cdot7205 \\ +0\cdot6398 \\ -0\cdot2631 \\ +0\cdot0490 \end{bmatrix}, \tag{12.6 c}$$

and

$$\mathbf{v}_4 = \begin{bmatrix} -0\cdot0009 \\ +1\cdot0000 \\ +0\cdot0001 \\ -0\cdot0001 \end{bmatrix}. \tag{12.6 d}$$

In accordance with eqn. (4.8), the longitudinal mode-controllability matrix is

$$\mathbf{P}_1 = \begin{bmatrix} \mathbf{v}_1' \\ \mathbf{v}_2' \\ \mathbf{v}_3' \\ \mathbf{v}_4' \end{bmatrix} \mathbf{B}_{11} = \begin{bmatrix} -0\cdot2767-1\cdot1087i, & -0\cdot0812+0\cdot0450i \\ -0\cdot2767+1\cdot1087i, & -0\cdot0812-0\cdot0450i \\ +1\cdot6541, & -0\cdot4492 \\ -0\cdot0010, & -0\cdot4239 \end{bmatrix}. \tag{12.7}$$

The eigenvalues of $\mathbf{A}_{22}$ are

$$\left.\begin{aligned} \lambda_5 &= +0\cdot0315+0\cdot4137i, \\ \lambda_6 &= +0\cdot0315-0\cdot4137i, \\ \lambda_7 &= -2\cdot0565, \\ \lambda_8 &= -0\cdot3024, \\ \lambda_9 &= 0, \end{aligned}\right\} \tag{12.8}$$

and the corresponding eigenvectors of $\mathbf{A}_{22}'$ are

$$\mathbf{v}_5 = \begin{bmatrix} +1\cdot0000 \\ -0\cdot0088-0\cdot0585i \\ +0\cdot0001-0\cdot0019i \\ +0\cdot0091-0\cdot1202i \\ 0 \end{bmatrix}, \tag{12.9 a}$$

$$\mathbf{v}_6 = \begin{bmatrix} +1\cdot000 \\ -0\cdot0088+0\cdot0585i \\ +0\cdot0001+0\cdot0019i \\ +0\cdot0091+0\cdot1202i \\ 0 \end{bmatrix}, \tag{12.9 b}$$

$$\mathbf{v}_7 = \begin{bmatrix} -0{\cdot}9633 \\ -0{\cdot}2673 \\ +0{\cdot}0020 \\ +0{\cdot}0234 \\ 0 \end{bmatrix}, \tag{12.9 c}$$

$$\mathbf{v}_8 = \begin{bmatrix} +0{\cdot}9670 \\ -0{\cdot}0975 \\ -0{\cdot}1730 \\ -0{\cdot}1599 \\ 0 \end{bmatrix}, \tag{12.9 d}$$

and

$$\mathbf{v}_9 = \begin{bmatrix} 0 \\ +0{\cdot}3866 \\ +0{\cdot}5013 \\ +0{\cdot}7598 \\ +0{\cdot}1480 \end{bmatrix}. \tag{12.9 e}$$

In accordance with eqn. (4.8), the lateral mode-controllability matrix is

$$\mathbf{P}_2 = \begin{bmatrix} \mathbf{v}_5' \\ \mathbf{v}_6' \\ \mathbf{v}_7' \\ \mathbf{v}_8' \\ \mathbf{v}_9' \end{bmatrix} \mathbf{B}_{22} = \begin{bmatrix} -0{\cdot}1422-1{\cdot}2764i, & +0{\cdot}0182-0{\cdot}0063i \\ -0{\cdot}1422+1{\cdot}2764i, & +0{\cdot}0182+0{\cdot}0063i \\ -5{\cdot}8781, & -0{\cdot}1292 \\ -2{\cdot}1074, & +1{\cdot}2815 \\ +8{\cdot}5180, & -3{\cdot}6158 \end{bmatrix}. \tag{12.10}$$

12.3. Design of single-input modal controllers

12.3.1. *Longitudinal mode control*

The eigenvalues given by (12.5) indicate that the uncontrolled helicopter has an unstable oscillatory mode (λ_1, λ_2) and two asymptotically stable non-oscillatory modes (λ_3, λ_4). It is evident from the mode-controllability matrix (12.7) that the longitudinal modes are all controllable and that, in the light of § 6.2.4, the first three modes (λ_1, λ_2, λ_3) would normally be controlled by U_P and the fourth mode (λ_4) by U_C.

If it is now desired to design a modal controller such that $\lambda_1 \to \rho_1$ $(= -0{\cdot}8+1{\cdot}2i)$, $\lambda_2 \to \rho_2$ $(= -0{\cdot}8-1{\cdot}2i)$, and $\lambda_3 \to \rho_3$ $(= -1{\cdot}0)$ using the control input U_P, then the formula (5.33) implies that the required control law is

$$U_P(t) = -6{\cdot}441[u(t)/V_R] - 1{\cdot}434[w(t)/V_R] + 0{\cdot}280q(t) + 0{\cdot}586\theta(t). \tag{12.11}$$

12.3.2. *Lateral mode control*

The eigenvalues given by (12.8) indicate that the uncontrolled helicopter has an unstable oscillatory mode (λ_5, λ_6), two asymptotically stable non-oscillatory modes (λ_7, λ_8), and a neutrally stable non-oscillatory mode (λ_9).

It is evident from the mode-controllability matrix (12.10) that the lateral modes are all controllable and that, in the light of § 6.2.4, all the first three modes (λ_5, λ_6, λ_7) would normally be controlled by U_R whilst the last two modes (λ_8, λ_9) could be controlled by either U_R or U_T.

If it is now desired to design a modal controller such that $\lambda_5 \to \rho_5$ $(= -0{\cdot}5 + 1{\cdot}0i)$, $\lambda_6 \to \rho_6$ $(= -0{\cdot}5 - 1{\cdot}0i)$, $\lambda_8 \to \rho_8$ $(= -1{\cdot}0)$ and $\lambda_9 \to \rho_9$ $(= -0{\cdot}6)$ using the control input U_R, then the formula (5.33) implies that the required control law is

$$U_R(t) = -6{\cdot}223[v(t)/V_R] - 0{\cdot}0889p(t) - 0{\cdot}628r(t) - 0{\cdot}311\phi(t) - 0{\cdot}250\psi(t). \quad (12.12)$$

12.4. Multi-stage design of multi-input modal controllers

12.4.1. *Longitudinal mode control*

If it is now desired to design a multi-input modal controller such that $\lambda_1 \to \rho_1$ $(= -0{\cdot}8 + 1{\cdot}2i)$, $\lambda_2 \to \rho_2$ $(= -0{\cdot}8 - 1{\cdot}2i)$ and $\lambda_3 \to \rho_3$ $(= -1{\cdot}0)$ using the control U_P, and $\lambda_4 \to \rho_4$ $(= -0{\cdot}6)$ using the control U_C, then the formula (5.33) and the procedure described in § 6.2.4 imply that the required control laws using the sequence $\{U_P, U_C\}$ are

$$U_P(t) = -6{\cdot}441[u(t)/V_R] - 1{\cdot}434[w(t)/V_R] + 0{\cdot}280q(t) + 0{\cdot}586\theta(t), \quad (12.13\ a)$$

and

$$U_C(t) = -0{\cdot}0014[u(t)/V_R] + 0{\cdot}651[w(t)/V_R] - 0{\cdot}000094\theta(t). \quad (12.13\ b)$$

Alternatively, the required control laws using the sequence $\{U_C, U_P\}$ are

$$U_C(t) = -0{\cdot}00061[u(t)/V_R] + 0{\cdot}651[w(t)/V_R) + 0{\cdot}000099q(t) - 0{\cdot}000094\theta(t), \quad (12.14\ a)$$

and

$$U_P(t) = -6{\cdot}441[u(t)/V_R] - 1{\cdot}592[w(t)/V_R] + 0{\cdot}279q(t) + 0{\cdot}586\theta(t). \quad (12.14\ b)$$

12.4.2. *Lateral mode control*

If it is now desired to design a multi-input modal controller such that $\lambda_5 \to \rho_5$ $(= -0{\cdot}5 + 1{\cdot}0i)$ and $\lambda_6 \to \rho_6$ $(= -0{\cdot}5 - 1{\cdot}0i)$, using the control U_R, and $\lambda_8 \to \rho_8$ $(= -1{\cdot}0)$ and $\lambda_9 \to \rho_9$ $(= -0{\cdot}6)$ using the control U_T, then the formula (5.33) and the procedure described in § 6.2.4 imply that the required control laws using the sequence $\{U_R, U_T\}$ are

$$U_R(t) = -1{\cdot}987[v(t)/V_R] - 0{\cdot}0441p(t) - 0{\cdot}0022r(t) - 0{\cdot}145\phi(t), \quad (12.15\ a)$$

and

$$U_T(t) = 1{\cdot}184[v(t)/V_R] + 0{\cdot}0111p(t) + 0{\cdot}177r(t) + 0{\cdot}0439\phi(t) + 0{\cdot}0819\psi(t). \quad (12.15\ b)$$

Alternatively, the required control laws using the sequence $\{U_T, U_R\}$ are

$$U_T(t) = 0{\cdot}518[v(t)/V_R] + 0{\cdot}160p(t) + 0{\cdot}182r(t) + 0{\cdot}331\phi(t) + 0{\cdot}0812\psi(t), \quad (12.16\ a)$$

and

$$U_R(t) = -1{\cdot}999[v(t)/V_R] - 0{\cdot}0441p(t) - 0{\cdot}0079r(t) - 0{\cdot}145\phi(t) - 0{\cdot}0022\psi(t). \quad (12.16\ b)$$

12.5. Design of multi-input modal controllers with gain constraints

The design procedure described in § 6.2.3 can be conveniently illustrated by designing a lateral modal controller such that $\lambda_8 \to \rho_8$ $(= -0{\cdot}6)$, $\lambda_9 \to \rho_9$ $(= -0{\cdot}1)$ using the controls U_R and U_T.

If the state variable $x_6(t)[=p(t)]$ is not available for feedback, then, as discussed in § 6.2.3, it is possible to determine a unique set of proportional-controller gains : the resulting control laws are

$$U_R(t) = -0{\cdot}507[v(t)/V_R] + 0{\cdot}0244r(t) - 0{\cdot}0166\phi(t) - 0{\cdot}0194\psi(t), \quad (12.17\ a)$$

and

$$U_T(t) = -0{\cdot}984[v(t)/V_R] + 0{\cdot}0474r(t) - 0{\cdot}0323\phi(t) - 0{\cdot}0380\psi(t). \quad (12.17\ b)$$

Similarly, if the state variable $x_7(t)[=r(t)]$ is not available for feedback, the required control laws are

$$U_R(t) = -0{\cdot}427[v(t)/V_R] - 0{\cdot}0159p(t) - 0{\cdot}0452\phi(t) - 0{\cdot}0226\psi(t), \quad (12.18\ a)$$

and

$$U_T(t) = -0{\cdot}853[v(t)/V_R] - 0{\cdot}0317p(t) - 0{\cdot}0903\phi(t) - 0{\cdot}0451\psi(t). \quad (12.18\ b)$$

Finally, if the state feedback $x_8(t)[=\phi(t)]$ is not available for feedback, the required control laws are

$$U_R(t) = -0{\cdot}566[v(t)/V_R] + 0{\cdot}0094p(t) + 0{\cdot}0395r(t) - 0{\cdot}0182\psi(t), \quad (12.19\ a)$$

and

$$U_T(t) = -1{\cdot}081[v(t)/V_R] + 0{\cdot}0180p(t) + 0{\cdot}0755r(t) + 0{\cdot}0348\psi(t). \quad (12.19\ b)$$

It is not possible to use this procedure if the state variables $x_5(t)[=v(t)/V_R]$ and $x_9(t)[=\psi(t)]$ are not available for feedback since the appropriate prescribed-gain equations (see § 6.2.3) are linearly dependent in these cases due to the fact that the last element of $\mathbf{v}_8$ and the first element of $\mathbf{v}_9$ are both zero.

Reference

[1] Murphy, R. D., and Narendra, K. S., 1969, " Design of helicopter stabilisation systems using optimal control theory ", *J. Aircraft*, **6**, 129.

Sensitivity analysis of an aircraft autostabilization system

CHAPTER 13

13.1. Introduction

In this chapter, the sensitivity theories developed in Chapter 9 are illustrated by determining the sensitivity characteristics of a controller for the modes of a VSTOL aircraft. Typical aerodynamic data for a VSTOL aircraft flying at an equivalent airspeed of 80 knots at sea level are used in the calculations.

13.2. Mathematical model of uncontrolled system

The dynamics of the uncontrolled aircraft considered in this chapter can be modelled by a state equation of the form (5.1 *a*) where the state vector, $\mathbf{x}(t)$, the input vector, $\mathbf{z}(t)$, the plant matrix, $\mathbf{A}$, and the input matrix, $\mathbf{B}$, are given by the respective equations

$$\mathbf{x}(t)=\begin{bmatrix}\beta(t)\\ p(t)\\ r(t)\\ \phi(t)\end{bmatrix}, \tag{13.1 a}$$

$$\mathbf{z}(t)=\begin{bmatrix}\zeta(t)\\ \\ \xi(t)\end{bmatrix}, \tag{13.1 b}$$

$$\mathbf{A}=\begin{bmatrix}-0{\cdot}0506, & 0, & -1, & +0{\cdot}2380\\ -0{\cdot}7374, & -1{\cdot}3345, & +0{\cdot}3696, & 0\\ +0{\cdot}0100, & +0{\cdot}1074, & -0{\cdot}3320, & 0\\ 0, & +1, & 0, & 0\end{bmatrix} \tag{13.1 c}$$

and

$$\mathbf{B}=\begin{bmatrix}+0{\cdot}0409, & 0\\ +1{\cdot}2714, & -20{\cdot}3106\\ -2{\cdot}0625, & +1{\cdot}3350\\ 0, & 0\end{bmatrix}. \tag{13.1 d}$$

In eqns. (13.1), $\beta(t)$ is the sideslip angle, $p(t)$ is the roll rate, $r(t)$ is the yaw rate, $\phi(t)$ is the bank angle, $\zeta(t)$ is the rudder angle, and $\xi(t)$ is the aileron angle.

The eigenvalues of $\mathbf{A}$ are

$$\left.\begin{aligned} \lambda_1 &= +0{\cdot}0470 + 0{\cdot}3147i, \\ \lambda_2 &= +0{\cdot}0470 - 0{\cdot}3147i, \\ \lambda_3 &= -1{\cdot}4088, \\ \lambda_4 &= -0{\cdot}4023, \end{aligned}\right\} \tag{13.2}$$

and the corresponding eigenvectors of $\mathbf{A}'$ are

$$\mathbf{v}_1 = \begin{bmatrix} -0{\cdot}3928 - 0{\cdot}2409i \\ -0{\cdot}0373 + 0{\cdot}1995i \\ +1{\cdot}0000 \\ -0{\cdot}2217 + 0{\cdot}2639i \end{bmatrix}, \tag{13.3 a}$$

$$\mathbf{v}_2 = \begin{bmatrix} -0{\cdot}3928 + 0{\cdot}2409i \\ -0{\cdot}0373 - 0{\cdot}1995i \\ +1{\cdot}0000 \\ -0{\cdot}2217 - 0{\cdot}2639i \end{bmatrix}, \tag{13.3 b}$$

$$\mathbf{v}_3 = \begin{bmatrix} -0{\cdot}4702 \\ -0{\cdot}8680 \\ -0{\cdot}1387 \\ +0{\cdot}0794 \end{bmatrix}, \tag{13.3 c}$$

and

$$\mathbf{v}_4 = \begin{bmatrix} +0{\cdot}0908 \\ +0{\cdot}0568 \\ +0{\cdot}9928 \\ -0{\cdot}0537 \end{bmatrix}. \tag{13.3 d}$$

In accordance with eqn. (4.8), the mode-controllability matrix is

$$\mathbf{P} = \begin{bmatrix} \mathbf{v}_1' \\ \mathbf{v}_2' \\ \mathbf{v}_3' \\ \mathbf{v}_4' \end{bmatrix} \mathbf{B} = \begin{bmatrix} -2{\cdot}1259 + 0{\cdot}2438i, & +2{\cdot}0918 - 4{\cdot}0527i \\ -2{\cdot}1259 - 0{\cdot}2438i, & +2{\cdot}0918 + 4{\cdot}0527i \\ -0{\cdot}8366, & +17{\cdot}4440 \\ -1{\cdot}9718, & +0{\cdot}1724 \end{bmatrix}. \tag{13.4}$$

13.3. Design of modal controller

The set of eigenvalues given in eqns. (13.2) indicates that the uncontrolled aircraft has an unstable oscillatory Dutch-roll mode (λ_1, λ_2) and two asymptotically stable non-oscillatory modes (λ_3, λ_4) : the latter two modes are the rolling and spiral modes of the aircraft, respectively. The mode-controllability matrix (13.4) indicates that, in accordance with the principles of § 6.2.4. the first three modes (λ_1, λ_2, λ_3) would normally be controlled by $\xi(t)$ and the fourth mode (λ_4) by $\zeta(t)$. However, for the purposes of the present sensitivity analysis, it will be assumed that the control law

$$\zeta(t) = -2{\cdot}35\beta(t) + 1{\cdot}33p(t) + 1{\cdot}69r(t) + 1{\cdot}47\phi(t) = \mathbf{g}'\mathbf{x}(t) \tag{13.5}$$

is used to control the Dutch-roll mode such that $\lambda_1 \to \rho_1$ $(-0{\cdot}9 + 2i)$ and $\lambda_2 \to \rho_2$ $(-0{\cdot}9 - 2i)$: this control law is derived by substituting the appropriate quantities from eqns. (13.2), (13.3), and (13.4) into eqn. (5.33).

13.4. Sensitivity analysis

The sensitivity characteristics of the closed-loop system defined by eqns. (13.1) and (13.5) may be calculated using the results presented in Chapter 9. In these calculations, the following selection of sensitivity vectors and coefficients are determined :

(i) $\partial \mathbf{g}/\partial a_{21}$, $\partial^2 \mathbf{g}/\partial {a_{21}}^2$;

(ii) $\partial \mathbf{g}/\partial b_3$, $\partial^2 \mathbf{g}/\partial {b_3}^2$;

(iii) $\partial \mathbf{g}/\partial \rho_1$, $\partial \mathbf{g}/\partial \rho_2$, $\partial^2 \mathbf{g}/\partial {\rho_1}^2$, $\partial^2 \mathbf{g}/\partial {\rho_2}^2$, $\partial^2 \mathbf{g}/\partial \rho_1 \rho_2$;

(iv) $\partial \rho_i/\partial g_1$, $\partial^2 \rho_i/\partial {g_1}^2$ $(i=1, 2, 3, 4)$;

(v) $\partial \rho_i/\partial a_{21}$, $\partial^2 \rho_i/\partial {a_{21}}^2$ $(i=1, 2, 3, 4)$;

(vi) $\partial \rho_i/\partial b_3$, $\partial^2 \rho_i/\partial {b_3}^2$ $(i=1, 2, 3, 4)$.

The plant matrix of the closed-loop system is

$$\mathbf{C}=\begin{bmatrix} -0{\cdot}1465, & +0{\cdot}0544, & -0{\cdot}9308, & +0{\cdot}2980 \\ -3{\cdot}7195, & +0{\cdot}3555, & +2{\cdot}5198, & +1{\cdot}8648 \\ +4{\cdot}8476, & -2{\cdot}6341, & -3{\cdot}8201, & -3{\cdot}0251 \\ 0, & +1, & 0, & 0 \end{bmatrix}. \tag{13.6}$$

It may be verified that the eigenvalue matrix of $\mathbf{C}$ is

$$\mathbf{R}=\text{diag}\,\{-0{\cdot}9000+2{\cdot}0000i,\ -0{\cdot}9000-2{\cdot}0000i,\ -1{\cdot}4088,\ -0{\cdot}4023\} \tag{13.7 a}$$

and that the corresponding modal matrices as defined in eqs. (9.26) are respectively

$$\mathbf{\Upsilon}=\begin{bmatrix} +0{\cdot}1996+0{\cdot}4154i, & +0{\cdot}1996-0{\cdot}4154i, & -0{\cdot}0411, & +0{\cdot}5391 \\ -0{\cdot}6920-0{\cdot}4143i, & -0{\cdot}6920+0{\cdot}4143i, & -0{\cdot}8121, & -0{\cdot}2845 \\ +1{\cdot}0000 & +1{\cdot}0000, & +0{\cdot}0814, & +0{\cdot}3580 \\ -0{\cdot}0428+0{\cdot}3653i, & -0{\cdot}0428-0{\cdot}3653i, & +0{\cdot}5764, & +0{\cdot}7073 \end{bmatrix} \tag{13.7 b}$$

and

$$\mathbf{\Xi}=\begin{bmatrix} +1{\cdot}0000, & +1{\cdot}0000, & -0{\cdot}7902, & -0{\cdot}0755 \\ -0{\cdot}6038-0{\cdot}1170i, & -0{\cdot}6038+0{\cdot}1170i, & -0{\cdot}3574, & +0{\cdot}5520 \\ -0{\cdot}6187+0{\cdot}3228i, & -0{\cdot}6187-0{\cdot}3228i, & -0{\cdot}0685, & +0{\cdot}4275 \\ -0{\cdot}6920-0{\cdot}2104i, & -0{\cdot}6920+0{\cdot}2104i, & +0{\cdot}4932, & +0{\cdot}7119 \end{bmatrix}. \tag{13.7 c}$$

It may also be verified that

$$\left.\begin{aligned} \partial \mathbf{g}/\partial a_{21} &= \begin{bmatrix} 0{\cdot}3113 \\ 1{\cdot}4773 \\ 0{\cdot}8517 \\ 1{\cdot}9151 \end{bmatrix}, \\ \partial^2 \mathbf{g}/\partial {a_{21}}^2 &= \begin{bmatrix} 0{\cdot}6265 \\ 3{\cdot}6499 \\ 2{\cdot}2088 \\ 4{\cdot}7274 \end{bmatrix}, \end{aligned}\right\} \tag{13.8}$$

$$\left.\begin{aligned}
\partial \mathbf{g}/\partial b_3 &= \begin{bmatrix} -1{\cdot}2132 \\ 0{\cdot}5853 \\ 1{\cdot}1568 \\ 0{\cdot}5748 \end{bmatrix}, \\
\partial^2 \mathbf{g}/\partial {b_3}^2 &= \begin{bmatrix} -1{\cdot}2286 \\ 0{\cdot}5064 \\ 1{\cdot}4096 \\ 0{\cdot}4269 \end{bmatrix},
\end{aligned}\right\} \qquad (13.9)$$

$$\left.\begin{aligned}
\partial \mathbf{g}/\partial \rho_1 &= \begin{bmatrix} 0{\cdot}5691 + 0{\cdot}8439i \\ -0{\cdot}2449 - 0{\cdot}5762i \\ -0{\cdot}6245 - 0{\cdot}3384i \\ -0{\cdot}2163 - 0{\cdot}7038i \end{bmatrix}, \\
\partial \mathbf{g}/\partial \rho_2 &= \begin{bmatrix} 0{\cdot}5691 - 0{\cdot}8439i \\ -0{\cdot}2449 + 0{\cdot}5762i \\ -0{\cdot}6245 + 0{\cdot}3384i \\ -0{\cdot}2163 + 0{\cdot}7038i \end{bmatrix}, \\
\partial^2 \mathbf{g}/\partial {\rho_1}^2 &= \begin{bmatrix} 0 \\ 0 \\ 0 \\ 0 \end{bmatrix}, \\
\partial^2 \mathbf{g}/\partial {\rho_2}^2 &= \begin{bmatrix} 0 \\ 0 \\ 0 \\ 0 \end{bmatrix}, \\
\partial^2 \mathbf{g}/\partial \rho_1 \partial \rho_2 &= \begin{bmatrix} -0{\cdot}4219 \\ 0{\cdot}2881 \\ 0{\cdot}1692 \\ 0{\cdot}3519 \end{bmatrix},
\end{aligned}\right\} \qquad (13.10)$$

$$\left.\begin{aligned}
\partial \rho_1/\partial g_1 &= 0{\cdot}1162 - 0{\cdot}5347i, \\
\partial^2 \rho_1/\partial {g_1}^2 &= 0{\cdot}1030 - 0{\cdot}1587i, \\
\partial \rho_2/\partial g_1 &= 0{\cdot}1162 + 0{\cdot}5347i, \\
\partial^2 \rho_2/\partial {g_1}^2 &= 0{\cdot}1030 + 0{\cdot}1587i, \\
\partial \rho_3/\partial g_1 &= 0{\cdot}0236, \\
\partial^2 \rho_3/\partial {g_1}^2 &= 0{\cdot}0325, \\
\partial \rho_4/\partial g_1 &= -0{\cdot}2151, \\
\partial^2 \rho_4/\partial {g_1}^2 &= -0{\cdot}2384,
\end{aligned}\right\} \qquad (13.11)$$

$$\left.\begin{aligned}
\partial \rho_1/\partial a_{21} &= -0{\cdot}3365 + 0{\cdot}0640i, \\
\partial^2 \rho_1/\partial {a_{21}}^2 &= 0{\cdot}0897 + 0{\cdot}1452i, \\
\partial \rho_2/\partial a_{21} &= -0{\cdot}3365 - 0{\cdot}0640i, \\
\partial^2 \rho_2/\partial {a_{21}}^2 &= 0{\cdot}0897 - 0{\cdot}1452i, \\
\partial \rho_3/\partial a_{21} &= 0{\cdot}0244, \\
\partial^2 \rho_3/\partial {a_{21}}^2 &= -0{\cdot}0305, \\
\partial \rho_4/\partial a_{21} &= 0{\cdot}6486, \\
\partial^2 \rho_4/\partial {a_{21}}^2 &= -0{\cdot}1490,
\end{aligned}\right\} \qquad (13.12)$$

$$\left.\begin{aligned}
\partial\rho_1/\partial b_3 &= 0{\cdot}8456 - 0{\cdot}1485i,\\
\partial^2\rho_1/\partial {b_3}^2 &= -0{\cdot}3686i,\\
\partial\rho_2/\partial b_3 &= 0{\cdot}8456 + 0{\cdot}1485i,\\
\partial^2\rho_2/\partial {b_3}^2 &= 0{\cdot}3686i,\\
\partial\rho_3/\partial b_3 = \partial\rho_4/\partial b_3 &= \partial^2\rho_3/\partial {b_3}^2 = \partial^2\rho_4/\partial {b_3}^2 = 0.
\end{aligned}\right\} \quad (13.13)$$

The expression given in eqn. (9.37) and the numerical values given in eqns. (13.8) to (13.10) may be used to obtain first- and second-order estimates of **g** resulting from changes in the system parameters a_{21}, b_3, ρ_1, and ρ_2. Similarly, the expression given in eqn. (9.38) and the numerical values given in eqns. (13.11) to (13.13) may be used to obtain first- and second-order estimates of ρ_1, ρ_2, ρ_3, and ρ_4, resulting from changes in the system parameters g_1, a_{21}, and b_3. These first- and second-order estimates, together with their directly-calculated values, are shown in figs. (13-1) to (13-6). These figures indicate that, in the design of practical modal control systems, a sensitivity analysis of the kind described in this chapter gives a clear indication as to the relative importance of the various design parameters involved.

Fig. 13-1

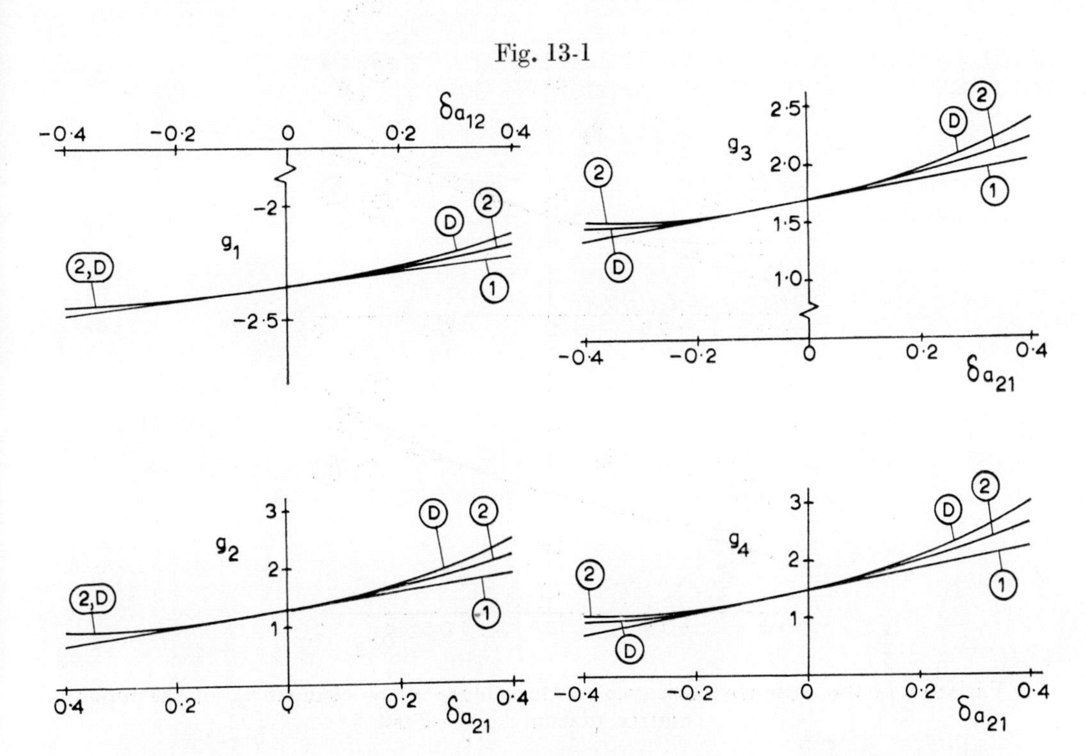

Variation of the state-feedback gains with changes in element a_{21} of the plant matrix (datum $a_{21} = -0{\cdot}737$).

Fig. 13-2

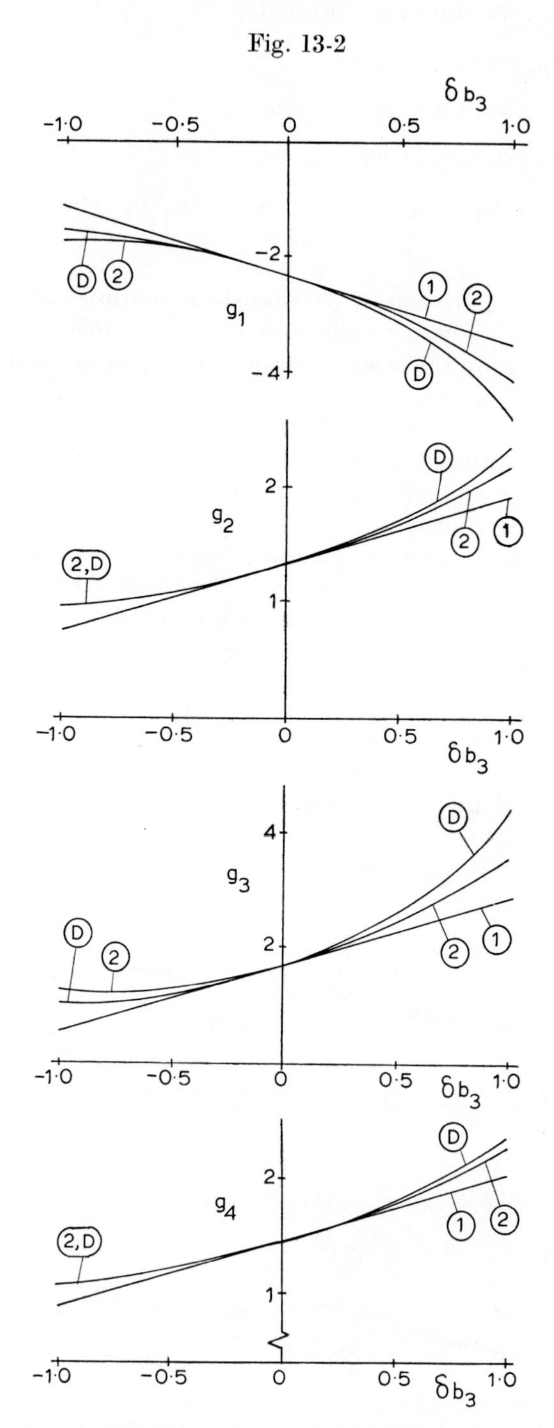

Variation of the state-feedback gains with changes in the element b_{31} of the input matrix (datum $b_{31} = -2{\cdot}06$).

Fig. 13-3

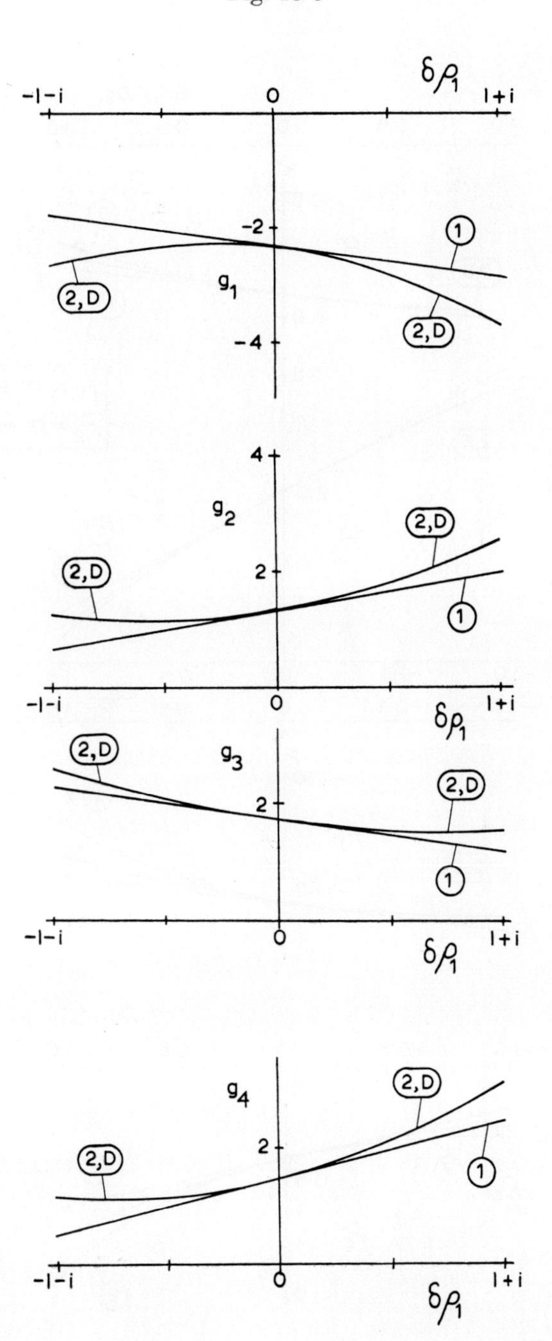

Variation of the state-feedback gains with changes in the required dutch-roll eigenvalue (datum $p_1 = -0{\cdot}9 + 2{\cdot}0\, i$).

Fig. 13-4

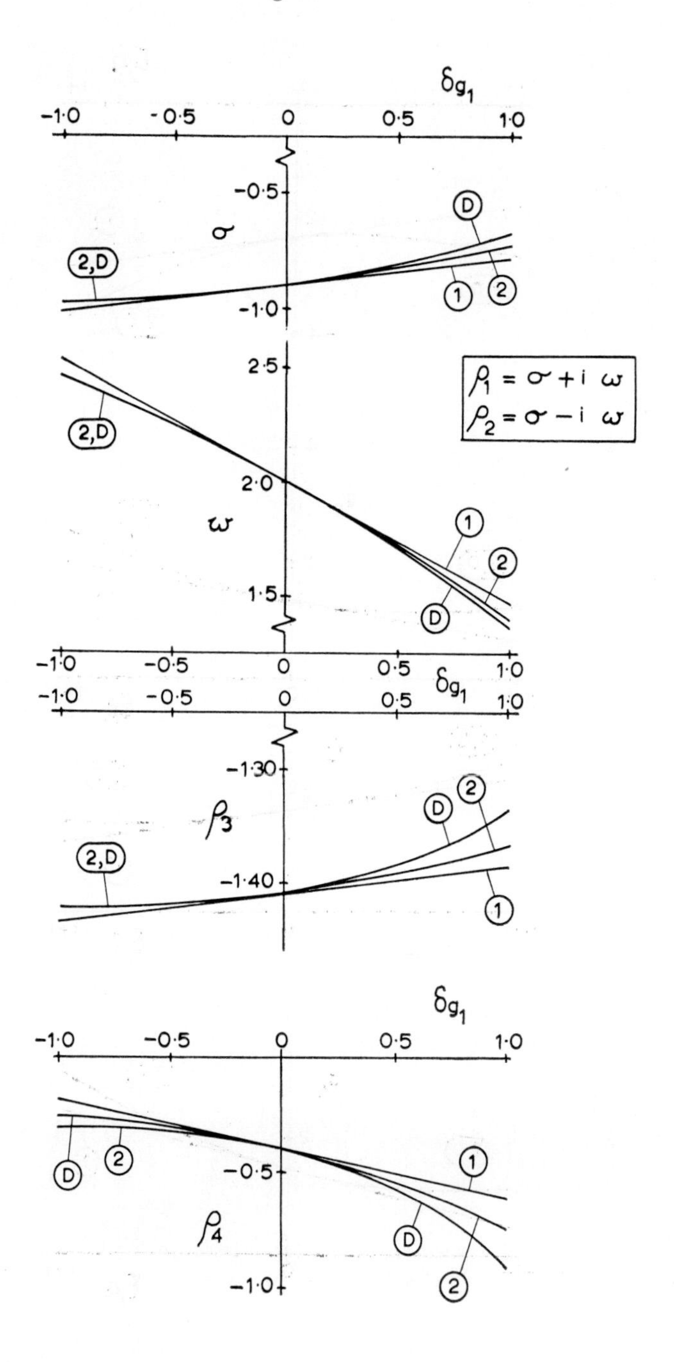

Variation of the closed-loop eigenvalues with changes in the element g_1 of the feedback-gain matrix (datum $g_1 = -2{\cdot}34$).

Fig. 13-5

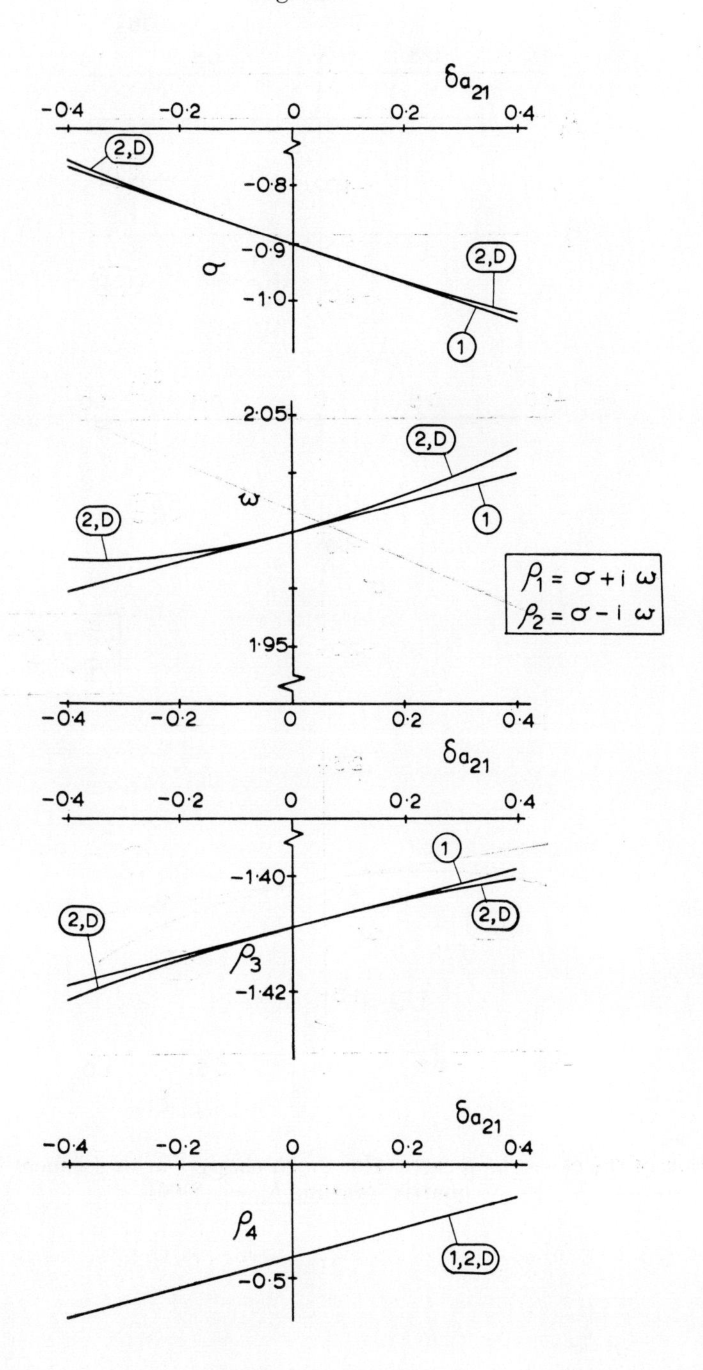

Variation of the closed-loop eigenvalues with changes in element a_{21} of the plant matrix (datum $a_{21} = -0{\cdot}737$).

Fig. 13-6

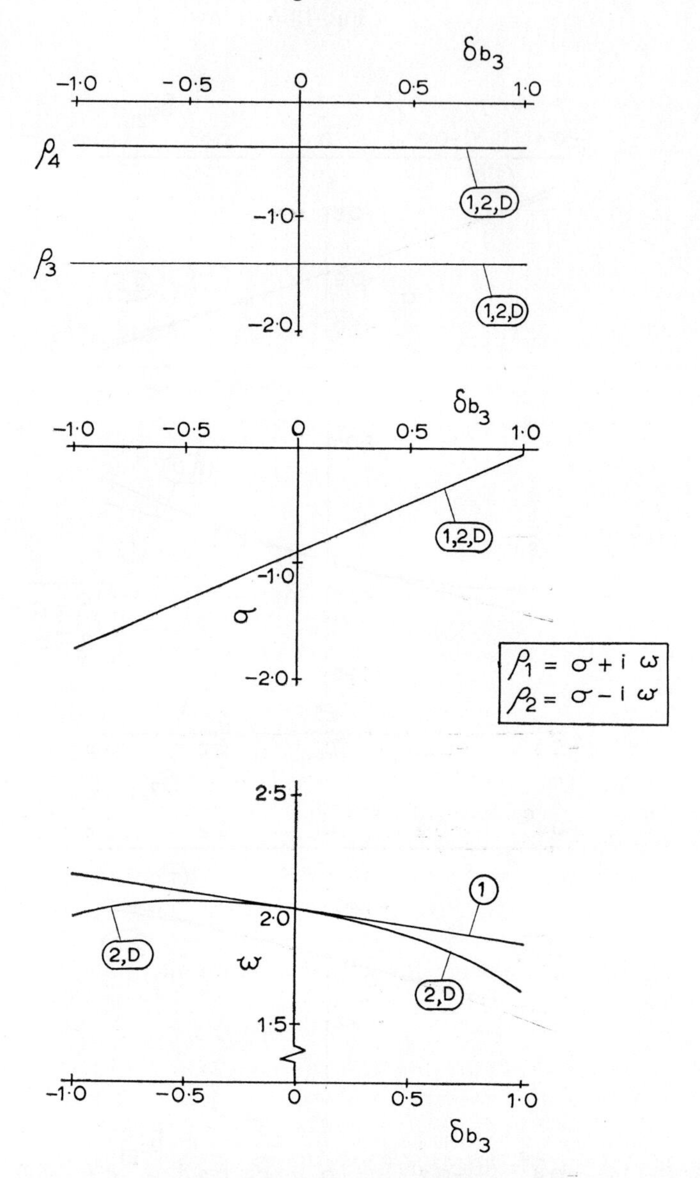

Variation of the closed-loop eigenvalues with changes in the element b_{31} of the input matrix (datum $b_{31} = -2{\cdot}06$).

Synthesis of modal controllers for cascaded-vehicle systems

CHAPTER 14

14.1. Introduction

Transportation systems incorporating a number of cascaded high-speed vehicles are currently being appraised in several parts of the world. The problem of controlling such systems is obviously of crucial significance from various points of view, although safety and passenger-comfort requirements are clearly of paramount importance. The problem of controlling the constituent vehicles of transportation systems of this type has been studied as a problem in continuous-time optimal control theory by Levine and Athans [1], and as a problem in discrete-time optimal control theory by Powner *et al.* [2].

In this chapter, however, control laws for cascaded-vehicle systems are derived by the application of both continuous-time and discrete-time modal control theory: the former type of control law would, of course, be implemented by an analogue controller, whereas the latter would enable an on-line digital controller to be used. The implementation of such control laws makes it possible to ensure that the eigenvalues of the plant matrix of the complete closed-loop system are such that the system is not only asymptotically stable (and, therefore, that it returns to a desired equilibrium state), but that the system possesses more specific time-domain properties characterized by the actual magnitudes of the closed-loop eigenvalues. It is shown that, using modal control theory, suitable control laws can be obtained as closed-form general formulae whose evaluation in any particular case involves only simple arithmetic computations. Furthermore, the control law for a given vehicle incorporates only terms representing the velocity of the vehicle concerned, its relative position with respect to the vehicle immediately in front, and the velocity of the latter vehicle. The control laws derived by modal control theory are therefore very much simpler than the corresponding laws derived by Levine and Athans [1] and by Powner *et al.* [2] since, for a given vehicle, laws of each of the latter types incorporate terms representing the velocities and relative positions of *all* the constituent vehicles of a cascade.

14.2. Continuous-time modal controllers

14.2.1. *Mathematical model of uncontrolled system*

It is assumed that the system to be controlled consists of n vehicles of masses $m_1, m_2, \ldots, m_n$ which are moving in a rectilinear cascade with absolute velocities $v_1, v_2, \ldots, v_n$ and absolute displacements $d_1, d_2, \ldots, d_n$. If the driving force applied to the jth vehicle is $F_j(t)$ and the corresponding drag force is $G_j\{v_j(t)\}$, then the equations of motion of the cascaded vehicles are

$$\left.\begin{aligned} \dot{v}_j(t) &= [F_j(t) - G_j\{v_j(t)\}]/m_j, \\ \dot{d}_j(t) &= v_j(t), \end{aligned}\right\} \quad (j=1, 2, \ldots, n). \tag{14.1}$$

The control problem consists in designing a controller which will generate appropriate variations, $\phi_j(t)$, in the forces $F_j(t)$ $(j=1, 2, \ldots, n)$ which are applied to the n vehicles of the cascade : specifically, the task of the controller is to maintain within acceptable bounds any deviations, $\delta_j(t)$, of the positions of the vehicles from specified positions, $\Delta_j(t)$, and similarly to maintain within acceptable bounds any deviations, $\nu_j(t)$, of the velocities of the vehicles from the unperturbed nominal velocity, N, of the cascade.

In order to solve this problem it is convenient to write eqns. (14.1) in the linearized forms

$$\left.\begin{aligned} \dot{\nu}_j(t) &= [\phi_j(t) - \alpha_j \nu_j(t)]/m_j, \\ \dot{\delta}_j(t) &= \nu_j(t), \end{aligned}\right\} (j=1, 2, \ldots, n) \tag{14.2}$$

where only first-order terms in the Taylor-series expansion of $G_j\{v_j(t)\}$ have been retained on the assumption that the $\nu_j(t)$ are of small modulus : also, the linearized drag coefficients are given by

$$\alpha_j = \left.\frac{\partial G_j\{v_j(t)\}}{\partial v_j(t)}\right|_{v_j(t)=N} \quad (j=1, 2, \ldots, n). \tag{14.3}$$

Equations (14.2) can clearly be expressed in the forms

$$\dot{\nu}_j(t) = [\phi_j(t) - \alpha_j \nu_j(t)]/m_j \quad (j=1, 2, \ldots, n), \tag{14.4 a}$$

and

$$\dot{\sigma}_j(t) = \nu_j(t) - \nu_{j+1}(t) \quad (j=1, 2, \ldots, n-1), \tag{14.4 b}$$

where

$$\sigma_j(t) = \delta_j(t) - \delta_{j+1}(t) \quad (j=1, 2, \ldots, n-1). \tag{14.4 c}$$

The $(2n-1)$ scalar differential eqns. (14.4 a) and (14.4 b) can be written as the single matrix equation

$$\dot{\mathbf{x}}(t) = \mathbf{A}\mathbf{x}(t) + \mathbf{B}\mathbf{z}(t), \tag{14.5}$$

where $\mathbf{A}$ is a constant $(2n-1)\times(2n-1)$ matrix, $\mathbf{B}$ is a constant $(2n-1)\times n$ matrix, $\mathbf{x}(t)$ is a $(2n-1)\times 1$ state vector of the system, and $\mathbf{z}(t)$ is an $n\times 1$ input vector.

It is evident from eqns. (14.4) that the plant matrix $\mathbf{A}$ in eqn. (14.5) has the partitioned form

$$\mathbf{A}=\begin{bmatrix}\mathbf{A}_{11}, & \mathbf{A}_{12}\\ \mathbf{A}_{21}, & \mathbf{A}_{22}\end{bmatrix}, \tag{14.6}$$

where $\mathbf{A}_{11}$, $\mathbf{A}_{12}$, $\mathbf{A}_{21}$, and $\mathbf{A}_{22}$ are the following $n\times n$, $n\times(n-1)$, $(n-1)\times n$, and $(n-1)\times(n-1)$ matrices :

$$\mathbf{A}_{11}=\operatorname{diag}\{-\alpha_1/m_1,\ -\alpha_2/m_2,\ \ldots,\ -\alpha_n/m_n\}, \tag{14.7 a}$$

$$\mathbf{A}_{12}=\mathbf{0}, \tag{14.7 b}$$

$$\mathbf{A}_{21}=\begin{bmatrix}1, & -1, & 0, & 0, \ldots, 0, & 0\\ 0, & 1, & -1, & 0, \ldots, 0, & 0\\ \cdots & \cdots & \cdots & \cdots & \cdots\\ \cdots & \cdots & \cdots & \cdots & \cdots\\ 0, & 0, & 0, & 0, \ldots, 1, & -1\end{bmatrix}, \tag{14.7 c}$$

$$\mathbf{A}_{22}=\mathbf{0}. \tag{14.7 d}$$

Similarly, the input matrix $\mathbf{B}$ has the partitioned form

$$\mathbf{B}=\begin{bmatrix}\mathbf{B}_1\\ \mathbf{B}_2\end{bmatrix}, \tag{14.8}$$

where $\mathbf{B}_1$ and $\mathbf{B}_2$ are the following $n\times n$ and $(n-1)\times n$ matrices :

$$\mathbf{B}_1=\operatorname{diag}\{1/m_1,\ 1/m_2,\ \ldots,\ 1/m_n\}, \tag{14.9 a}$$

$$\mathbf{B}_2=\mathbf{0}. \tag{14.9 b}$$

Finally, the $n\times 1$ input vector $\mathbf{z}(t)$ is given by the equation

$$\mathbf{z}'(t)=[\phi_1(t),\ \phi_2(t),\ \ldots,\ \phi_n(t)], \tag{14.10}$$

and the state vector $\mathbf{x}(t)$ has the partitioned form

$$\mathbf{x}(t)=\begin{bmatrix}\boldsymbol{\nu}(t)\\ \boldsymbol{\sigma}(t)\end{bmatrix}, \tag{14.11 a}$$

where $\boldsymbol{\nu}'(t)$ and $\boldsymbol{\sigma}'(t)$ are the following $1\times n$ and $1\times(n-1)$ vectors :

$$\boldsymbol{\nu}'(t)=[\nu_1(t),\ \nu_2(t),\ \ldots,\ \nu_n(t)],$$

$$\boldsymbol{\sigma}'(t)=[\sigma_1(t),\ \sigma_2(t),\ \ldots,\ \sigma_{n-1}(t)]. \tag{14.11 b}$$

It may readily be verified that the Jordan canonical form of the plant matrix (14.6) is given by the direct sum

$$\mathbf{J}=\mathbf{J}_1(-\alpha_1/m_1)\oplus\mathbf{J}_1(-\alpha_2/m_2)\oplus\ldots\oplus\mathbf{J}_1(-\alpha_n/m_n)\oplus\mathbf{J}_1(0)\oplus\ldots\oplus\mathbf{J}_1(0). \tag{14.12}$$

In addition, it is convenient to express the generalized modal matrices of $\mathbf{A}$ and $\mathbf{A}'$ in the partitioned forms

$$\mathbf{U}=\begin{bmatrix}\mathbf{U}_{11}, & \mathbf{U}_{12}\\ \mathbf{U}_{21}, & \mathbf{U}_{22}\end{bmatrix}, \tag{14.13}$$

and

$$\mathbf{V}=\begin{bmatrix}\mathbf{V}_{11}, & \mathbf{V}_{12}\\ \mathbf{V}_{21}, & \mathbf{V}_{22}\end{bmatrix}, \tag{14.14}$$

respectively, and to choose these matrices such that

$$\mathbf{AU}=\mathbf{UJ} \tag{14.15 a}$$

and

$$\mathbf{A}'\mathbf{V}=\mathbf{VJ}' \tag{14.15 b}$$

where

$$\mathbf{V}'\mathbf{U}=\mathbf{I}_{2n-1}. \tag{14.15 c}$$

In the case of eqn. (14.13) it can be shown that $\mathbf{U}_{11}$, $\mathbf{U}_{12}$, $\mathbf{U}_{21}$, and $\mathbf{U}_{22}$ are the following $n\times n$, $n\times(n-1)$, $(n-1)\times n$, and $(n-1)\times(n-1)$ matrices :

$$\mathbf{U}_{11}=\mathbf{I}_n, \tag{14.16 a}$$

$$\mathbf{U}_{12}=\mathbf{0}, \tag{14.16 b}$$

$$\mathbf{U}_{21}=\begin{bmatrix}-m_1/\alpha_1, & m_2/\alpha_2, & 0 & \ldots, & 0, & 0\\ 0, & -m_2/\alpha_2, & m_3/\alpha_3, & \ldots, & 0, & 0\\ 0, & 0, & -m_3/\alpha_3, & \ldots, & 0, & 0\\ \cdots & \cdots & \cdots & \cdots & \cdots & \cdots\\ \cdots & \cdots & \cdots & \cdots & \cdots & \cdots\\ 0, & 0, & 0, & \ldots, & m_{n-1}/\alpha_{n-1}, & 0\\ 0, & 0, & 0, & \ldots, & -m_{n-1}/\alpha_{n-1}, & m_n/\alpha_n\end{bmatrix} \tag{14.16 c}$$

and

$$\mathbf{U}_{22}=\mathbf{I}_{n-1}. \tag{14.16 d}$$

Similarly, in the case of eqn. (14.14) it can be shown that $\mathbf{V}_{11}$, $\mathbf{V}_{12}$, $\mathbf{V}_{21}$, and $\mathbf{V}_{22}$ are the following $n\times n$, $n\times(n-1)$, $(n-1)\times n$, and $(n-1)\times(n-1)$ matrices :

$$\mathbf{V}_{11}=\mathbf{I}_n, \tag{14.17 a}$$

$$\mathbf{V}_{12}=\begin{bmatrix}m_1/\alpha_1, & 0, & 0, & \ldots, & 0, & 0\\ -m_2/\alpha_2, & m_2/\alpha_2, & 0, & \ldots, & 0, & 0\\ 0, & -m_3/\alpha_3, & m_3/\alpha_3, & \ldots, & 0, & 0\\ \cdots & \cdots & \cdots & \cdots & \cdots & \cdots\\ \cdots & \cdots & \cdots & \cdots & \cdots & \cdots\\ 0, & 0, & 0, & \ldots, & -m_{n-1}/\alpha_{n-1}, & m_{n-1}/\alpha_{n-1}\\ 0, & 0, & 0, & \ldots, & 0, & -m_n/\alpha_n\end{bmatrix} \tag{14.17 b}$$

$$\mathbf{V}_{21}=\mathbf{0}, \tag{14.17 c}$$

and

$$\mathbf{V}_{22}=\mathbf{I}_{n-1}. \tag{14.17 d}$$

The corresponding partitioned form of the Jordan matrix (14.12) is

$$\mathbf{J} = \begin{bmatrix} \mathbf{J}_{11}, & \mathbf{J}_{12} \\ \mathbf{J}_{21}, & \mathbf{J}_{22} \end{bmatrix}, \tag{14.18}$$

where $\mathbf{J}_{11}$, $\mathbf{J}_{12}$, $\mathbf{J}_{21}$, and $\mathbf{J}_{22}$ are the following $n \times n$, $n \times (n-1)$, $(n-1) \times n$, and $(n-1) \times (n-1)$ matrices :

$$\mathbf{J}_{11} = \operatorname{diag}\{-\alpha_1/m_1, -\alpha_2/m_2, \ldots, -\alpha_n/m_n\}, \tag{14.19 a}$$

$$\mathbf{J}_{12} = \mathbf{0}, \tag{14.19 b}$$

$$\mathbf{J}_{21} = \mathbf{0}, \tag{14.19 c}$$

and

$$\mathbf{J}_{22} = \mathbf{0}. \tag{14.19 d}$$

It may be deduced from eqns. (14.8) and (14.15) that the mode-controllability matrix, $\mathbf{P}$, of the cascaded-vehicle system (14.5) is given by the expression

$$\mathbf{P} = \mathbf{V}'\mathbf{B} = \begin{bmatrix} \mathbf{P}_1 \\ \mathbf{P}_2 \end{bmatrix}, \tag{14.20}$$

where $\mathbf{P}_1$ and $\mathbf{P}_2$ are the following $n \times n$ and $(n-1) \times n$ matrices :

$$\mathbf{P}_1 = \operatorname{diag}\{1/m_1, 1/m_2, \ldots, 1/m_n\}, \tag{14.21 a}$$

and

$$\mathbf{P}_2 = \begin{bmatrix} 1/\alpha_1, & -1/\alpha_2, & 0, & \ldots, & 0, & 0 \\ 0, & 1/\alpha_2, & -1/\alpha_3, & \ldots, & 0, & 0 \\ 0, & 0, & 1/\alpha_3, & \ldots, & 0, & 0 \\ \cdots & \cdots & \cdots & \cdots & \cdots & \cdots \\ \cdots & \cdots & \cdots & \cdots & \cdots & \cdots \\ 0, & 0, & 0, & \ldots, & -1/\alpha_{n-1}, & 0 \\ 0, & 0, & 0, & \ldots, & 1/\alpha_{n-1}, & -1/\alpha_n \end{bmatrix}. \tag{14.21 b}$$

Since the rows of $\mathbf{P}$ are linearly independent, it may be inferred that the cascaded-vehicle system is controllable.

14.2.2. *Design of continuous-time modal controllers*

It will be recalled from Chapters 5 and 6 that the elements of the matrix $\mathbf{P}$ defined in eqn. (14.20) play a central role in the design of modal control systems : it is therefore to be expected that the special form of $\mathbf{P}$ described in eqns. (14.21) will lead to a special form of modal controller. Accordingly, it is the object of this section to determine the characteristics of the closed-loop system if the n input variables are chosen to be

$$z_1(t) = K_1 \mathbf{v}_1{}' \mathbf{x}(t), \tag{14.22 a}$$

and

$$z_{i+1}(t) = K_{i+1} \mathbf{v}_{i+1}{}' \mathbf{x}(t) + K_{n+i} \mathbf{v}_{n+i}{}' \mathbf{x}(t) \quad (i = 1, 2, \ldots, n-1), \tag{14.22 b}$$

where the vector $\mathbf{v}_j$ is the jth column of the generalized modal matrix $\mathbf{V}$. This form of modal control implies that, apart from the first vehicle in the cascade, the force applied to a given vehicle controls the two modes of the complete system which are introduced by the presence of that vehicle in the cascade.

It follows by substituting the expressions for $z_i(t)$ $(i=1, 2, \ldots, n)$ given in eqns. (14.22) into eqn. (14.5) that the governing equation for the resulting closed-loop system is

$$\dot{\mathbf{x}}(t)=\mathbf{A}\mathbf{x}(t)+K_1\mathbf{b}_1\mathbf{v}_1'\mathbf{x}(t)+\sum_{i=1}^{n-1}[K_{i+1}\mathbf{b}_{i+1}\mathbf{v}_{i+1}'\mathbf{x}(t) + K_{n+i}\mathbf{b}_{i+1}\mathbf{v}_{n+i}'\mathbf{x}(t)]=\mathbf{C}\mathbf{x}(t), \qquad (14.23)$$

where the $\mathbf{b}_j$ are the columns of $\mathbf{B}$. Equation (14.23) indicates that the effect of the input variables defined in eqn. (14.22) is to change the plant matrix $\mathbf{A}$ to a new matrix $\mathbf{C}$ given by

$$\mathbf{C}=\mathbf{A}+K_1\mathbf{b}_1\mathbf{v}_1'+\sum_{i=1}^{n-1}(K_{i+1}\mathbf{b}_{i+1}\mathbf{v}_{i+1}'+K_{n+i}\mathbf{b}_{i+1}\mathbf{v}_{n+i}'). \qquad (14.24)$$

The results expressed by eqns. (14.15) and (14.24) indicate that

$$\mathbf{C}\mathbf{u}_1=\lambda_1\mathbf{u}_1+K_1\mathbf{b}_1, \qquad (14.25\ a)$$

$$\mathbf{C}\mathbf{u}_{i+1}=\lambda_{i+1}\mathbf{u}_{i+1}+K_{i+1}\mathbf{b}_{i+1} \quad (i=1, 2, \ldots, n-1), \qquad (14.25\ b)$$

and

$$\mathbf{C}\mathbf{u}_{n+i}=\lambda_{n+i}\mathbf{u}_{n+i}+K_{n+i}\mathbf{b}_{i+1} \quad (i=1, 2, \ldots, n-1). \qquad (14.25\ c)$$

Equations (14.25) imply that λ_j $(j=1, 2, \ldots, 2n-1)$ is not an eigenvalue of $\mathbf{C}$ and also that $\mathbf{u}_j$ is no longer the corresponding eigenvector if $K_j \neq 0$. The effect of implementing the control laws (14.22) is thus to change the eigenvalues, λ_j, to new values, ρ_j, and the eigenvectors, $\mathbf{u}_j$, to corresponding new vectors, $\mathbf{w}_j$ $(j=1, 2, \ldots, 2n-1)$.

In order to calculate the values of the proportional-controller gains, K_j, in terms of the eigenvalues of $\mathbf{A}$ and the required eigenvalues of $\mathbf{C}$, it is convenient to express the $\mathbf{w}_j$ and $\mathbf{b}_j$ in the respective forms

$$\mathbf{w}_j=\sum_{k=1}^{2n-1} q_{jk}\mathbf{u}_k \quad (j=1, 2, \ldots, 2n-1), \qquad (14.26)$$

and

$$\mathbf{b}_j=\sum_{k=1}^{2n-1} p_{kj}\mathbf{u}_k \quad (j=1, 2, \ldots, n). \qquad (14.27)$$

It then follows from eqns. (14.13) and (14.27) that

$$p_{ij}=\mathbf{v}_i'\mathbf{b}_j \quad (i=1, 2, \ldots, 2n-1\ ;\ j=1, 2, \ldots, n). \qquad (14.28)$$

In addition, the special form of eqns. (14.21) and of the expressions given in eqns. (14.27) and (14.28) imply that

$$\mathbf{b}_1=p_{11}\mathbf{u}_1+p_{n+1,1}\mathbf{u}_{n+1}, \qquad (14.29\ a)$$

$$\mathbf{b}_{i+1}=p_{i+1,i+1}\mathbf{u}_{i+1}+p_{n+i,i+1}\mathbf{u}_{n+i}+p_{n+i+1,i+1}\mathbf{u}_{n+i+1},\quad (i=1,\,2,\,\ldots,\,n-2), \tag{14.29 b}$$

and

$$\mathbf{b}_n=p_{nn}\mathbf{u}_n+p_{2n-1,n}\mathbf{u}_{2n-1}. \tag{14.29 c}$$

Since, by definition, the eigenvalues, ρ_j, and the linearly independent eigenvectors, $\mathbf{w}_j$, satisfy the equation

$$\mathbf{C}\mathbf{w}_j=\rho_j\mathbf{w}_j \quad (j=1,\,2,\,\ldots,\,2n-1) \tag{14.30}$$

substituting the expressions for $\mathbf{C}$ and $\mathbf{w}_j$ from eqns. (14.24) and (14.26) into eqn. (14.30) yields the equations

$$\left\{\mathbf{A}+K_1\mathbf{b}_1\mathbf{v}_1'+\sum_{i=1}^{n-1}(K_{i+1}\mathbf{b}_{i+1}\mathbf{v}_{i+1}'+K_{n+i}\mathbf{b}_{i+1}\mathbf{v}_{n+i}')\right\}\sum_{k=1}^{2n-1}q_{jk}\mathbf{u}_k=\rho_j\sum_{k=1}^{2n-1}q_{jk}\mathbf{u}_k \quad (j=1,\,2,\,\ldots,\,2n-1). \tag{14.31}$$

If the relationships expressed by eqns. (14.13) are then used in eqn. (14.31) this equation becomes

$$\sum_{k=1}^{2n-1}\lambda_k q_{jk}\mathbf{u}_k+K_1\mathbf{b}_1q_{j1}+\sum_{i=1}^{n-1}(K_{i+1}\mathbf{b}_{i+1}q_{j,i+1}+K_{n+i}\mathbf{b}_{i+1}q_{j,n+i})=\rho_j\sum_{k=1}^{2n-1}q_{jk}\mathbf{u}_k \quad (j=1,\,2,\,\ldots,\,2n-1). \tag{14.32}$$

It is thus evident that if the expressions for the vectors given in eqns. (14.29) are substituted into eqns. (14.32), then the latter equations assume the form

$$\begin{aligned}\sum_{k=1}^{2n-1}\lambda_k q_{jk}\mathbf{u}_k&+K_1q_{j1}(p_{11}\mathbf{u}_1+p_{n+1,1}\mathbf{u}_{n+1})+\sum_{i=1}^{n-2}(K_{i+1}q_{j,i+1}+K_{n+i}q_{j,n+i})\\&\times(p_{i+1,i+1}\mathbf{u}_{i+1}+p_{n+i,i+1}\mathbf{u}_{n+i}+p_{n+i+1,i+1}\mathbf{u}_{n+i+1})\\&+(K_nq_{jn}+K_{2n-1}q_{j,2n-1})(p_{nn}\mathbf{u}_n+p_{2n-1,n}\mathbf{u}_{2n-1})=\rho_j\sum_{k=1}^{2n-1}q_{jk}\mathbf{u}_k\\&(j=1,\,2,\,\ldots,\,2n-1).\end{aligned} \tag{14.33}$$

Equation (14.33) is a vector equation in the linearly independent vectors $\mathbf{u}_k$ and is thus equivalent to the following $(2n-1)^2$ scalar equations :

$$[(\rho_1-\lambda_1-K_1p_{11})q_{j1}]=0 \quad (j=1,\,2,\,\ldots,\,2n-1), \tag{14.34 a}$$

$$[(\rho_j-\lambda_{i+1}-K_{i+1}p_{i+1,i+1})q_{j,i+1}-K_{n+i}p_{i+1,i+1}q_{j,n+i}]=0 \quad (i=1,\,2,\,\ldots,\,n-1;\ j=1,\,2,\,\ldots,\,2n-1), \tag{14.34 b}$$

$$\begin{aligned}[(\rho_j-\lambda_{n+i}-K_{n+i}p_{n+i,i+1})q_{j,n+i}&-K_{i+1}p_{n+i,i+1}q_{j,i+1}\\-K_ip_{n+i,i}q_{ji}&-K_{n+i-1}p_{n+i,i}q_{j,n+i-1}]=0\\&(i=2,\,3,\,\ldots,\,n-1\ ;\ j=1,\,2,\,\ldots,\,2n-1),\end{aligned} \tag{14.34 c}$$

and

$$(\rho_j - \lambda_{n+1} - K_{n+1}p_{n+1,2})q_{j,n+1} - K_2 p_{n+1,2}q_{j2} - K_1 p_{n+1,1}q_{j1} = 0. \quad (14.34\ d)$$

It may be verified that eqns. (14.34) can be written as the $(2n-1)$ matrix equations

$$\begin{bmatrix} \mathbf{F}_{11}^{(j)}, & \mathbf{F}_{12}^{(j)} \\ \mathbf{F}_{21}^{(j)}, & \mathbf{F}_{22}^{(j)} \end{bmatrix} \begin{bmatrix} \mathbf{q}_1^{(j)} \\ \mathbf{q}_2^{(j)} \end{bmatrix} = \mathbf{0} \quad (j = 1, 2, \ldots, 2n-1),, \quad (14.35)$$

where $\mathbf{F}_{11}^{(j)}$, $\mathbf{F}_{12}^{(j)}$, $\mathbf{F}_{21}^{(j)}$, and $\mathbf{F}_{22}^{(j)}$ are respectively $n \times n$, $n \times (n-1)$, $(n-1) \times n$ and $(n-1) \times (n-1)$ matrices, and where $\mathbf{q}_1^{(j)}$ and $\mathbf{q}_2^{(j)}$ are respectively $n \times 1$ and $(n-1) \times 1$ vectors. These matrices and vectors are given by the explicit formulae

$$\mathbf{F}_{11}^{(j)} = \text{diag}\,\{\rho_j - \lambda_1 - K_1 p_{11}, \rho_j - \lambda_2 - K_2 p_{22}, \ldots, \rho_j - \lambda_n - K_n p_{nn}\}, \quad (14.36\ a)$$

$$\mathbf{F}_{12}^{(j)} = \begin{bmatrix} 0, & 0, & 0, & 0, \ldots, & 0 \\ -K_{n+1}p_{22}, & 0, & 0, & 0, \ldots, & 0 \\ 0, & -K_{n+2}p_{33}, & 0, & 0, \ldots, & 0 \\ 0, & 0, & -K_{n+3}p_{44}, & 0, \ldots, & 0 \\ \cdots & \cdots & \cdots & \cdots & \cdots \\ \cdots & \cdots & \cdots & \cdots & \cdots \\ 0, & 0, & 0, & 0, \ldots, & -K_{2n-1}p_{nn} \end{bmatrix}, \quad (14.36\ b)$$

$$\mathbf{q}_1^{(j)\prime} = [q_{j1}, q_{j2}, \ldots, q_{jn}], \quad (14.36\ e)$$

and

$$\mathbf{q}_2^{(j)\prime} = [q_{j,n+1}, q_{j,n+2}, \ldots, q_{j,2n-1}]. \quad (14.36\ f)$$

Since

$$\mathbf{q}_j = \begin{bmatrix} \mathbf{q}_1^{(j)} \\ \mathbf{q}_2^{(j)} \end{bmatrix} \neq \mathbf{0},$$

it follows from eqns. (14.35) that

$$\begin{vmatrix} \mathbf{F}_{11}^{(j)}, & \mathbf{F}_{12}^{(j)} \\ \mathbf{F}_{21}^{(j)}, & \mathbf{F}_{22}^{(j)} \end{vmatrix} = 0 \quad (j = 1, 2, \ldots, 2n-1). \quad (14.37)$$

Expansion of the determinants in eqns. (14.37) yields the equations

$$(\rho_j - \lambda_1 - K_1 p_{11}) \prod_{i=1}^{n-1} [(\rho_j - \lambda_{i+1})(\rho_j - \lambda_{n+i}) - K_{i+1}p_{i+1,i+1}(\rho_j - \lambda_{n+i}) - K_{n+i}p_{n+i,i+1}(\rho_j - \lambda_{i+1})] = 0 \quad (j = 1, 2, \ldots, 2n-1). \quad (14.38)$$

The $(2n-1)$ eqns. (14.38) will be identically satisfied if

$$\rho_1 - \lambda_1 - K_1 p_{11} = 0, \quad (14.39\ a)$$

$$(\rho_k - \lambda_k)(\rho_k - \lambda_{n+k-1}) - K_k p_{kk}(\rho_k - \lambda_{n+k-1}) - K_{n+k-1}p_{n+k-1,k}(\rho_k - \lambda_k) = 0, \quad (14.39\ b)$$

$$\mathbf{F}_{21}{}^{(j)}=\begin{bmatrix} -K_1p_{n+1,1}, & -K_2p_{n+1,2}, & 0, & 0, & \ldots, & 0, & 0 \\ 0, & -K_2p_{n+2,2}, & -K_3p_{n+2,3}, & 0, & \ldots, & 0, & 0 \\ 0, & 0, & -K_3p_{n+3,3}, & -K_4p_{n+3,4}, & \ldots, & 0, & 0 \\ 0, & 0, & 0, & -K_4p_{n+4,4}, & \ldots, & 0, & 0 \\ \cdots & \cdots & \cdots & \cdots & \cdots & \cdots & \cdots \\ \cdots & \cdots & \cdots & \cdots & \cdots & \cdots & \cdots \\ 0, & 0, & 0, & 0, & \ldots, & -K_{n-1}p_{2n-2,n-1}, & 0 \\ 0, & 0, & 0, & 0, & \ldots, & -K_{n-1}p_{2n-1,n-1}, & -K_np_{2n-1,n} \end{bmatrix}, \quad (14.36\,c)$$

$$\mathbf{F}_{22}{}^{(j)}=\begin{bmatrix} \rho_j-\lambda_{n+1}-K_{n+1}p_{n+1,2}, & 0, & 0, & \ldots, & 0, & 0 \\ -K_{n+1}p_{n+2,2}, & \rho_j-\lambda_{n+2}-K_{n+2}p_{n+2,3}, & 0, & \ldots, & 0, & 0 \\ 0, & -K_{n+2}p_{n+3,3}, & \rho_j-\lambda_{n+3}-K_{n+3}p_{n+3,4}, & \ldots, & 0, & 0 \\ 0, & 0, & -K_{n+3}p_{n+4,4} & \ldots, & 0, & 0 \\ \cdots & \cdots & \cdots & \cdots & \cdots & \cdots \\ \cdots & \cdots & \cdots & \cdots & \cdots & \cdots \\ 0, & 0, & 0, & \ldots, & \rho_j-\lambda_{2n-2}-K_{2n-2}p_{2n-2,n-1}, & 0 \\ 0, & 0, & 0, & \ldots, & -K_{2n-2}p_{2n-1,n-1}, & \rho_j-\lambda_{2n-1}-K_{2n-1}p_{2n-1,n} \end{bmatrix}, \quad (14.36\,d)$$

and

$$(\rho_{n+k-1}-\lambda_k)(\rho_{n+k-1}-\lambda_{n+k-1})-K_k p_{kk}(\rho_{n+k-1}-\lambda_{n+k-1})$$
$$-K_{n+k-1}p_{n+k-1,k}(\rho_{n+k-1}-\lambda_k)=0 \quad (k=2, 3, \ldots, n). \qquad (14.39\ c)$$

Equations (14.39) can be solved for the proportional-controller gains, K_1, K_k, and K_{n+k-1}, thus yielding the values

$$K_1=(\rho_1-\lambda_1)/p_{11}, \qquad (14.40\ a)$$

$$K_k=(\rho_k-\lambda_k)(\rho_{n+k-1}-\lambda_k)/p_{kk}(\lambda_{n+k-1}-\lambda_k) \quad (k=2, 3, \ldots, n), \qquad (14.40\ b)$$

and

$$K_{n+k-1}=(\rho_k-\lambda_{n+k-1})(\rho_{n+k-1}-\lambda_{n+k-1})/p_{n+k-1,k}(\lambda_k-\lambda_{n+k-1})$$
$$(k=2, 3, \ldots, n). \qquad (14.40\ c)$$

Equations (14.40) indicate that the K_j $(j=1, 2, \ldots, 2n-1)$ will always be calculable since

$$p_{kk}\neq 0 \quad (k=1, 2, \ldots, n),$$

$$p_{n+k-1,k}\neq 0 \quad (k=2, 3, \ldots, n),$$

and

$$\lambda_{n+k-1}-\lambda_k\neq 0 \quad (k=2, 3, \ldots, n).$$

The foregoing analysis indicates that the control laws postulated in eqns. (14.22) lead to a realizable modal control system : the explicit forms of the control laws may be obtained by substituting from eqns. (14.17), (14.19), (14.28), and (14.40) into eqn. (14.22). Thus, it is found that

$$z_1(t)=m_1[\rho_1+(\alpha_1/m_1)]x_1(t), \qquad (14.41\ a)$$

and

$$z_k(t)=m_k[\{\rho_k+\rho_{n+k-1}+(\alpha_k/m_k)\}x_k(t)$$
$$+\rho_k\rho_{n+k-1}\{(m_{k-1}/\alpha_{k-1})x_{k-1}(t)+x_{n+k-1}(t)\}] \quad (k=2, 3, \ldots, n). \qquad (14.41\ b)$$

It is thus evident from eqns. (14.10), (14.11), and (14.41) that the force variations to be applied to each vehicle, expressed in terms of the vehicle velocity and relative displacement deviations, are

$$\phi_1(t)=m_1[\rho_1+(\alpha_1/m_1)]\nu_1(t) \qquad (14.42\ a)$$

and

$$\phi_k(t)=m_k[\{\rho_k+\rho_{n+k-1}+(\alpha_k/m_k)\}\nu_k(t)$$
$$+\rho_k\rho_{n+k-1}\{m_{k-1}/\alpha_{k-1})\nu_{k-1}(t)+\sigma_{k-1}(t)\}] \quad (k=2, 3, \ldots, n). \qquad (14.24\ b)$$

Any desired set of closed-loop eigenvalues ρ_j $(j=1, 2, \ldots, 2n-1)$ can now be selected and used to determine a corresponding set of control laws (14.42) provided that, corresponding to any specified initial state, $\mathbf{x}(0)$, the resulting motion of the cascade is such that no vehicles collide. Thus, consider that a choice of the ρ_j is made such that the closed-loop system has a transition matrix

$$\mathbf{\Phi}(t)=\exp(\mathbf{C}t)=\mathbf{Y}\exp(\mathbf{R}t)\mathbf{Y}^{-1}, \qquad (14.43)$$

where $\mathbf{Y}$ is the modal matrix of $\mathbf{C}$ and

$$\mathbf{R} = \text{diag}\,\{\rho_1, \rho_2, \ldots, \rho_{2n-1}\}. \tag{14.44}$$

This transition matrix, $\mathbf{\Phi}(t)$, can clearly be computed in terms of the eigenproperties of $\mathbf{C}$ and may then be written in the partitioned form

$$\mathbf{\Phi}(t) = \begin{bmatrix} \mathbf{\Phi}_{11}(t), & \mathbf{\Phi}_{12}(t) \\ \mathbf{\Phi}_{21}(t), & \mathbf{\Phi}_{22}(t) \end{bmatrix}. \tag{14.45}$$

If the state vector (14.10) is written in the partitioned form

$$\mathbf{x}(t) = \begin{bmatrix} \mathbf{v}(t) \\ \boldsymbol{\sigma}(t) \end{bmatrix}, \tag{14.46}$$

then substituting the expressions for $\mathbf{\Phi}(t)$ and $\mathbf{x}(t)$ from eqns. (14.45) and (14.46) into the eqn.

$$\mathbf{x}(t) = \mathbf{\Phi}(t)\mathbf{x}(0) \tag{14.47}$$

indicates that

$$\mathbf{v}(t) = \mathbf{\Phi}_{11}(t)\mathbf{v}(0) + \mathbf{\Phi}_{12}(t)\boldsymbol{\sigma}(0), \tag{14.48 a}$$

and

$$\boldsymbol{\sigma}(t) = \mathbf{\Phi}_{21}(t)\mathbf{v}(0) + \mathbf{\Phi}_{22}(t)\boldsymbol{\sigma}(0). \tag{14.48 b}$$

It is clear that a sufficient condition for no vehicles of the cascade to collide is that the ρ_j be selected so that all the elements of the vector defined by the right-hand member of eqn. (14.48 *b*) remain positive for all t. Furthermore, if it is deemed desirable that no absolute velocity reversals should occur during the motion of the cascade, then the closed-loop eigenvalues must also be chosen such that the elements of the vector defined by the right-hand member of eqn. (14.48 *a*) remain greater than $-N$ for all $t \geqslant 0$, where N is the unperturbed nominal velocity of the cascade. On the assumption that these conditions are satisfied, the modes corresponding to the open-loop eigenvalues $\lambda_j = -\alpha_j/m_j$ $(j = 1, 2, \ldots, n)$ will normally be assigned closed-loop eigenvalues, ρ_j, which are such that

$$\rho_j < -\frac{\alpha_j}{m_j} \quad (j = 1, 2, \ldots, n),$$

where, in view of eqn. (14.30), the ρ_j must be chosen to be distinct. If this is the case, it is apparent from eqns. (14.42) that the control law for the kth vehicle $(k = 2, 3, \ldots, n)$ embodies negative feedback of the velocity error of the kth vehicle, positive feedback of the velocity deviation of $(k-1)$*th* vehicle, and positive feedback of the relative positional deviation between the kth and $(k-1)$th vehicles. The control law for the first vehicle is the obvious special case of this general control-law structure.

The control laws given by eqns. (14.40) are obviously very simple to compute since they are defined by closed-form formulae. In addition, the control force acting on each vehicle is a function of only one state variable for the first vehicle and of only three state variables for each of the remaining vehicles : thus, for example, the control of a cascade consisting of 50 vehicles will involve the use of only 148 control links. This simplicity is in sharp contrast to the complexity of the corresponding control system which is obtained using optimal control theory : thus, the optimal control of a 50 vehicle cascade will, in general, require the use of 4950 control links, since there are 99 state variables associated with this system [1].

14.2.3. *Numerical example*

In this section, the control laws for the continuous-time modal control of a cascaded-vehicle system comprising three vehicles are determined, and typical transient-response characteristics of the controlled system are presented. The numerical data used in the calculations is as follows :

$$m_1 = m_2 = m_3 = 1, \tag{14.49 a}$$

$$\alpha_1 = \alpha_2 = \alpha_3 = 1, \tag{14.49 b}$$

$$\rho_1 = -1{\cdot}90, \tag{14.49 c}$$

$$\rho_2 = -2{\cdot}00, \tag{14.49 d}$$

$$\rho_3 = -2{\cdot}10, \tag{14.49 e}$$

$$\rho_4 = -0{\cdot}45, \tag{14.49 f}$$

$$\rho_5 = -0{\cdot}55. \tag{14.49 g}$$

Equations (14.42) indicate that the appropriate force variations to be applied to each vehicle, expressed in terms of the vehicle velocity and relative displacement variations, are

$$\phi_1(t) = -\nu_1(t), \tag{14.50 a}$$

$$\phi_2(t) = 0{\cdot}945\nu_1(t) - 1{\cdot}55\nu_2(t) + 0{\cdot}945\sigma_1(t), \tag{14.50 b}$$

and

$$\phi_3(t) = 1{\cdot}045\nu_2(t) - 1{\cdot}45\nu_3(t) + 1{\cdot}045\sigma_2(t). \tag{14.50 c}$$

Figure 14-1 shows the respective transient-response characteristics of the cascaded-vehicle system when controlled by a continuous-time modal controller in the case when the initial conditions of the system are

$$\nu_1(0) = -0{\cdot}5, \tag{14.51 a}$$

$$\nu_2(0) = 0{\cdot}5, \tag{14.51 b}$$

$$\nu_3(0) = 1{\cdot}0, \tag{14.51 c}$$

$$\sigma_1(0) = -1{\cdot}0, \tag{14.51 d}$$

and

$$\sigma_2(0) = -1{\cdot}5. \tag{14.51 e}$$

It is interesting to note that, despite the adverse initial conditions (14.51), the cascaded-vehicle system is rapidly brought under control with no overshoots in the relative displacement variations.

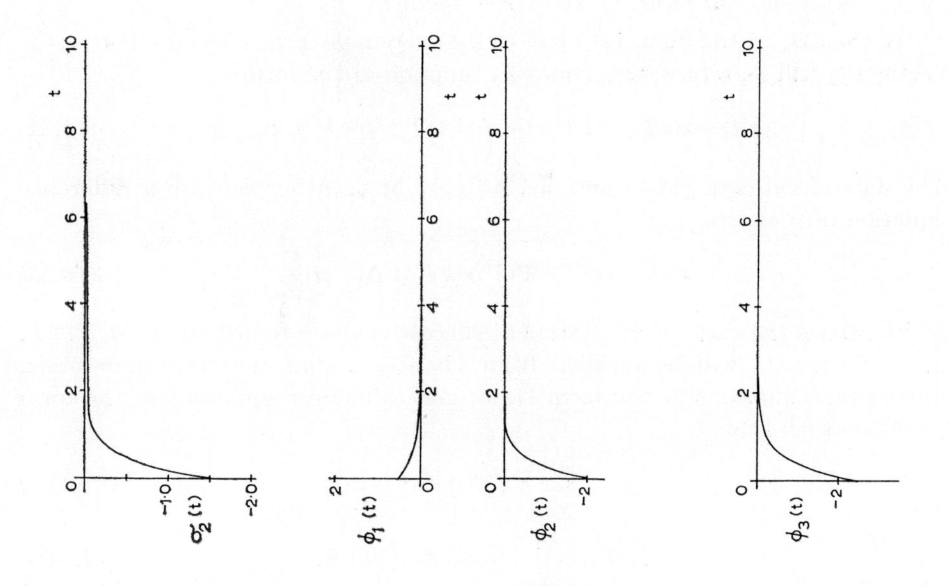

σ2 (t)
−1·0
−2·0
ϕ1 (t)
2
ϕ2 (t)
−2
ϕ3 (t)
−2
t
0
2
4
6
8
10

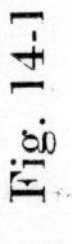

Fig. 14-1

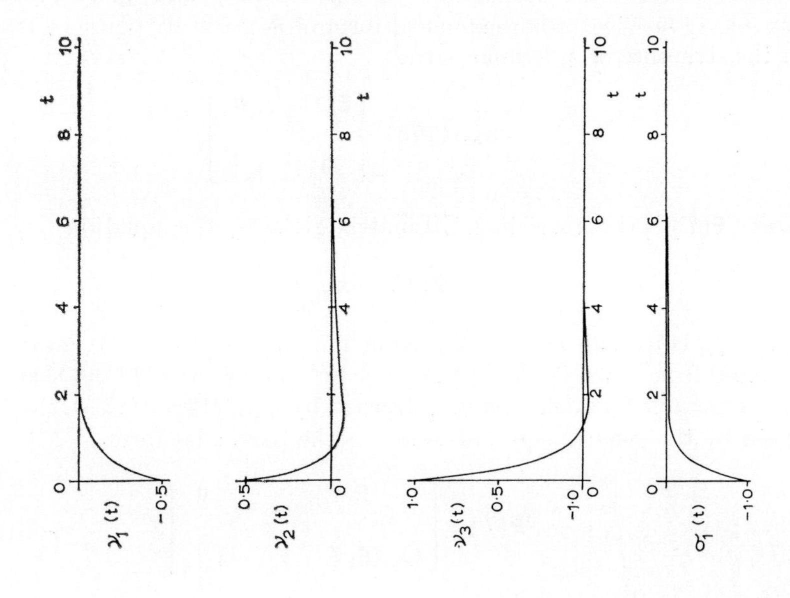

ν1 (t)
−0·5
ν2 (t)
0·5
ν3 (t)
1·0
0·5
−1·0
σ1 (t)
−1·0
t
0
2
4
6
8
10

14.3. Discrete-time modal controllers

14.3.1. *Mathematical model of uncontrolled system*

In the case of the digital control of the system governed by eqn. (14.5), the vector $\mathbf{z}(t)$ will be a piecewise-constant function of the form

$$\mathbf{z}(t)=\mathbf{z}(kT) \quad (kT \leqslant t<(k+1)T\ ;\ k=0,\ 1,\ 2,\ \ldots). \tag{14.52}$$

The differential eqn. (14.5) may accordingly be transformed into a difference equation of the form

$$\mathbf{x}\{(k+1)T\}=\mathbf{\Psi}(T)\mathbf{x}(kT)+\mathbf{\Delta}(T)\mathbf{z}(kT) \tag{14.53}$$

which relates the state of the system at successive discrete instants $t=0,\ T,\ 2T,\ \ldots,\ kT,\ \ldots$. It will be recalled from Chapter 2 that the relation between differential equations of the form (14.5) and difference equations of the form (14.53) is such that

$$\mathbf{\Psi}(T)=\exp(\mathbf{A}T) \tag{14.54 a}$$

and

$$\mathbf{\Delta}(T)=\left\{\int_0^T [\exp(\mathbf{A}t)]\,dt\right\}\mathbf{B}. \tag{14.54 b}$$

It can be readily shown that

$$\mathbf{\Psi}(T)=\mathbf{U}[\exp(\mathbf{J}T)]\mathbf{V}', \tag{14.55}$$

where $\mathbf{U}$ and $\mathbf{V}$ are respectively the $(2n-1)\times(2n-1)$ generalized modal matrices of $\mathbf{A}$ and $\mathbf{A}'$ given by eqns. (14.3) and (14.4), and $\mathbf{J}$ is the $(2n-1)\times(2n-1)$ Jordan canonical form of $\mathbf{A}$ given by eqn. (14.18). In view of the structure of $\mathbf{J}$, it follows that

$$\exp(\mathbf{J}T)=\begin{bmatrix}\mathbf{E}(T), & \mathbf{0}\\ \mathbf{0}, & \mathbf{I}_{n-1}\end{bmatrix}, \tag{14.56}$$

where $\mathbf{E}(T)$ is the $n\times n$ diagonal matrix given by the equation

$$\mathbf{E}(T)=\exp(\mathbf{J}_{11}T), \tag{14.57}$$

where $\mathbf{J}_{11}$ is the sub-matrix of $\mathbf{J}$ defined in eqn. (14.19 *a*). It may therefore be deduced from eqns. (14.3), (14.14), (14.18), (14.55), and (14.56) that, in the case of the cascaded-vehicle system governed by eqn. (14.53), the transition matrix given by the general eqn. (14.54 *a*) has the particular form

$$\mathbf{\Psi}(T)=\begin{bmatrix}\mathbf{E}(T), & \mathbf{0}\\ \mathbf{U}_{21}\{\mathbf{E}(T)-\mathbf{I}_n\}, & \mathbf{I}_{n-1}\end{bmatrix}; \tag{14.58}$$

where $\mathbf{U}_{21}$ is the sub-matrix of $\mathbf{U}$ given by eqn. (14.16 *c*) : it is evident from (14.58) that the eigenvalues of $\mathbf{\Psi}(T)$ are the members of the set

$$S=\{\mu_j\}=S_1\cup S_2,$$

where S_1 consists of the n non-zero elements of $\mathbf{E}(T)$ and S_2 consists of $(n-1)$ unit elements. Equation (14.55) indicates that the generalized modal matrices of $\mathbf{\Psi}(T)$ and $\mathbf{\Psi}'(T)$ are respectively $\mathbf{U}$ and $\mathbf{V}$.

Similarly, it may be shown by performing the integration indicated in the general eqn. (14.54 *b*) that the matrix $\mathbf{\Delta}(T)$ for the cascaded-vehicle system has the particular form

$$\mathbf{\Delta}(T)=\begin{bmatrix} \mathbf{F}(T)\mathbf{B}_1 \\ \\ \mathbf{U}_{21}\mathbf{B}_1\{\mathbf{F}(T)-T\mathbf{I}_n\} \end{bmatrix}, \tag{14.59}$$

where

$$\mathbf{F}(T)=\operatorname{diag}\,[\mathbf{J}_{11}^{-1}\{\mathbf{E}(T)-\mathbf{I}_n\}], \tag{14.60}$$

$\mathbf{J}_{11}$ is the sub-matrix of $\mathbf{J}$ given by eqn. (14.19 *a*), $\mathbf{B}_1$ is the sub-matrix of $\mathbf{B}$ given by eqn. (14.9 *a*), and $\mathbf{U}_{21}$ is the sub-matrix of $\mathbf{U}$ given by eqn. (14.16 *c*).

Finally, the mode-controllability matrix, $\mathbf{P}(T)$, associated with the system governed by the difference eqn. (14.53) is given by the expression

$$\mathbf{P}(T)=\mathbf{V}'\mathbf{\Delta}(T)=\begin{bmatrix} \mathbf{F}(T)\mathbf{B}_1 \\ \\ -T\mathbf{U}_{21}\mathbf{B}_1 \end{bmatrix}, \tag{14.61}$$

where the various matrices in eqn. (14.61) are as previously defined.

14.3.2. *Design of discrete-time modal controllers*

In view of the essential isomorphism which can now be seen to exist between the eigenstructure of the continuous-time modal control problem discussed in § 14.2 and the corresponding discrete-time problem, it may be inferred that the elements of the piecewise-constant control vector, $\mathbf{z}(kT)$, defined in eqn. (14.52) have the same form as the corresponding elements of the continuous-time modal control vector. Thus, it may be inferred that

$$z_1(kT)=K_1\mathbf{v}_1{}'\mathbf{x}(kT), \tag{14.62 a}$$

and

$$z_{i+1}(kT)=K_{i+1}\mathbf{v}_{i+1}{}'\mathbf{x}(kT)+K_{n+i}\mathbf{v}_{n+i}{}'\mathbf{x}(kT) \qquad (i=1,\,2,\,\ldots,\,n-1), \tag{14.62 b}$$

where $\mathbf{x}(kT)$ is the value of the state vector at the beginning of an interval $kT \leqslant t < (k+1)T$ during which each element of $\mathbf{z}(kT)$ has a constant value. In eqn. (14.62), $\mathbf{v}_j$ is the jth column of the generalized modal matrix, $\mathbf{V}$, and

$$K_1=(\rho_1-\mu_1)/p_{11}(T), \tag{14.63 a}$$

$$K_j=(\rho_j-\mu_j)(\rho_{n+j-1}-\mu_j)/p_{jj}(T)(\mu_{n+j-1}-\mu_j) \quad (j=2,\,3,\,\ldots,\,n), \tag{14.63 b}$$

and

$$K_{n+j-1}=(\rho_j-\mu_{n+j-1})(\rho_{n+j-1}-\mu_{n+j-1})/p_{n+j-1,j}(T)(\mu_j-\mu_{n+j-1}) \qquad (j=2,\,3,\,\ldots,\,n). \tag{14.63 c}$$

In eqns. (14.63), the ρ_j ($j=1, 2, \ldots, 2n-1$) are the eigenvalues of the plant matrix of the closed-loop system defined by eqns. (14.53), (14.58), (14.59), (14.62), and (14.63), and $p_{ij}(T)$ is the element in the ith row and jth column of the mode-controllability matrix, $\mathbf{P}(T)$, given by eqn. (14.61).

If the appropriate substitutions are made in eqns. (14.62), it is readily found that the modal control laws for the digital control of the cascaded-vehicle system are given by the explicit formulae

$$\phi_1(kT)=\frac{\alpha_1\{\rho_1-\exp(-\alpha_1 T/m_1)\}}{\{1-\exp(-\lambda_1 T/m_1)\}}\nu_1(kT), \tag{14.64 a}$$

and

$$\phi_j(kT)=\frac{\alpha_j\{\rho_j-\exp(-\alpha_j T/m_j)\}\{\rho_{n+j-1}-\exp(-\alpha_j T/m_j)\}}{\{1-\exp(-\alpha_j T/m_j)\}^2}\nu_j(kT)$$

$$+\frac{\alpha_j(\rho_j-1)(\rho_{n+j-1}-1)}{T\{1-\exp(-\lambda_j T/m_j)\}}\{(m_{j-1}/\alpha_{j-1})\nu_{j-1}(kT)$$

$$-(m_j/\alpha_j)\nu_j(kT)+\sigma_{j-1}(kT)\}\quad (j=2, 3, \ldots, n). \tag{14.64 b}$$

The quantities given in these formulae are, of course, the n elements of the piecewise-constant force-variation vector, $\boldsymbol{\phi}(kT)$, which is the vector $\mathbf{z}(kT)$ defined in eqn. (14.52) : the $\nu_j(kT)$ and $\sigma_i(kT)$ are the values of the elements of the system state vector (14.11) at the beginning of the associated time interval $kT \leqslant t < (k+1)T$.

The control laws defined in eqns. (14.64) are the desired discrete-time algorithms for the direct digital control of the cascaded-vehicle system : these laws are the precise analogues of the continuous-time algorithms (14.42).

14.3.3. *Numerical example*

In this section, the control laws for the discrete-time modal control of a cascaded-vehicle system comprising three vehicles are determined, and typical transient-response characteristics of the controlled system are presented. The numerical data used in the calculations is as follows :

$$m_1=m_2=m_3=1, \tag{14.65 a}$$

$$\alpha_1=\alpha_2=\alpha_3=1, \tag{14.65 b}$$

$$T=0{\cdot}693, \tag{14.65 c}$$

$$\rho_1=0{\cdot}05, \tag{14.65 d}$$

$$\rho_2=0{\cdot}10, \tag{14.65 e}$$

$$\rho_3=0{\cdot}15, \tag{14.65 f}$$

$$\rho_4=0{\cdot}80, \tag{14.65 g}$$

$$\rho_5=0{\cdot}90. \tag{14.65 h}$$

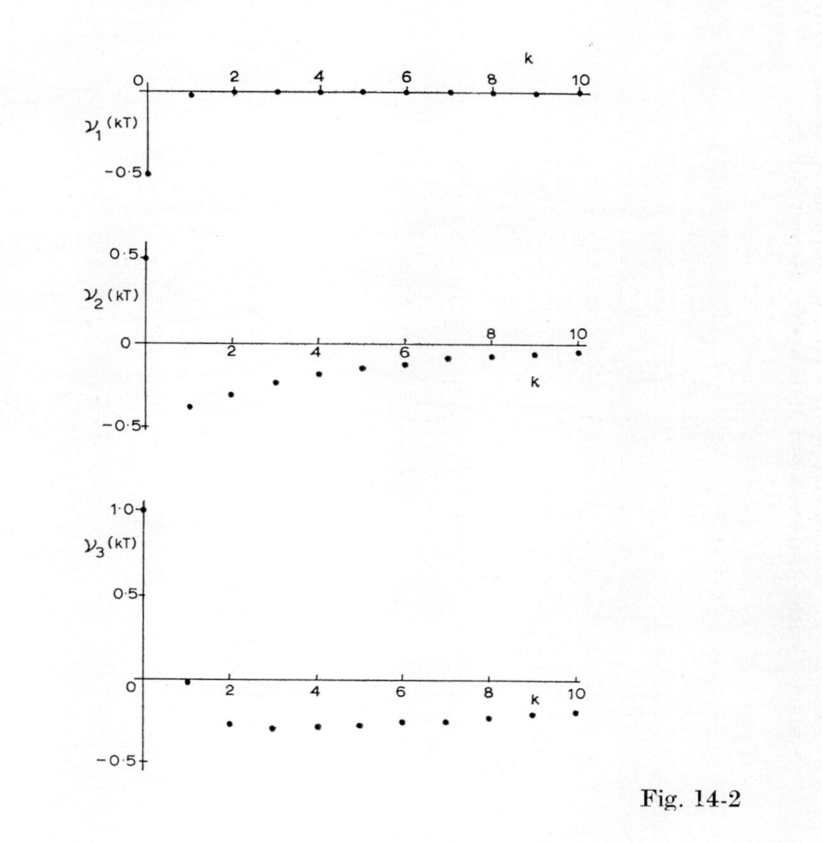

$\nu_1(kT)$
$\nu_2(kT)$
$\nu_3(kT)$
k

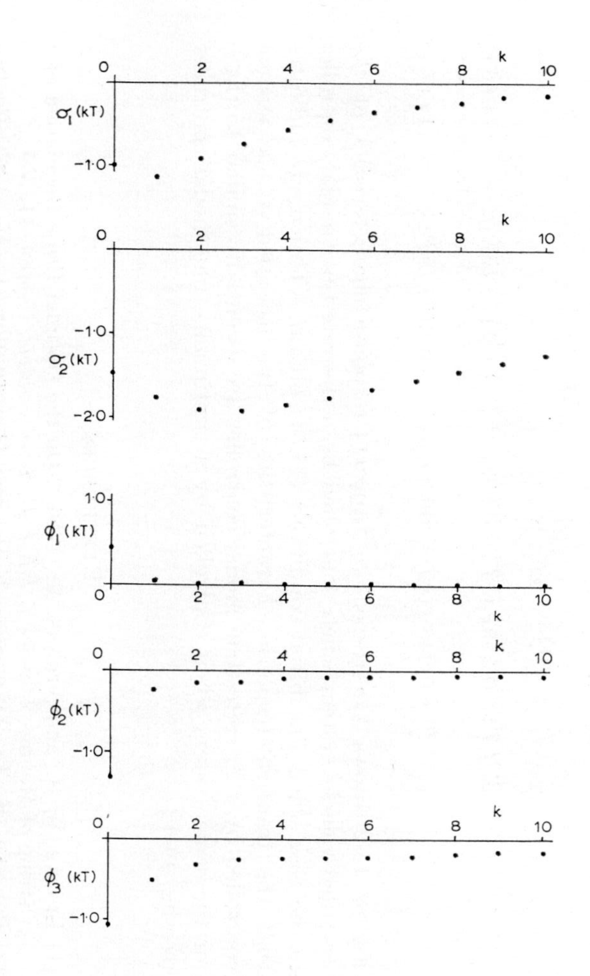

$\sigma_1(kT)$
$\sigma_2(kT)$
$\phi_1(kT)$
$\phi_2(kT)$
$\phi_3(kT)$
k

Fig. 14-2

Equations (14.64) indicate that the elements of the piecewise-constant force-variations vector, $\boldsymbol{\phi}(kT)$, are

$$\phi_1(kT) = -0{\cdot}900\nu_1(kT), \tag{14.66 a}$$

$$\phi_2(kT) = 0{\cdot}519\nu_1(kT) - 0{\cdot}999\nu_2(kT) + 0{\cdot}519\sigma_1(kT), \tag{14.66 b}$$

and

$$\phi_3(kT) = 0{\cdot}245\nu_2(kT) - 0{\cdot}805\nu_3(kT) + 0{\cdot}245\sigma_2(kT). \tag{14.66 c}$$

Figure 14-2 shows the respective transient-response characteristics of the cascaded-vehicle system when controlled by a discrete-time modal controller in the case when the initial conditions (14.51) obtain. This figure indicates that the transient-response characteristics of the cascaded-vehicle system controlled by a discrete-time modal controller are essentially similar to those for the same system when controlled by a continuous-time modal controller (see fig. 14-1).

References

[1] Levine, W. S., and Athans, M., 1966, " On the Optimal Error Regulation of a string of Moving Vehicles ", *I.E.E.E. Trans. autom. Control,* **11,** 355.

[2] Powner, E. T., Anderson, J. H., and Qualtrough, G. H., 1967, Control Systems Centre, University of Manchester, Report No. 25.

Synthesis of modal control policies for economic systems

CHAPTER 15

15.1. Introduction

In this chapter it is shown that modal control theory provides a very effective means of synthesizing control policies for economies which can be adequately modelled by linear state equations of the form (1.1) or (1.3). In order to make this demonstration as clear as possible, only simple economic models, namely, the continuous-time and discrete-time multiplier models [1], [2], [3] are discussed : the same approach can, however, be used in connection with more complicated models. The methods presented in this chapter may be regarded as inversions of certain classical methods [1], [2], [3] for determining control policies for such economic models in that the latter methods are *analytic* : thus, the analytic methods express the eigenproperties of an economic model in terms of the system parameters, whereas the synthetic methods express the system parameters in terms of the eigenproperties of the economic model under consideration. The difficulties inherent in the analytic approach are expressed in Allen [1] where it is stated that the equations to be solved involve ' a considerable number of parameters. General solutions are, therefore, not to be expected. Rather it is a matter of solving numerical equations in a series of particular cases '. The results obtained by the synthetic approach used in this chapter are clearly general and are not obtained by solving numerical equations in a series of particular cases.

15.2. Continuous-time modal control policies

15.2.1. *Mathematical model of uncontrolled system*

It is convenient to use the open-loop multiplier model in the form described by Allen [1], [2] and also to adopt the associated nomenclature initially. Thus, the model is governed by the equations

$$Z(t) = C(t) - A + G(t) \tag{15.1}$$

and

$$T_y \dot{Y}(t) + Y(t) = Z(t), \tag{15.2}$$

where $Z(t)$ is the total demand, $C(t)$ is the consumption, $-A$ is a negative step function change in private demand, $G(t)$ is the official demand, $Y(t)$ is the output, and T_y is the time-constant relating the speed of response of $Y(t)$ to changes in $Z(t)$. In addition,

$$C(t)=cY(t) \tag{15.3}$$

and

$$T_g\dot{G}(t)+G(t)=P(t), \tag{15.4}$$

where c is the marginal propensity to consume, $P(t)$ is the formulated control policy, and T_g is the time-constant relating the speed of response of $G(t)$ to changes in $P(t)$.

It may be readily deduced from eqns. (15.1), (15.2), (15.3), and (15.4) that the scalar differential equation relating $Y(t)$ and $P(t)$ is

$$T_gT_y\ddot{Y}(t)+(T_y+sT_g)\dot{Y}(t)+sY(t)=P(t)-A, \tag{15.5}$$

where

$$s=1-c \tag{15.6}$$

is the marginal propensity to save. Equation (15.5) can be readily put into the standard vector-matrix form (1.1) by choosing $x_1(t)=Y(t)$ and $x_2(t)=\dot{Y}(t)$ as state variables and by writing $P(t)$ as $z(t)$. The resulting state equation is

$$\dot{\mathbf{x}}(t)=\begin{bmatrix} 0, & 1 \\ -s/T_gT_y, & -(T_y+sT_g)/T_gT_y \end{bmatrix}\mathbf{x}(t)$$

$$+\begin{bmatrix} 0 \\ 1/T_gT_y \end{bmatrix}z(t)+\begin{bmatrix} 0 \\ -A/T_gT_y \end{bmatrix}, \tag{15.7}$$

where the state vector $\mathbf{x}(t)$ is equal to

$$\begin{bmatrix} x_1(t) \\ x_2(t) \end{bmatrix}.$$

Equation (15.7) indicates that the behaviour of the uncontrolled model is characterized by the eigenproperties of the plant matrix

$$\mathbf{A}=\begin{bmatrix} 0, & 1 \\ -s/T_gT_y, & -(T_y+sT_g)/T_gT_y \end{bmatrix}. \tag{15.8}$$

In fact, it can be readily shown that the eigenvalues of $\mathbf{A}$ are

$$\left.\begin{aligned} \lambda_1&=-s/T_y, \\ \lambda_2&=-1/T_g, \end{aligned}\right\} \tag{15.9}$$

and that a pair of corresponding eigenvectors of $\mathbf{A}'$ are

$$\left.\begin{aligned} \mathbf{v}_1 &= \begin{bmatrix} 1 \\ T_g \end{bmatrix}, \\ \mathbf{v}_2 &= \begin{bmatrix} s \\ T_y \end{bmatrix}. \end{aligned}\right\} \tag{15.10}$$

The mode-controllability matrix, $\mathbf{p}$, of the system can now be calculated and is found to be

$$\mathbf{p} = \mathbf{V}'\mathbf{b} = \begin{bmatrix} 1, & T_g \\ s, & T_y \end{bmatrix} \begin{bmatrix} 0 \\ 1/T_gT_y \end{bmatrix} = \begin{bmatrix} 1/T_y \\ 1/T_g \end{bmatrix} = \begin{bmatrix} p_1 \\ p_2 \end{bmatrix}, \tag{15.11}$$

where

$$\mathbf{V}' = \begin{bmatrix} \mathbf{v}_1' \\ \mathbf{v}_2' \end{bmatrix}$$

and $\mathbf{b}$ is the input matrix in eqn. (15.7). Since eqn. (15.11) indicates that both p_1 and p_2 are non-zero, it follows that the multiplier model (15.7) is controllable.

15.2.2. *Synthesis of control policy*

It is evident from eqn. (15.7) that, in the absence of control (i.e., when $z(t) = 0$),

$$\left.\begin{aligned} \dot{x}_1(t) &= x_2(t), \\ \dot{x}_2(t) &= -(s/T_gT_y)x_1(t) - \{(T_y + sT_g)/T_gT_y\}x_2(t) - (A/T_gT_y). \end{aligned}\right\} \tag{15.12}$$

Now, since s, T_y, and T_g are assumed to be real positive parameters, the open-loop eigenvalues given in equations (15.9) are negative. It may therefore be concluded that $\dot{x}_1 \to 0$ and $\dot{x}_2 \to 0$ in eqns. (15.12) as $t \to \infty$, which implies that the uncontrolled economy approaches the steady state

$$\left.\begin{aligned} \hat{x}_1 &= -A/s, \\ \hat{x}_2 &= 0, \end{aligned}\right\} \tag{15.13}$$

with a speed which depends upon the values of λ_1 and λ_2. In the words of Allen [2], ‘ there are two things wrong with such an economy ’: these are that $x_1 \neq 0$ in the steady state and that the speed of response may not be satisfactory. The problem is thus to determine a policy (i.e., to determine $z(t)$ in eqn. (15.7)) such that $x_1 = 0$ in the steady state, and that the speed of response is satisfactory.

Since the multiplier model is controllable, the eigenvalues which determine the speed of response of the two dynamical modes of this model can be assigned any desired values by using the results presented in Chapter 5 to determine an appropriate policy, $z(t)$. Thus, let $z(t)$ be generated by linear feedback of state according to a control law of the form

$$z(t) = g_1 x_1(t) + g_2 x_2(t), \tag{15.14}$$

where

$$[g_1, g_2] = K_1 \mathbf{v}_1' + K_2 \mathbf{v}_2'. \tag{15.15}$$

It is evident from eqns. (15.7) and (15.14) that the effect of introducing $z(t)$ in accordance with eqn. (15.14) is to change eqn. (15.7) to an equation of the form

$$\dot{\mathbf{x}}(t) = \mathbf{C}\mathbf{x}(t) + \begin{bmatrix} 0 \\ -A/T_g T_y \end{bmatrix}, \tag{15.16}$$

where the plant matrix $\mathbf{C}$ of the resulting closed-loop system involves g_1 and g_2. The behaviour of the closed-loop system is determined by the eigen-structure of $\mathbf{C}$: in particular, the eigenvalues of $\mathbf{C}$ determine the speed of response of the controlled economy.

Now the formula (5.33) indicates that the quantities K_1 and K_2 in eqns. (15.15) are given by

$$K_1 = \frac{(\rho_1 - \lambda_1)(\rho_2 - \lambda_1)}{p_1(\lambda_2 - \lambda_1)} \tag{15.17 a}$$

and

$$K_2 = \frac{(\rho_1 - \lambda_2)(\rho_2 - \lambda_2)}{p_2(\lambda_1 - \lambda_2)}, \tag{15.17 b}$$

where ρ_1 and ρ_2 are the desired closed-loop eigenvalues, p_1 and p_2 are given by eqn. (15.11), and the open-loop eigenvalues λ_1 and λ_2 are given in eqns. (15.9). It is obvious that substitution of the various known quantities into eqns. (15.17) yields the explicit values

$$K_1 = \frac{T_g(T_y\rho_1 + s)(T_y\rho_2 + s)}{(T_g s - T_y)} \tag{15.18 a}$$

and

$$K_2 = \frac{-T_y(T_g\rho_1 + 1)(T_g\rho_2 + 1)}{(T_g s - T_y)} \tag{15.18 b}$$

for K_1 and K_2. It thus follows from eqns. (15.10), (15.15), and (15.18) that the feedback gains in the control law (15.15) are

$$g_1 = s - T_g T_y \rho_1 \rho_2 \tag{15.19 a}$$

and

$$g_2 = (T_y + T_g s) + T_g T_y(\rho_1 + \rho_2). \tag{15.19 b}$$

The control law (15.14) accordingly has the explicit form

$$z(t) = (s - T_g T_y \rho_1 \rho_2) x_1(t) + \{(T_y + T_g s) + T_g T_y(\rho_1 + \rho_2)\} x_2(t) \tag{15.20}$$

which clearly defines a control policy of the proportional-plus-derivative type since $x\ (t) = Y(t)$ and $x_2(t) = \dot{Y}(t)$.

If this expression for $z(t)$ is substituted into eqn. (15.7), the resulting state equation for the controlled economy is found to be

$$\dot{\mathbf{x}}(t)=\begin{bmatrix} 0, & 1 \\ -\rho_1\rho_2, & \rho_1+\rho_2 \end{bmatrix}\mathbf{x}(t)+\begin{bmatrix} 0 \\ -A/T_gT_y \end{bmatrix}. \tag{15.21}$$

Equation (15.21) is of the form (15.16) and indicates that the behaviour of the controlled economy is characterized by the eigenproperties of the matrix

$$\mathbf{C}=\begin{bmatrix} 0, & 1 \\ -\rho_1\rho_2, & \rho_1+\rho_2 \end{bmatrix}. \tag{15.22}$$

Thus, the eigenvalues of $\mathbf{C}$ are clearly ρ_1 and ρ_2, as desired, and the solution of eqn. (15.21) is given by the scalar equations

$$\left.\begin{aligned} x_1(t)&=-(A/\rho_1\rho_2T_gT_y)+c_1\exp(\rho_1 t)+c_2\exp(\rho_2 t), \\ x_2(t)&=c_1\rho_1\exp(\rho_1 t)+c_2\rho_2\exp(\rho_2 t), \end{aligned}\right\} \tag{15.23}$$

where $x_1(t)=Y(t)$, $x_2(t)=\dot{Y}(t)$, and c_1 and c_2 are constants which depend upon the initial conditions.

It is evident that the speed and nature (e.g., oscillatory or non-oscillatory) of the response of the controlled economy to the step change, $-A$, in private demand can be selected as desired by choosing appropriate values for the eigenvalues ρ_1 and ρ_2 : the corresponding control policy is then obtained by substituting these quantities into the control-law eqn. (15.20). It is thus clear that, of the ‘ two things wrong with such an economy ’ [2] mentioned at the beginning of this section, the use of the control policy defined by eqn. (15.20) can certainly make one of these ‘ things ’ right. However, the other ‘ thing ’ cannot be put wholly right by the use of this control policy since it is clear from eqns. (15.23) that the controlled economy approaches the steady state

$$\left.\begin{aligned} \hat{x}_1&=-A/\rho_1\rho_2T_gT_y, \\ \hat{x}_2&=0, \end{aligned}\right\} \tag{15.24}$$

as $t\to\infty$.† Eqns. (15.24) indicate that x_1 is still non-zero in the steady state, although it is clear that this steady-state value of x_1 for the controlled economy will usually be less than the corresponding value for the uncontrolled economy given in (15.13). However, the only way of making $x_1\to 0$ as $t\to\infty$ (without assuming *a priori* knowledge of the magnitude of A) is to introduce a new state variable into the multiplier model, as is shown in the next section. Typical response characteristics of the systems governed by eqns. (15.7) and (15.21) are shown in figs. 15-1 and 15-2, respectively.

---

† Provided, of course, that ρ_1 and ρ_2 are chosen so that Re $\rho_1<0$ and Re $\rho_2<0$.

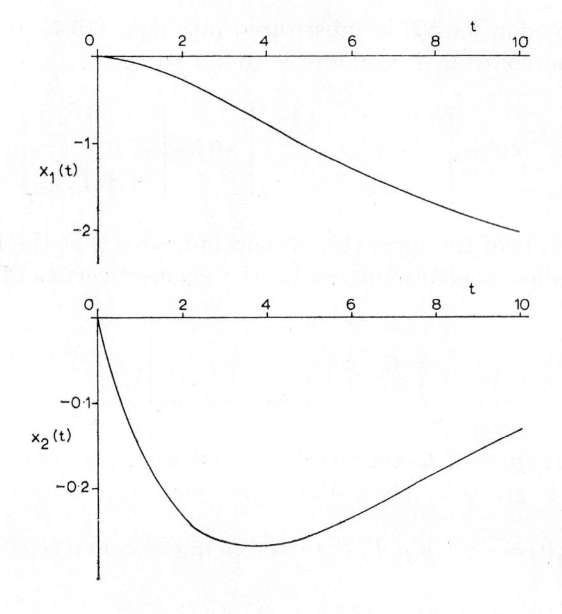

S = 0·36 , T_y = 1,
T_g = 5 , A = 1.

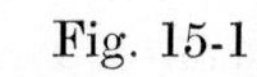
Fig. 15-1

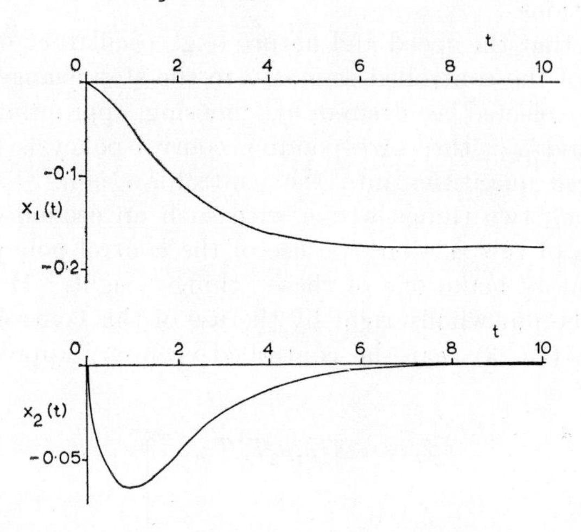

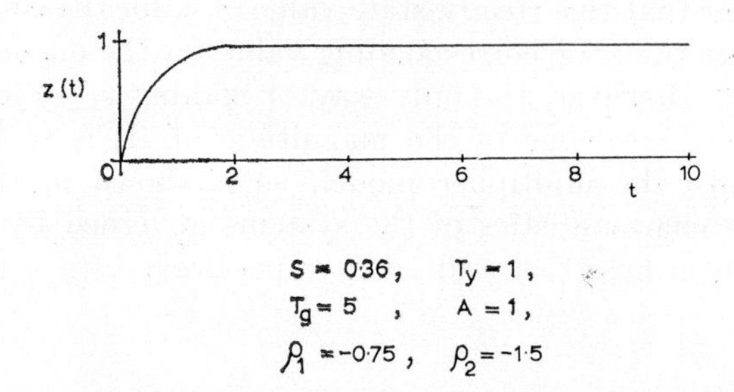

S = 0·36, T_y = 1,
T_g = 5 , A = 1,
ρ_1 = -0·75 , ρ_2 = -1·5

Fig. 15-2

15.2.3. *Synthesis of control policy for modified model*

Thus, if a new state variable, $x_3(t)$, defined by the equation

$$\dot{x}_3(t) = x_1(t) \tag{15.25}$$

is introduced into the multiplier model, the vector-matrix eqn. (15.7) assumes the form

$$\dot{\mathbf{x}}(t) = \begin{bmatrix} 0, & 1, & 0 \\ -s/T_gT_y, & -(T_y+T_gs)/T_gT_y, & 0 \\ 1, & 0, & 0 \end{bmatrix} \mathbf{x}(t) + \begin{bmatrix} 0 \\ 1/T_gT_y \\ 0 \end{bmatrix} z(t) + \begin{bmatrix} 0 \\ -A/T_gT_y \\ 0 \end{bmatrix}, \tag{15.26}$$

where

$$\mathbf{x}(t) = \begin{bmatrix} x_1(t) \\ x_2(t) \\ x_3(t) \end{bmatrix}.$$

Equation (15.26) indicates that the behaviour of the uncontrolled modified multiplier model (in which $x_3(t)$ is obviously obtained by integrating $x_1(t)$) is characterized by the eigenproperties of the plant matrix

$$\mathbf{A} = \begin{bmatrix} 0, & 1, & 0 \\ -s/T_gT_y, & -(T_y+T_gs)/T_gT_y, & 0 \\ 1, & 0, & 0 \end{bmatrix}. \tag{15.27}$$

The eigenvalues of this matrix are

$$\left.\begin{aligned} \lambda_1 &= 0, \\ \lambda_2 &= -s/T_y, \\ \lambda_3 &= -1/T_g, \end{aligned}\right\} \tag{15.28}$$

and the corresponding eigenvectors of $\mathbf{A}'$ are

$$\mathbf{v}_1 = \begin{bmatrix} T_y+T_gs \\ T_gT_y \\ s \end{bmatrix}, \quad \mathbf{v}_2 = \begin{bmatrix} 1 \\ T_g \\ 0 \end{bmatrix}, \quad \mathbf{v}_3 = \begin{bmatrix} s \\ T_y \\ 0 \end{bmatrix}. \tag{15.29}$$

The mode-controllability matrix, $\mathbf{p}$, of the system (15.26) can now be calculated and is found to be

$$\mathbf{p} = \begin{bmatrix} T_y+T_gs, & T_gT_y, & s \\ 1, & T_g, & 0 \\ s, & T_y, & 0 \end{bmatrix} \begin{bmatrix} 0 \\ 1/T_gT_y \\ 0 \end{bmatrix} = \begin{bmatrix} 1 \\ 1/T_y \\ 1/T_g \end{bmatrix} = \begin{bmatrix} p_1 \\ p_2 \\ p_3 \end{bmatrix}. \tag{15.30}$$

Since it is evident from (15.30) that p_1, p_2, and p_3 are non-zero, it follows that the modified multiplier model is controllable.

The formula (5.33) can therefore be used to determine a control policy, $z(t)$, of the form

$$z(t) = g_1x_1(t) + g_2x_2(t) + g_3x_3(t) \tag{15.31}$$

which makes it possible to assign any desired values to the eigenvalues ρ_1, ρ_2, and ρ_3 of the matrix, $\mathbf{C}$, of the controlled system defined by eqns. (15.26) and (15.31). In fact

$$[g_1, g_2, g_3] = K_1\mathbf{v}_1' + K_2\mathbf{v}_2' + K_3\mathbf{v}_3', \tag{15.32}$$

where

$$K_1 = \frac{(\rho_1 - \lambda_1)(\rho_2 - \lambda_1)(\rho_3 - \lambda_1)}{p_1(\lambda_2 - \lambda_1)(\lambda_3 - \lambda_1)}, \tag{15.33 a}$$

$$K_2 = \frac{(\rho_1 - \lambda_2)(\rho_2 - \lambda_2)(\rho_3 - \lambda_2)}{p_2(\lambda_1 - \lambda_2)(\lambda_3 - \lambda_2)}, \tag{15.33 b}$$

and

$$K_3 = \frac{(\rho_1 - \lambda_3)(\rho_2 - \lambda_3)(\rho_3 - \lambda_3)}{p_3(\lambda_1 - \lambda_3)(\lambda_2 - \lambda_3)}. \tag{15.33 c}$$

In the case of the modified multiplier model, it is evident that substitution of the various known quantities into eqns. (15.33) yields the explicit values

$$K_1 = \frac{\rho_1\rho_2\rho_3}{(s/T_gT_y)}, \tag{15.34 a}$$

$$K_2 = \frac{T_g(T_y\rho_1 + s)(T_y\rho_2 + s)(T_y\rho_3 + s)}{s(T_gs - T_y)}, \tag{15.34 b}$$

and

$$K_3 = \frac{-T_y(T_g\rho_1 + 1)(T_g\rho_2 + 1)(T_g\rho_3 + 1)}{(T_gs - T_y)}, \tag{15.34 c}$$

for K_1, K_2, and K_3. It thus follows from eqns. (15.29), (15.32), and (15.34) that the feedback gains defined in (15.15) are

$$g_1 = s - T_gT_y(\rho_1\rho_2 + \rho_2\rho_3 + \rho_3\rho_1), \tag{15.35 a}$$

$$g_2 = (T_y + T_gs) + T_gT_y(\rho_1 + \rho_2 + \rho_3), \tag{15.35 b}$$

and

$$g_3 = T_gT_y\rho_1\rho_2\rho_3. \tag{15.35 c}$$

The control law (15.31) therefore has the explicit form

$$\begin{aligned} z(t) = \{s - T_gT_y(\rho_1\rho_2 + \rho_2\rho_3 + \rho_3\rho_1)\}x_1(t) \\ + \{(T_y + T_gs) + T_gT_y(\rho_1 + \rho_2 + \rho_3)\}x_2(t) \\ + \{T_gT_y\rho_1\rho_2\rho_3\}x_3(t) \end{aligned} \tag{15.36}$$

which is clearly of the proportional-plus-derivative-plus-integral type since $x_1(t) = Y(t)$, $x_2(t) = \dot{Y}(t)$, and $x_3(t) = \int_{t_0}^{t} Y(t)\,dt$.

If this expression for the control policy is substituted for $z(t)$ in eqn. (15.26), the resulting state equation for the controlled economy is found to be

$$\dot{\mathbf{x}}(t) = \begin{bmatrix} 0, & 1, & 0 \\ -(\rho_1\rho_2 + \rho_2\rho_3 + \rho_3\rho_1), & (\rho_1 + \rho_2 + \rho_3), & \rho_1\rho_2\rho_3 \\ 1, & 0, & 0 \end{bmatrix} \mathbf{x}(t) + \begin{bmatrix} 0 \\ -A/T_gT_y \\ 0 \end{bmatrix}. \tag{15.37}$$

Equation (15.37) indicates that the behaviour of the controlled economy is characterized by the eigenproperties of the matrix

$$\mathbf{C}=\begin{bmatrix} 0, & 1, & 0 \\ -(\rho_1\rho_2+\rho_2\rho_3+\rho_3\rho_1), & (\rho_1+\rho_2+\rho_3), & \rho_1\rho_2\rho_3 \\ 1, & 0, & 0 \end{bmatrix}, \tag{15.38}$$

The eigenvalues of $\mathbf{C}$ are clearly ρ_1, ρ_2, and ρ_3 as desired, and the solution of eqn. (15.37) is given by the scalar equations

$$\left.\begin{aligned} x_1(t) &= c_1 \exp(\rho_1 t)+c_2 \exp(\rho_2 t)+c_3 \exp(\rho_3 t), \\ x_2(t) &= c_1\rho_1 \exp(\rho_1 t)+c_2\rho_2 \exp(\rho_2 t)+c_3\rho_3 \exp(\rho_3 t), \\ x_3(t) &= (-A/\rho_1\rho_2\rho_3 T_g T_y)+(c_1/\rho_1)\exp(\rho_1 t) \\ &\qquad +(c_2/\rho_2)\exp(\rho_2 t)+(c_3/\rho_3)\exp(\rho_3 t), \end{aligned}\right\} \tag{15.39}$$

where $x(t)=Y(t)$, $x_2(t)=\dot{Y}(t)$, $\dot{x}_3=Y(t)$, and c_1, c_2, and c_3 are constants which depend upon the initial conditions.

It is now evident that (as in the case of the unmodified multiplier model) the speed and nature of the response of the economy to the negative step change, $-A$, in private demand can be selected as required by choosing appropriate values for the closed-loop eigenvalues ρ_1, ρ_2, and ρ_3 and substituting these quantities into the control policy equation (15.36). However, in this case the '*two* things wrong with such an economy' [2] can both be put right by the use of the control policy equation (15.36) since it may be inferred from eqns. (15.39) that the controlled economy approaches the steady state

$$\left.\begin{aligned} \hat{x}_1 &= 0, \\ \hat{x}_2 &= 0, \\ \hat{x}_3 &= -A/\rho_1\rho_2\rho_3 T_g T_y, \end{aligned}\right\}$$

as $t\to\infty$.† Thus, the policy (15.36) controls the economy so that $x_1(t)=Y(t)\to 0$ as $t\to\infty$, as required. Typical response characteristics of the system governed by eqn. (15.37) are shown in fig. 15-3.

15.3. Discrete-time modal control policies

15.3.1. *Mathematical model of uncontrolled system*

In the case of the discrete-time multiplier model [1], [2], the appropriate governing equations are

$$Z(kT)=C(kT)+G(kT)-A \quad (k=0, 1, 2, \ldots) \tag{15.41}$$

and

$$Y\{(k+1)T\}=Z(kT) \quad (k=0, 1, 2, \ldots), \tag{15.42}$$

where, for the period corresponding to a particular value of k, $Z(kT)$ is the total demand, $C(kT)$ is the consumption, $-A$ is a negative step function change in private demand, $G(kT)$ is the official demand, and $Y(kT)$ is the output: eqn. (15.42) indicates the assumed presence of a Lundbergian lag

† Provided, of course, that ρ_1, ρ_2, and ρ_3 are chosen so that Re $\rho_1<0$, Re $\rho_2<0$, and Re $\rho_3<0$.

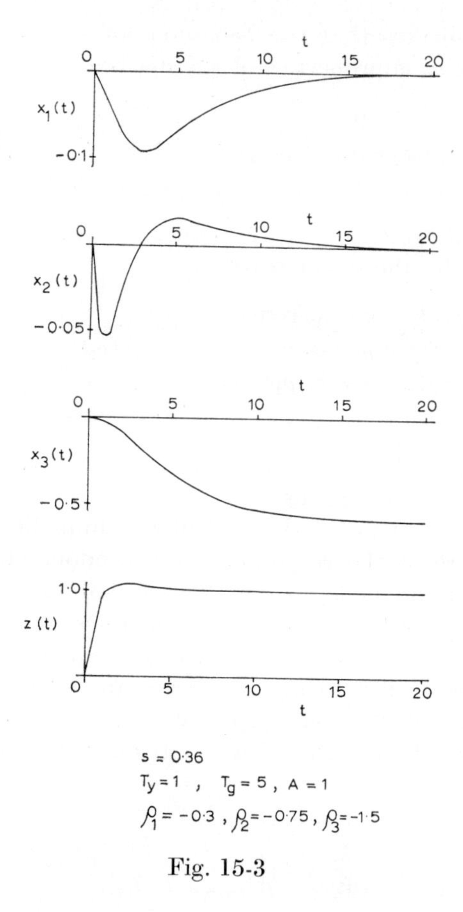

Fig. 15-3

[1], [2], of one period, T, between output and demand. The remaining governing equations of the model are

$$C\{(k+1)T\} = cY(kT) \quad (k=0, 1, 2, \ldots) \tag{15.43}$$

and

$$G\{(k+1)T\} = P(kT) \quad (k=0, 1, 2, \ldots), \tag{15.44}$$

where c is the marginal propensity to consume, $P(kT)$ is the formulated control policy for the period corresponding to a particular value of k : equation (15.43) indicates the assumed presence of a Robertsonian lag [1], [2] of one period, T, between consumption and output. It follows from eqns. (15.41) and (15.42) that

$$Y\{(k+1)T\} = C(kT) + G(kT) - A. \tag{15.45}$$

Equations (15.43), (15.44), and (15.45) can be readily expressed in standard vector-matrix form by choosing $x_1(kT) = G(kT)$, $x_2(kT) = C(kT)$ and $x_3(kT) = Y(kT$ as state variables and by writing $P(kT)$ as $z(kT)$. Thus, the state equation assumes the form

$$\mathbf{x}\{(k+1)T\} = \begin{bmatrix} 0, & 0, & 0 \\ 0, & 0, & c \\ 1, & 1, & 0 \end{bmatrix} \mathbf{x}(kT) + \begin{bmatrix} 1 \\ 0 \\ 0 \end{bmatrix} z(kT) + \begin{bmatrix} 0 \\ 0 \\ -A \end{bmatrix}, \tag{15.46}$$

where the state vector $\mathbf{x}(kT)$ is

$$\begin{bmatrix} x_1(kT) \\ x_2(kT) \\ x_3(kT) \end{bmatrix}.$$

It is evident from eqn. (15.46) that the behaviour of the uncontrolled model is determined by the eigenproperties of the plant matrix

$$\mathbf{\Psi}(T) = \begin{bmatrix} 0, & 0, & 0 \\ 0, & 0, & c \\ 1, & 1, & 0 \end{bmatrix}. \tag{15.47}$$

It is a simple matter to show that the eigenvalues of $\mathbf{\Psi}(T)$ are

$$\left.\begin{aligned} \lambda_1 &= -\xi, \\ \lambda_2 &= \xi, \\ \lambda_3 &= 0, \end{aligned}\right\} \quad (\xi^2 = c) \tag{15.48}$$

and that a corresponding set of eigenvectors of $\mathbf{\Psi}'(T)$ are

$$\mathbf{v}_1 = \begin{bmatrix} -1/\xi \\ -1/\xi \\ 1 \end{bmatrix}, \quad \mathbf{v}_2 = \begin{bmatrix} 1/\xi \\ 1/\xi \\ 1 \end{bmatrix}, \quad \mathbf{v}_3 = \begin{bmatrix} 1 \\ 0 \\ 0 \end{bmatrix}. \tag{15.49}$$

The mode controllability matrix, $\mathbf{p}(T)$, of the system (15.46) is accordingly

$$\mathbf{p}(T) = \mathbf{V}'\mathbf{\delta}(T) = \begin{bmatrix} -1/\xi, & -1/\xi, & 1 \\ 1/\xi, & 1/\xi, & 1 \\ 1, & 0, & 0 \end{bmatrix} \begin{bmatrix} 1 \\ 0 \\ 0 \end{bmatrix} = \begin{bmatrix} -1/\xi \\ 1/\xi \\ 1 \end{bmatrix} = \begin{bmatrix} p_1(T) \\ p_2(T) \\ p_3(T) \end{bmatrix}, \tag{15.50}$$

where

$$\mathbf{V}' = \begin{bmatrix} \mathbf{v}_1' \\ \mathbf{v}_2' \\ \mathbf{v}_3' \end{bmatrix}$$

and $\mathbf{\delta}(T)$ is the input matrix in eqn. (15.46). Since it is evident from (15.50) that $p_1(T)$, $p_2(T)$, and $p_3(T)$ are non-zero, it may be concluded that the discrete-time multiplier model (15.46) is controllable.

15.3.2. *Synthesis of control policy*

In the absence of control (i.e., when $z(kT) = 0$), eqn. (15.46) indicates that

$$\left.\begin{aligned} x_1\{(k+1)T\} &= 0, \\ x_2\{(k+1)T\} &= cx_3(kT), \\ x_3\{(k+1)T\} &= x_1(kT) + x_2(kT) - A. \end{aligned}\right\} \tag{15.51}$$

Now, provided that $c < 1$, the open-loop eigenvalues given in eqns. (15.48) will lie within the unit circle centred at the origin of the λ plane. It therefore follows that $x_2(kT) \to \hat{x}_2$ and $x_3(kT) \to \hat{x}_3$ (where $\hat{x}_2$ and $\hat{x}_3$ are constants) as $k \to \infty$. Thus, as $k \to \infty$, the uncontrolled economy approaches the steady state

$$\left.\begin{aligned} \hat{x}_2 &= -Ac/s, \\ \hat{x}_3 &= -A/s, \end{aligned}\right\} \quad (s = 1 - c) \tag{15.52}$$

with a speed which depends upon the eigenvalues of the uncontrolled system : in (15.52), s is, of course, the marginal propensity to save.

This behaviour of the uncontrolled economy is defective in that $x_3(kT)$ $(=Y(kT))$ does not return to the desired equilibrium state $Y(kT)=0$ after the step change in private demand, and also in that the approach to the equilibrium state, $-A/s$, may well be too sluggish in view of the values of λ_1 and λ_2. Because of the controllability of the system (15.46), it is a simple matter to rectify the latter defect by synthesizing an appropriate feedback policy, $z(kT)$: it transpires, however, that the former defect can only be rectified by incorporating an additional state variable into the system.

Thus, initially, let $z(kT)$ be generated by linear feedback of state according to a control law of the form

$$z(kT)=g_1x_1(kT)+g_2x_2(kT)+g_3x_3(kT), \tag{15.53}$$

where

$$[g_1, g_2, g_3]=K_1\mathbf{v}_1'+K_2\mathbf{v}_2'+K_3\mathbf{v}_3' \tag{15.54}$$

and, in accordance with the formula (5.33),

$$K_1=\frac{(\rho_1-\lambda_1)(\rho_2-\lambda_1)(\rho_3-\lambda_1)}{p_1(T)(\lambda_2-\lambda_1)(\lambda_3-\lambda_1)}, \tag{15.55 a}$$

$$K_2=\frac{(\rho_1-\lambda_2)(\rho_2-\lambda_2)(\rho_3-\lambda_2)}{p_2(T)(\lambda_1-\lambda_2)(\lambda_3-\lambda_2)}, \tag{15.55 b}$$

and

$$K_3=\frac{(\rho_1-\lambda_3)(\rho_2-\lambda_3)(\rho_3-\lambda_3)}{p_3(T)(\lambda_1-\lambda_3)(\lambda_2-\lambda_3)}. \tag{15.55 c}$$

In eqns. (15.55), ρ_1, ρ_2, and ρ_3 are the desired eigenvalues of the closed-loop system defined by eqns. (15.46) and (15.52) ; $p_1(T)$, $p_2(T)$, and $p_3(T)$ are given by eqn. (15.50) ; and the open-loop eigenvalues λ_1, λ_2, and λ_3 are given in equations (15.48). Equations (15.55) yield the explicit values

$$K_1=-(\rho_1+\xi)(\rho_2+\xi)(\rho_3+\xi)/2\xi, \tag{15.56 a}$$

$$K_2=(\rho_1-\xi)(\rho_2-\xi)(\rho_3-\xi)/2\xi, \tag{15.56 b}$$

and

$$K_3=\rho_1\rho_2\rho_3/\xi^2 \tag{15.56 c}$$

for K_1, K_2, and K_3 upon substitution of the various known quantities into these equations. It may thus be deduced from eqns. (15.49), (15.54), and (15.55) that the feedback gains in the control law (15.53) are

$$g_1=\sigma_1, \tag{15.57 a}$$

$$g_2=\sigma_1+(\sigma_3/c), \tag{15.57 b}$$

and

$$g_3=-\sigma_2-c, \tag{15.57 c}$$

where

$$\sigma_1=\rho_1+\rho_2+\rho_3, \tag{15.58 a}$$

$$\sigma_2=\rho_1\rho_2+\rho_2\rho_3+\rho_3\rho_1, \tag{15.58 b}$$

and

$$\sigma_3=\rho_1\rho_2\rho_3. \tag{15.58 c}$$

The control law (15.53) therefore has the explicit form

$$z(kT) = \sigma_1 x_1(kT) + \{\sigma_1 + (\sigma_3/c)\} x_2(kT) - (\sigma_2 + c) x_3(kT). \tag{15.59}$$

If the expression for $z(kT)$ given in eqn. (15.58) is substituted into eqn. (15.46) the resulting state equation for the controlled economy is found to be

$$\mathbf{x}\{(k+1)T\} = \begin{bmatrix} \sigma_1, & \sigma_1 + (\sigma_3/c), & -\sigma_2 - c \\ 0, & 0, & c \\ 1, & 1, & 0 \end{bmatrix} \mathbf{x}(kT) + \begin{bmatrix} 0 \\ 0 \\ -A \end{bmatrix}. \tag{15.60}$$

Equation (15.60) indicates that the behaviour of the controlled economy is determined essentially by the eigenproperties of the closed-loop plant matrix

$$\mathbf{C} = \begin{bmatrix} \sigma_1, & \sigma_1 + (\sigma_3/c), & -\sigma_2 - c \\ 0, & 0, & c \\ 1, & 1, & 0 \end{bmatrix}. \tag{15.61}$$

Thus, the eigenvalues of $\mathbf{C}$ are clearly ρ_1, ρ_2, and ρ_3, as desired, and the solution of eqn. (15.60) is given by the scalar equations

$$\left.\begin{aligned} x_1(kT) &= \left[\frac{(1-\sigma_1)c + \sigma_2 - \sigma_3}{1 - \sigma_1 + \sigma_2 - \sigma_3}\right] A + c_{11}\rho_1{}^k + c_{12}\rho_2{}^k + c_{13}\rho_3{}^k, \\ x_2(kT) &= \left[\frac{\sigma_1 - 1}{1 - \sigma_1 + \sigma_2 - \sigma_3}\right] Ac + c_{21}\rho_1{}^k + c_{22}\rho_2{}^k + c_{23}\rho_3{}^k, \\ x_3(kT) &= \left[\frac{\sigma_1 - 1}{1 - \sigma_1 + \sigma_2 - \sigma_3}\right] A + c_{31}\rho_1{}^k + c_{32}\rho_2{}^k + c_{33}\rho_3{}^k, \end{aligned}\right\} \tag{15.62}$$

provided that ρ_1, ρ_2, and ρ_3 are chosen such that†

$$1 - \sigma_1 + \sigma_2 - \sigma_3 \neq 0.$$

In eqns. (15.61), $x_1(kT) = G(kT)$, $x_2(kT) = C(kT)$, and $x_3(kT) = Y(kT)$, and the c_{ij} ($i, j = 1, 2, 3$) are constants which depend upon the initial conditions.

These results indicate that the speed and nature of the response of the economy to the step change, $-A$, in private demand can be selected as required by choosing appropriate values for the closed-loop eigenvalues ρ_1, ρ_2, and ρ_3: the corresponding control policy is then obtained by substituting these quantities into the control-law eqn. (15.59). However, eqns. (15.62) indicate that, for any set of initial conditions, the output, $Y(kT)$, of the controlled economy approaches the steady state

$$\hat{Y} = \hat{x}_3 = \left[\frac{\sigma_1 - 1}{1 - \sigma_1 + \sigma_2 - \sigma_3}\right] A \tag{15.64}$$

as $k \to \infty$. It is evident from eqn. (15.64) that, by choosing $\sigma_1 = 1$, it is possible to make $\hat{Y} = 0$ as desired. However, such a specific choice of σ_1 constrains the freedom of choice of the eigenvalues ρ_1, ρ_2, ρ_3, and thus imposes

† This inequality will always be satisfied in practice since ρ_1, ρ_2, and ρ_3 will always be chosen so that $|\rho_1| < 1$, $|\rho_2| < 1$, and $|\rho_3| < 1$.

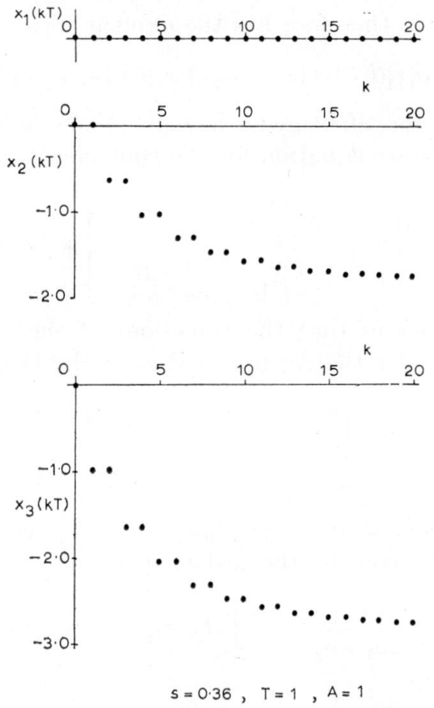

Fig. 15-4

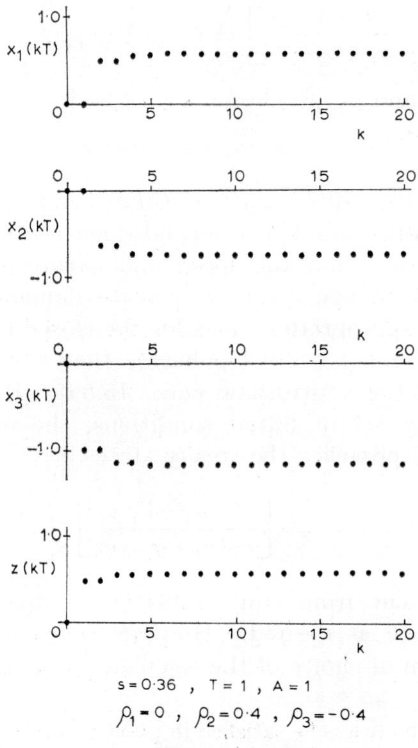

Fig. 15-5

bounds on the speed of response of the system. The only means of making $\hat{Y}=0$ without imposing such bounds is to incorporate a new state variable into the model (15.46), as is shown in the next section. Typical response characteristics of the systems governed by eqns. (15.46) and (15.64) are shown in the figs. 15-4 and 15-5, respectively.

15.3.3. *Synthesis of control policy for modified model*

As in the case of the continuous-time multiplier model, the desired final value of $x_3(kT)$ can be achieved by introducing a new state variable which corresponds to integral action into the model. Thus, in the present case, if a new state variable, $x_4(kT)$, defined by the equation

$$x_4(kT)=x_4\{(k-1)T\}+T[x_3(kT)+x_3\{(k-1)T\}]/2 \tag{15.65}$$

is introduced into the multiplier model, the vector-matrix eqn. (15.46) becomes

$$\begin{bmatrix}1, & 0, & 0, & 0\\ 0, & 1, & 0, & 0\\ 0, & 0, & 1, & 0\\ 0, & 0, & -T/2, & 1\end{bmatrix}\mathbf{x}\{(k+1)T\}=\begin{bmatrix}0, & 0, & 0, & 0\\ 0, & 0, & c, & 0\\ 1, & 1, & 0, & 0\\ 0, & 0, & T/2, & 1\end{bmatrix}\mathbf{x}(kT)$$

$$+\begin{bmatrix}1\\0\\0\\0\end{bmatrix}z(kT)+\begin{bmatrix}0\\0\\-A\\0\end{bmatrix}. \tag{15.66}$$

Before it is possible to synthesize the desired control policy for the modified model, it is necessary to pre-multiply eqn. (15.66) by the matrix

$$\begin{bmatrix}1, & 0, & 0, & 0\\ 0, & 1, & 0, & 0\\ 0, & 0, & 1, & 0\\ 0, & 0, & -T/2, & 1\end{bmatrix}^{-1}.$$

The resulting equation is found to be

$$\mathbf{x}\{(k+1)T\}=\begin{bmatrix}0, & 0, & 0, & 0\\ 0, & 0, & c, & 0\\ 1, & 1, & 0, & 0\\ T/2, & T/2, & T/2, & 1\end{bmatrix}\mathbf{x}(kT)+\begin{bmatrix}1\\0\\0\\0\end{bmatrix}z(kT)+\begin{bmatrix}0\\0\\-A\\-AT/2\end{bmatrix} \tag{15.67}$$

which indicates that the behaviour of the uncontrolled modified multiplier model is characterized by the eigenproperties of the plant matrix

$$\mathbf{\Psi}(T)=\begin{bmatrix}0, & 0, & 0, & 0\\ 0, & 0, & c, & 0\\ 1, & 1, & 0, & 0\\ T/2, & T/2, & T/2, & 1\end{bmatrix}. \tag{15.68}$$

The eigenvalues of this matrix are

$$\left.\begin{array}{l}\lambda_1=-\xi,\\ \lambda_2=\xi,\\ \lambda_3=0,\\ \lambda_4=1,\end{array}\right\} \quad (\xi^2=c) \tag{15.69}$$

and the corresponding eigenvectors of $\mathbf{\Psi}'(T)$ are

$$\mathbf{v}_1=\begin{bmatrix}-1/\xi\\-1/\xi\\1\\0\end{bmatrix},\quad \mathbf{v}_2=\begin{bmatrix}1/\xi\\1/\xi\\1\\0\end{bmatrix},\quad \mathbf{v}_3=\begin{bmatrix}1\\0\\0\\0\end{bmatrix},\quad \mathbf{v}_4=\begin{bmatrix}2\\2\\1+\xi^2\\2(1-\xi^2)/T\end{bmatrix}. \tag{15.70}$$

In terms of these quantities, the mode-controllability matrix, $\mathbf{p}(T)$, of the system (15.67) can now be calculated and is found to be

$$\mathbf{p}(T)=\begin{bmatrix}-1/\xi, & -1/\xi, & 1, & 0\\ 1/\xi, & 1/\xi, & 1, & 0\\ 1, & 0, & 0, & 0\\ 2, & 2, & 1+\xi^2, & 2(1-\xi^2)/T\end{bmatrix}\begin{bmatrix}1\\0\\0\\0\end{bmatrix}=\begin{bmatrix}-1/\xi\\1/\xi\\1\\2\end{bmatrix}=\begin{bmatrix}p_1(T)\\p_2(T)\\p_3(T)\\p_4(T)\end{bmatrix}. \tag{15.71}$$

Since $p_1(T)$, $p_2(T)$, $p_3(T)$, and $p_4(T)$ are all non-zero in (15.71), it follows that the modified multiplier model is controllable.

The formula (5.33) can therefore be used to compute a control policy, $z(kT)$, of the form

$$z(kT)=g_1x_1(kT)+g_2x_2(kT)+g_3x_3(kT)+g_4x_4(kT), \tag{15.72}$$

where

$$[g_1, g_2, g_3, g_4]=K_1\mathbf{v}_1'+K_2\mathbf{v}_2'+K_3\mathbf{v}_3'+K_4\mathbf{v}_4' \tag{15.73}$$

and

$$K_1=\frac{(\rho_1-\lambda_1)(\rho_2-\lambda_1)(\rho_3-\lambda_1)(\rho_4-\lambda_1)}{p_1(T)(\lambda_2-\lambda_1)(\lambda_3-\lambda_1)(\lambda_4-\lambda_1)}, \tag{15.74 a}$$

$$K_2=\frac{(\rho_1-\lambda_2)(\rho_2-\lambda_2)(\rho_3-\lambda_2)(\rho_4-\lambda_2)}{p_2(T)(\lambda_1-\lambda_2)(\lambda_3-\lambda_2)(\lambda_4-\lambda_2)}, \tag{15.74 b}$$

$$K_3=\frac{(\rho_1-\lambda_3)(\rho_2-\lambda_3)(\rho_3-\lambda_3)(\rho_4-\lambda_3)}{p_3(T)(\lambda_1-\lambda_3)(\lambda_2-\lambda_3)(\lambda_4-\lambda_3)}, \tag{15.74 c}$$

and

$$K_4=\frac{(\rho_1-\lambda_4)(\rho_2-\lambda_4)(\rho_3-\lambda_4)(\rho_4-\lambda_4)}{p_4(T)(\lambda_1-\lambda_4)(\lambda_2-\lambda_4)(\lambda_3-\lambda_4)}. \tag{15.74 d}$$

In the case of the present system, eqns. (15.74) yield the explicit values

$$K_1=-(\rho_1+\xi)(\rho_2+\xi)(\rho_3+\xi)(\rho_4+\xi)/2\xi(1+\xi), \tag{15.75 a}$$

$$K_2=(\rho_1-\xi)(\rho_2-\xi)(\rho_3-\xi)(\rho_4-\xi)/2\xi(1-\xi), \tag{15.75 b}$$

$$K_3=-\rho_1\rho_2\rho_3\rho_4/\xi^2, \tag{15.75 c}$$

and

$$K_4=-(\rho_1-1)(\rho_2-1)(\rho_3-1)(\rho_4-1)/2(1-\xi^2) \tag{15.75 d}$$

for K_1, K_2, K_3, and K_4. It may thus be deduced from eqns. (15.70), (15.73), and (15.75) that the feedback gains defined in (15.73) are

$$g_1 = -1 + \nu_1, \tag{15.76 a}$$

$$g_2 = -1 + \nu_1 + (\nu_4/c), \tag{15.76 b}$$

$$g_3 = (\nu_1 - \nu_2 - \nu_3 + \nu_4)/2 - (2c+1)/2, \tag{15.76 c}$$

and

$$g_4 = (-1 + \nu_1 - \nu_2 + \nu_3 - \nu_4)/T, \tag{15.76 d}$$

where

$$\nu_1 = \rho_1 + \rho_2 + \rho_3 + \rho_4, \tag{15.77 a}$$

$$\nu_2 = \rho_1\rho_2 + \rho_2\rho_3 + \rho_3\rho_4 + \rho_4\rho_1 + \rho_1\rho_3 + \rho_4\rho_2, \tag{15.77 b}$$

$$\nu_3 = \rho_1\rho_2\rho_3 + \rho_2\rho_3\rho_4 + \rho_3\rho_4\rho_1 + \rho_4\rho_1\rho_2, \tag{15.77 c}$$

and

$$\nu_4 = \rho_1\rho_2\rho_3\rho_4. \tag{15.77 d}$$

The control law (15.72) consequently has the explicit form

$$\begin{aligned} z(kT) = (-1+\nu_1)x_1(kT) &+ \{-1 + \nu_1 + (\nu_4/c)\}x_2(kT) \\ &+ \{(\nu_1 - \nu_2 - \nu_3 + \nu_4)/2 - (2c+1)/2\}x_3(kT) \\ &+ \{(-1 + \nu_1 - \nu_2 + \nu_3 - \nu_4)/T\}x_4(kT). \end{aligned} \tag{15.78}$$

If this expression for $z(kT)$ is substituted into eqn. (15.67), the resulting governing equation for the controlled economy is found to be

$$\mathbf{x}\{(k+1)T\} = \mathbf{C}\mathbf{x}(kT) + \begin{bmatrix} 0 \\ 0 \\ -A \\ -AT/2 \end{bmatrix} \tag{15.79}$$

where

$$\mathbf{C} = \begin{bmatrix} -1+\nu_1, & \nu_1 + (\nu_4/c) - 1, & (\nu_1 - \nu_2 - \nu_3 + \nu_4)/2 - (2c+1)/2, & (-1+\nu_1 - \nu_2 + \nu_3 - \nu_4)/T \\ 0, & 0, & c, & 0 \\ 1, & 1, & 0, & 1 \\ T/2, & T/2, & T/2, & 1 \end{bmatrix}. \tag{15.80}$$

The eigenvalues of the closed-loop plant $\mathbf{C}$ are ρ_1, ρ_2, ρ_3, and ρ_4, as desired, and the solution of eqn. (15.79) is given by the scalar equations

$$\left.\begin{aligned} x_1(kT) &= A + c_{11}\rho_1{}^k + c_{12}\rho_2{}^k + c_{13}\rho_3{}^k + c_{14}\rho_4{}^k, \\ x_2(kT) &= c_{21}\rho_1{}^k + c_{22}\rho_2{}^k + c_{23}\rho_3{}^k + c_{24}\rho_4{}^k, \\ x_3(kT) &= c_{31}\rho_1{}^k + c_{32}\rho_2{}^k + c_{33}\rho_3{}^k + c_{34}\rho_4{}^k, \\ x_4(kT) &= \left[\frac{AT(\nu_1 - 2)}{1 - \nu_1 + \nu_2 - \nu_3 + \nu_4}\right] + c_{41}\rho_1{}^k + c_{42}\rho_2{}^k + c_{43}\rho_3{}^k + c_{44}\rho_4{}^k, \end{aligned}\right\} \tag{15.81}$$

where $x_1(kT)=G(kT)$, $x_2(kT)=C(kT)$, $x_3(kT)=Y(kT)$, and $x_4(kT)$ is as defined in eqn. (15.65), and the c_{ij} $(i, j=1, 2, 3, 4)$ are constants which depend upon the initial conditions.

It may now be concluded that (as in the case of the unmodified multiplier model) any desired speed of response of the economy to the negative step change, $-A$, in private demand can be obtained by selecting a suitable set of values for the closed-loop eigenvalues ρ_1, ρ_2, ρ_3, and ρ_4 and substituting these quantities into the control policy eqn. (15.78). In addition, eqns. (15.81) imply that the controlled economy approaches the steady state

$$\begin{aligned}\hat{x}_1&=A,\\ \hat{x}_2&=0,\\ \hat{x}_3&=0,\\ \hat{x}_4&=AT(\nu_1-2)/(1-\nu_1+\nu_2-\nu_3+\nu_4),\end{aligned}$$

as $k\to\infty$: the policy (15.78) thus controls the economy so that $x_3(kT)=Y(kT)\to 0$ as $k\to\infty$, as required. Typical response characteristics of the system governed by eqns. (15.79) and (15.80) are shown in fig. 15-6.

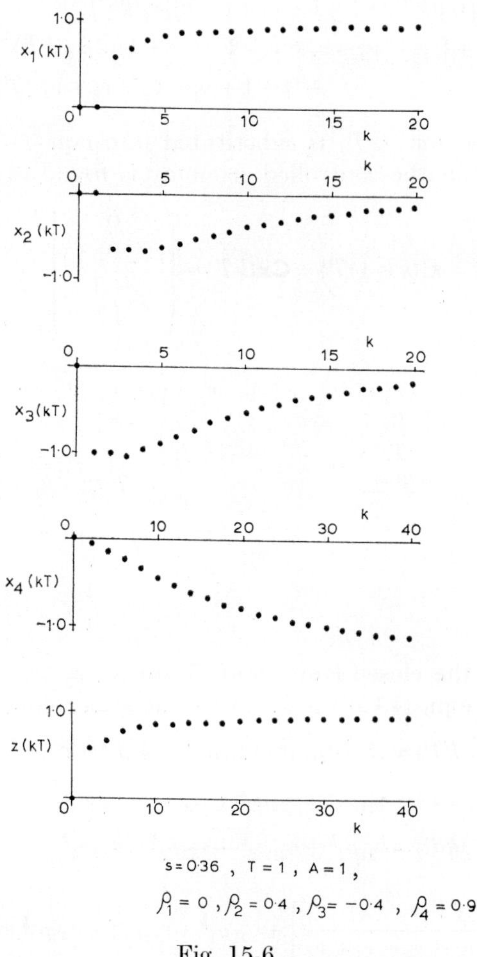

Fig. 15-6

References

[1] Allen, R. G. D., 1963, *Mathematical Economics* (Macmillan).
[2] Allen, R. G. D., 1967, *Macro-economic Theory* (Macmillan).
[3] Philips, A. W., 1954, " Stabilisation policy in a closed economy ", *Econ. J.*, **64,** 290.

Synthesis of modal control policies for manufacturing systems

CHAPTER 16

16.1. Introduction

In this chapter the modal control design procedures presented in Chapters 5 and 6 are illustrated by using these procedures to synthesize policies for the control of inventory and backlog levels in manufacturing systems, whilst maintaining balanced production lines.

16.2. Mathematical model of uncontrolled system

16.2.1. *Mathematical model of manufacturing system*

The mathematical model of the dynamics of the manufacturing system considered in this chapter is based upon the discrete-time model of a production process investigated using optimal control theory by Christensen and Brogan [1]. The block diagram shown in fig. 16-1 indicates that this system comprises six sub-systems, S_j $(j=1, 2, \ldots, 6)$, whose state equations are

$$\chi_1\{(k+1)T\}=\chi_1(kT)+\alpha\beta_1\mu_1(kT) + \alpha\beta_2\mu_2(kT)-2\alpha\beta_3\mu_3(kT) \quad (k=0, 1, 2, \ldots), \tag{16.1}$$

$$\chi_2\{(k+1)T\}=\beta_3\mu_3(kT) \quad (k=0, 1, 2, \ldots), \tag{16.2}$$

$$\left.\begin{aligned} \chi_3\{(k+1)T\}&=\chi_4(kT),\\ \chi_4\{(k+1)T\}&=\chi_5(kT),\\ \chi_5\{(k+1)T\}&=\chi_6(kT),\\ \chi_6\{(k+1)T\}&=\beta_4\mu_4(kT), \end{aligned}\right\} \quad (k=0, 1, 2, \ldots), \tag{16.3}$$

$$\chi_7\{(k+1)T\}=\chi_7(kT)+\alpha\chi_2(kT)+\alpha\chi_3(kT) + \alpha\beta_5\mu_5(kT)+\alpha\beta_6\mu_6(kT)-4\alpha\beta_7\mu_7(kT) \quad (k=0, 1, 2, \ldots), \tag{16.4}$$

$$\chi_8\{(k+1)T\}=\beta_7\mu_7(kT) \quad (k=0, 1, 2, \ldots), \tag{16.5}$$

and

$$\chi_9\{(k+1)T\}=\chi_9(kT)+\alpha\chi_8(kT)-\sigma \quad (k=0, 1, 2, \ldots), \tag{16.6}$$

where, in order to ensure balanced production lines, it is assumed that

$$\beta_1\mu_1(kT)=\beta_2\mu_2(kT)=\ldots=\beta_7\mu_7(kT) \quad (k=0,\ 1,\ 2,\ \ldots). \tag{16.7}$$

In eqns. (16.1) to (16.6), σ is the constant planned sales rate measured in parts per shift, the state variables $\chi_i(kT)$ $(i=1,\ 7)$ are backlog variables measured in parts, the state variables $\chi_i(kT)$ $(i=2,\ 3,\ 4,\ 5,\ 6,\ 8)$ are rates-of-flow of parts measured in parts per week, the state variable $\chi_9(kT)$ is the inventory level measured in parts, the $\mu_i(kT)$ $(i=1,\ 2,\ 3,\ 4,\ 5,\ 6,\ 7)$ are inputs measured in man-hours per week, the $\beta_i(kT)$ $(i=1,\ 2,\ 3,\ 4,\ 5,\ 6,\ 7)$ represents parts processed per man-hour, T is the shift duration in hours denoted by the operator E, and α is equal to (T/H) where H is the number of hours worked per week.

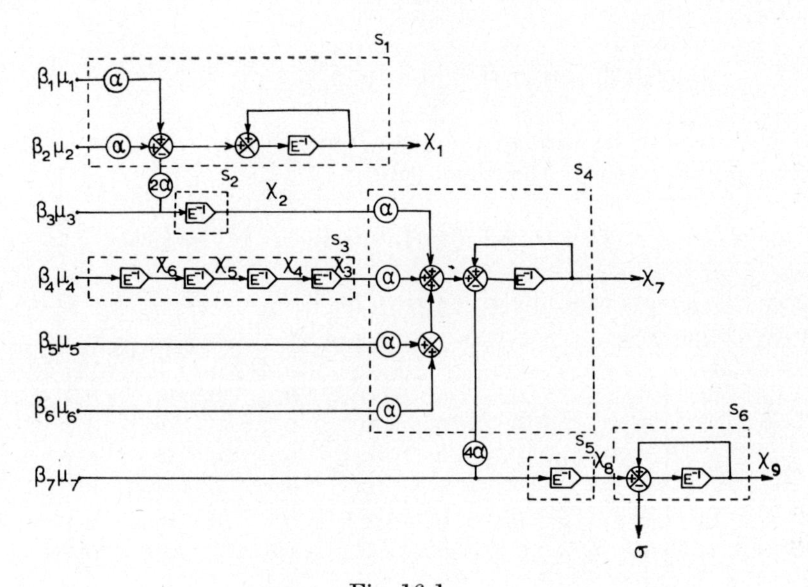

Fig. 16-1

Equations (16.1) to (16.6) can be written in the vector-matrix form

$$\boldsymbol{\chi}\{(k+1)T\}=\boldsymbol{\Psi}(T)\boldsymbol{\chi}(kT)+\boldsymbol{\Delta}(T)\boldsymbol{\zeta}(kT)+\boldsymbol{\nu}(T)\sigma \quad (k=0,\ 1,\ 2,\ \ldots), \tag{16.8}$$

where the plant matrix, $\boldsymbol{\Psi}(T)$, the input matrix, $\boldsymbol{\Delta}(T)$, the state vector, $\boldsymbol{\chi}(kT)$, the input vector, $\boldsymbol{\zeta}(kT)$, and the vector, $\boldsymbol{\nu}(T)$, are given by the respective equations

$$\boldsymbol{\Psi}(T)=\begin{bmatrix} 1, & 0, & 0, & 0, & 0, & 0, & 0, & 0, & 0 \\ 0, & 0, & 0, & 0, & 0, & 0, & 0, & 0, & 0 \\ 0, & 0, & 0, & 1, & 0, & 0, & 0, & 0, & 0 \\ 0, & 0, & 0, & 0, & 1, & 0, & 0, & 0, & 0 \\ 0, & 0, & 0, & 0, & 0, & 1, & 0, & 0, & 0 \\ 0, & 0, & 0, & 0, & 0, & 0, & 0, & 0, & 0 \\ 0, & \alpha, & \alpha, & 0, & 0, & 0, & 1, & 0, & 0 \\ 0, & 0, & 0, & 0, & 0, & 0, & 0, & 0, & 0 \\ 0, & 0, & 0, & 0, & 0, & 0, & 0, & \alpha, & 1 \end{bmatrix}, \tag{16.9}$$

$$\mathbf{\Delta}(T)=\begin{bmatrix} \alpha, & \alpha, & -2\alpha, & 0, & 0, & 0, & 0 \\ 0, & 0, & 1, & 0, & 0, & 0, & 0 \\ 0, & 0, & 0, & 0, & 0, & 0, & 0 \\ 0, & 0, & 0, & 0, & 0, & 0, & 0 \\ 0, & 0, & 0, & 0, & 0, & 0, & 0 \\ 0, & 0, & 0, & 1, & 0, & 0, & 0 \\ 0, & 0, & 0, & 0, & \alpha, & \alpha, & -4\alpha \\ 0, & 0, & 0, & 0, & 0, & 0, & 1 \\ 0, & 0, & 0, & 0, & 0, & 0, & 0 \end{bmatrix}, \tag{16.10}$$

$$\boldsymbol{\chi}'(kT)=[\chi_1(kT),\ \chi_2(kT),\ \dots,\ \chi_9(kT)], \tag{16.11}$$

$$\boldsymbol{\zeta}'(kT)=[\beta_1\mu_1(kT),\ \beta_2\mu_2(kT),\ \dots,\ \beta_7\mu_7(kT)], \tag{16.12}$$

and

$$\boldsymbol{\nu}'(T)=[0,\ 0,\ 0,\ 0,\ 0,\ 0,\ 0,\ 0,\ -1]. \tag{16.13}$$

It may readily be verified that the Jordan canonical form of the plant matrix (16.9) is given by the direct sum

$$\mathbf{J}=\mathbf{J}_1(1)\oplus\mathbf{J}_1(0)\oplus\mathbf{J}_4(0)\oplus\mathbf{J}_1(1)\oplus\mathbf{J}_1(0)\oplus\mathbf{J}_1(1), \tag{16.14}$$

and that the generalized modal matrices of $\mathbf{\Psi}(T)$ and $\mathbf{\Psi}'(T)$ are given by the respective equations

$$\mathbf{U}=\begin{bmatrix} 1, & 0, & 0, & 0, & 0, & 0, & 0, & 0, & 0 \\ 0, & 1, & 0, & 0, & 0, & 0, & 0, & 0, & 0 \\ 0, & 0, & 1, & 0, & 0, & 0, & 0, & 0, & 0 \\ 0, & 0, & 0, & 1, & 0, & 0, & 0, & 0, & 0 \\ 0, & 0, & 0, & 0, & 1, & 0, & 0, & 0, & 0 \\ 0, & 0, & 0, & 0, & 0, & 1, & 0, & 0, & 0 \\ 0, & -\alpha, & -\alpha, & -\alpha, & -\alpha, & -\alpha, & 1, & 0, & 0 \\ 0, & 0, & 0, & 0, & 0, & 0, & 0, & 1, & 0 \\ 0, & 0, & 0, & 0, & 0, & 0, & 0, & -\alpha, & 1 \end{bmatrix}, \tag{16.15}$$

and

$$\mathbf{V}=\begin{bmatrix} 1, & 0, & 0, & 0, & 0, & 0, & 0, & 0, & 0 \\ 0, & 1, & 0, & 0, & 0, & 0, & \alpha, & 0, & 0 \\ 0, & 0, & 1, & 0, & 0, & 0, & \alpha, & 0, & 0 \\ 0, & 0, & 0, & 1, & 0, & 0, & \alpha, & 0, & 0 \\ 0, & 0, & 0, & 0, & 1, & 0, & \alpha, & 0, & 0 \\ 0, & 0, & 0, & 0, & 0, & 1, & \alpha, & 0, & 0 \\ 0, & 0, & 0, & 0, & 0, & 0, & 1, & 0, & 0 \\ 0, & 0, & 0, & 0, & 0, & 0, & 0, & 1, & \alpha \\ 0, & 0, & 0, & 0, & 0, & 0, & 0, & 0, & 1 \end{bmatrix}. \tag{16.16}$$

It can be deduced from eqns. (16.10) and (16.16) that the mode-controllability matrix, $\mathbf{P}(T)$, of the system governed by the state eqn. (16.7 a) is given by the equation

$$\mathbf{P}(T)=\mathbf{V}'\mathbf{\Delta}(T)=\begin{bmatrix} \alpha, & \alpha, & -2\alpha, & 0, & 0, & 0, & 0 \\ 0, & 0, & 1, & 0, & 0, & 0, & 0 \\ 0, & 0, & 0, & 0, & 0, & 0, & 0 \\ 0, & 0, & 0, & 0, & 0, & 0, & 0 \\ 0, & 0, & 0, & 0, & 0, & 0, & 0 \\ 0, & 0, & 0, & 1, & 0, & 0, & 0 \\ 0, & 0, & \alpha, & \alpha, & \alpha, & \alpha, & -4\alpha \\ 0, & 0, & 0, & 0, & 0, & 0, & 1 \\ 0, & 0, & 0, & 0, & 0, & 0, & \alpha \end{bmatrix}. \qquad (16.17)$$

Since all the rows of $\mathbf{P}(T)$ which correspond to the last rows of the Jordan blocks in eqn. (16.14) which contain the same-valued eigenvalue are linearly independent, it follows that the system (16.8) is controllable.

It may be readily verified that the steady-state solution of eqn. (16.8) is given by the equations

$$\left.\begin{aligned} \hat{\boldsymbol{\chi}}' &= [\hat{\chi}_1, \sigma/\alpha, \sigma/\alpha, \sigma/\alpha, \sigma/\alpha, \sigma/\alpha, \hat{\chi}_7, \sigma/\alpha, \hat{\chi}_9], \\ \hat{\boldsymbol{\zeta}}' &= [\sigma/\alpha, \sigma/\alpha, \sigma/\alpha, \sigma/\alpha, \sigma/\alpha, \sigma/\alpha, \sigma/\alpha], \end{aligned}\right\} \qquad (16.18)$$

where $\hat{\chi}_1$, $\hat{\chi}_7$, and $\hat{\chi}_9$ are constants which depend upon the initial state of the system. If the system (16.8) is disturbed from the steady state defined by eqns. (16.18) as a result of a sales deviation from the planned sales rate, σ, to some new rate, $\sigma+s(kT)$, then the state eqn. (16.8) can be written in the form

$$\mathbf{x}\{(k+1)T\}=\mathbf{\Psi}(T)\mathbf{x}(kT)+\mathbf{\Delta}(T)\mathbf{z}(kT)+\boldsymbol{\nu}(T)s(kT) \qquad (k=0, 1, 2, \ldots). \qquad (16.19)$$

In eqn. (16.19), $\mathbf{x}(kT)$ and $\mathbf{z}(kT)$ are the respective perturbations of $\boldsymbol{\chi}(kT)$ and $\boldsymbol{\zeta}(kT)$ from the steady-state values, $\hat{\boldsymbol{\chi}}$ and $\hat{\boldsymbol{\zeta}}$, given in eqns. (16.18). It is evident that the introduction of $s(kT)$ will lead to consequential changes in the backlog variables, $\chi_1(kT)$ and $\chi_7(kT)$, and in the inventory level, $\chi_9(kT)$. The precise nature of these changes will clearly depend upon the time variation of $s(kT)$ and may lead, for example, to inventory levels which are either too low or too high : in extreme cases, of course, the inventory could be completely exhausted. However, this deficiency in the behaviour of the system can be overcome by introducing state-feedback control, in association with the introduction of a new state variable, $x_{10}(kT)$, which corresponds to local integral control of the inventory level.

16.2.2. *Mathematical model of modified manufacturing system*

Thus, if a new state variable, $x_{10}(kT)$, defined by the equation

$$x_{10}\{(k+1)T\}=x_{10}(kT)+x_9(kT) \qquad (16.20)$$

is introduced into the model, eqns. (16.1), (16.2), (16.3), (16.4), (16.5), (16.6), and (16.20) can be written in the form (16.19), where the plant matrix, $\mathbf{\Psi}(T)$, the input matrix, $\boldsymbol{\Delta}(T)$, the state vector, $\mathbf{x}(kT)$, the input vector, $\mathbf{z}(kT)$, and the vector, $\boldsymbol{\nu}(T)$, are now given by the respective equations

$$\boldsymbol{\Psi}(T)=\begin{bmatrix} 1, & 0, & 0, & 0, & 0, & 0, & 0, & 0, & 0, & 0 \\ 0, & 0, & 0, & 0, & 0, & 0, & 0, & 0, & 0, & 0 \\ 0, & 0, & 0, & 1, & 0, & 0, & 0, & 0, & 0, & 0 \\ 0, & 0, & 0, & 0, & 1, & 0, & 0, & 0, & 0, & 0 \\ 0, & 0, & 0, & 0, & 0, & 1, & 0, & 0, & 0, & 0 \\ 0, & 0, & 0, & 0, & 0, & 0, & 0, & 0, & 0, & 0 \\ 0, & \alpha, & \alpha, & 0, & 0, & 0, & 1, & 0, & 0, & 0 \\ 0, & 0, & 0, & 0, & 0, & 0, & 0, & 0, & 0, & 0 \\ 0, & 0, & 0, & 0, & 0, & 0, & 0, & \alpha, & 1, & 0 \\ 0, & 0, & 0, & 0, & 0, & 0, & 0, & 0, & 1, & 1 \end{bmatrix}, \tag{16.21}$$

$$\boldsymbol{\Delta}(T)=\begin{bmatrix} \alpha, & \alpha, & -2\alpha, & 0, & 0, & 0, & 0 \\ 0, & 0, & 1, & 0, & 0, & 0, & 0 \\ 0, & 0, & 0, & 0, & 0, & 0, & 0 \\ 0, & 0, & 0, & 0, & 0, & 0, & 0 \\ 0, & 0, & 0, & 0, & 0, & 0, & 0 \\ 0, & 0, & 0, & 1, & 0, & 0, & 0 \\ 0, & 0, & 0, & 0, & \alpha, & \alpha, & -4\alpha \\ 0, & 0, & 0, & 0, & 0, & 0, & 1 \\ 0, & 0, & 0, & 0, & 0, & 0, & 0 \\ 0, & 0, & 0, & 0, & 0, & 0, & 0 \end{bmatrix}, \tag{16.22}$$

$$\mathbf{x}'(kT)=[x_1(kT),\ x_2(kT),\ \ldots,\ x_{10}(kT)], \tag{16.23}$$

$$\mathbf{z}'(kT)=[z_1(kT),\ z_2(kT),\ \ldots,\ z_7(kT)], \tag{16.24}$$

and

$$\boldsymbol{\nu}'(T)=[0,\ 0,\ 0,\ 0,\ 0,\ 0,\ 0,\ 0,\ -1,\ 0]. \tag{16.25}$$

It may readily be verified that the Jordan canonical form of the plant matrix (16.21) is given by the direct sum

$$\mathbf{J}=\mathbf{J}_1(1)\oplus\mathbf{J}_1(0)\oplus\mathbf{J}_4(0)\oplus\mathbf{J}_1(1)\oplus\mathbf{J}_1(0)\oplus\mathbf{J}_2(1), \tag{16.26}$$

and that the generalized modal matrices of $\boldsymbol{\Psi}(T)$ and $\boldsymbol{\Psi}'(T)$ are given by the respective equations

$$\mathbf{U}=\begin{bmatrix} 1, & 0, & 0, & 0, & 0, & 0, & 0, & 0, & 0, & 0 \\ 0, & 1, & 0, & 0, & 0, & 0, & 0, & 0, & 0, & 0 \\ 0, & 0, & 1, & 0, & 0, & 0, & 0, & 0, & 0, & 0 \\ 0, & 0, & 0, & 1, & 0, & 0, & 0, & 0, & 0, & 0 \\ 0, & 0, & 0, & 0, & 1, & 0, & 0, & 0, & 0, & 0 \\ 0, & 0, & 0, & 0, & 0, & 1, & 0, & 0, & 0, & 0 \\ 0, & -\alpha, & -\alpha, & -\alpha, & -\alpha, & -\alpha, & 1, & 0, & 0, & 0 \\ 0, & 0, & 0, & 0, & 0, & 0, & 0, & 1, & 0, & 0 \\ 0, & 0, & 0, & 0, & 0, & 0, & 0, & -\alpha, & 0, & 1 \\ 0, & 0, & 0, & 0, & 0, & 0, & 0, & \alpha, & 1, & 0 \end{bmatrix} \tag{16.27}$$

and

$$\mathbf{V}=\begin{bmatrix} 1, & 0, & 0, & 0, & 0, & 0, & 0, & 0, & 0, & 0 \\ 0, & 1, & 0, & 0, & 0, & 0, & \alpha, & 0, & 0, & 0 \\ 0, & 0, & 1, & 0, & 0, & 0, & \alpha, & 0, & 0, & 0 \\ 0, & 0, & 0, & 1, & 0, & 0, & \alpha, & 0, & 0, & 0 \\ 0, & 0, & 0, & 0, & 1, & 0, & \alpha, & 0, & 0, & 0 \\ 0, & 0, & 0, & 0, & 0, & 1, & \alpha, & 0, & 0, & 0 \\ 0, & 0, & 0, & 0, & 0, & 0, & 1, & 0, & 0, & 0 \\ 0, & 0, & 0, & 0, & 0, & 0, & 0, & 1, & -\alpha, & \alpha \\ 0, & 0, & 0, & 0, & 0, & 0, & 0, & 0, & 0, & 1 \\ 0, & 0, & 0, & 0, & 0, & 0, & 0, & 0, & 1, & 0 \end{bmatrix}. \tag{16.28}$$

It can be deduced from eqns. (16.22) and (16.28) that the mode-controllability matrix of the modified system is given by the equation

$$\mathbf{P}(T)=\mathbf{V}'\mathbf{\Delta}(T)=\begin{bmatrix} \alpha, & \alpha, & -2\alpha, & 0, & 0, & 0, & 0 \\ 0, & 0, & 1, & 0, & 0, & 0, & 0 \\ 0, & 0, & 0, & 0, & 0, & 0, & 0 \\ 0, & 0, & 0, & 0, & 0, & 0, & 0 \\ 0, & 0, & 0, & 0, & 0, & 0, & 0 \\ 0, & 0, & 0, & 1, & 0, & 0, & 0 \\ 0, & 0, & \alpha, & \alpha, & \alpha, & \alpha, & -4\alpha \\ 0, & 0, & 0, & 0, & 0, & 0, & 1 \\ 0, & 0, & 0, & 0, & 0, & 0, & -\alpha \\ 0, & 0, & 0, & 0, & 0, & 0, & \alpha \end{bmatrix}. \tag{16.29}$$

Since all the rows of $\mathbf{P}(T)$ which correspond to the last rows of the Jordan blocks in the eqn. (16.26) which contain the same-valued eigenvalue are linearly independent, it follows that the modified system is controllable.

16.3. Synthesis of control policy for modified manufacturing system

Since the model of the modified manufacturing system described in § 16.2.2 is controllable, the eigenvalues which determine the speed of response of the manufacturing system can be assigned any desired values by using the results presented in Chapter 6 to determine an appropriate control policy, $\mathbf{z}(kT)$. Thus it is evident from the structure of the mode-controllability matrix (16.29) that, in order to control the inventory level $x_9(kT)$, it is necessary to use the control input $z_7(kT)$. Furthermore, if it is also desired to maintain balanced production lines in accordance with eqn. (16.7), then the control policy, $\mathbf{z}(kT)$, must have the form

$$\mathbf{z}(kT)=\begin{bmatrix} 1 \\ 1 \\ 1 \\ 1 \\ 1 \\ 1 \\ 1 \end{bmatrix} z_7(kT)=\mathbf{f}z_7(kT). \tag{16.30}$$

Eqns. (16.19), (16.21), (16.22), (16.23), (16.24), (16.25), and (16.30) then indicate that the state equation governing the modified manufacturing system has the form

$$\mathbf{x}\{(k+1)T\} = \mathbf{\Psi}(T)\mathbf{x}(kT) + \boldsymbol{\delta}(T)z_7(kT) + \boldsymbol{\nu}(T)s(kT), \tag{16.31}$$

where

$$\boldsymbol{\delta}(T) = \boldsymbol{\Delta}(T)\mathbf{f}. \tag{16.32}$$

The mode-controllability matrix, $\mathbf{p}(T)$, of the system (16.31) is given by the equation

$$\mathbf{p}(T) = \mathbf{V}'\boldsymbol{\delta}(T) = \mathbf{P}(T)\mathbf{f} = \begin{bmatrix} 0 \\ 1 \\ 0 \\ 0 \\ 0 \\ 1 \\ 0 \\ 1 \\ -\alpha \\ \alpha \end{bmatrix}, \tag{16.33}$$

where $\mathbf{V}$ and $\mathbf{P}(T)$ are given by eqns. (16.28) and (16.29), respectively. Since the element $p_{10}(T)$ associated with the last row of the Jordan block, $\mathbf{J}_2(1)$, in eqn. (16.26) is non-zero, and since this block is associated with the state variable $x_{10}(kT)$, it follows that the design procedure presented in § 5.3 is applicable.

Thus, eqns. (5.54), (5.87), (5.88), (5.90), (5.91), and (5.93) indicate that the effect of the control law

$$z_7(kT) = K_9\mathbf{v}_9'\mathbf{x}(kT) + K_{10}\mathbf{v}_{10}'\mathbf{x}(kT) \tag{16.34}$$

will be to change the eigenvalues $\lambda_9(=1)$ and $\lambda_{10}(=1)$ associated with the Jordan block, $\mathbf{J}_2(1)$, to desired new values, ρ_9 and ρ_{10}, if the proportional-controller gains K_9 and K_{10} are given by the equation

$$\begin{bmatrix} K_9 \\ K_{10} \end{bmatrix} = \begin{bmatrix} 1/(\rho_9-\lambda_9), & 1/(\rho_9-\lambda_9)^2 \\ 1/(\rho_{10}-\lambda_{10}), & 1/(\rho_{10}-\lambda_{10})^2 \end{bmatrix}^{-1} \begin{bmatrix} 1/p_9(T), & 1/p_{10}(T) \\ 1/p_{10}(T), & 0 \end{bmatrix}^{-1} \begin{bmatrix} 1 \\ 1 \end{bmatrix}. \tag{16.35}$$

In eqns. (16.34) and (16.35), $\mathbf{v}_9$ and $\mathbf{v}_{10}$ are the ninth and tenth columns of the matrix $\mathbf{V}$ given in eqn. (16.28), and $p_9(T)$ and $p_{10}(T)$ are the ninth and tenth elements of the vector $\mathbf{p}(T)$ defined by eqn. (16.33) : equation (16.35) has the solutions

$$K_9 = -(\rho_9-1)(\rho_{10}-1)/\alpha \tag{16.36 a}$$

and

$$K_{10} = \{(\rho_9-1)+(\rho_{10}-1)-(\rho_9-1)(\rho_{10}-1)\}/\alpha. \tag{16.36 b}$$

In view of eqns. (16.36), the control-law eqn. (16.34) therefore has the explicit form

$$z_7(kT)=\{(\rho_9-1)+(\rho_{10}-1)\}x_8(kT)$$
$$+[\{(\rho_9-1)+(\rho_{10}-1)-(\rho_9-1)(\rho_{10}-1)\}/\alpha]x_9(kT)$$
$$-\{(\rho_9-1)(\rho_{10}-1)/\alpha\}x_{10}(kT). \qquad (16.37)$$

In order to investigate the steady-state behaviour of the modified system governed by eqn. (16.31) when controlled in accordance with eqn. (16.37), it is convenient to assume that the system is subjected to a constant sales-rate perturbation

$$s(kT)=s \quad (k=0,\ 1,\ 2,\ \ldots). \qquad (16.38)$$

If this is the case, eqns. (16.31), (16.37), and (16.38) imply that the resulting steady-state perturbations are given by the vectors

$$\hat{\mathbf{x}}'=[0,\ s/\alpha,\ s/\alpha,\ s/\alpha,\ s/\alpha,\ s/\alpha,\ -2s/\alpha,\ s/\alpha,\ 0,$$
$$-(3-\rho_9-\rho_{10})s/(\rho_9-1)(\rho_{10}-1)] \qquad (16.39\ a)$$

and

$$\hat{\mathbf{z}}'=[s/\alpha,\ s/\alpha,\ s/\alpha,\ s/\alpha,\ s/\alpha,\ s/\alpha,\ s/\alpha]. \qquad (16.39\ b)$$

The significant feature of the behaviour of the modified system, as indicated by eqns. (16.39), is that the inventory-level perturbation, $x_9(kT)$, and the backlog-level perturbation, $x_1(kT)$, both tend to zero in the steady state: it will be noted that the backlog-level perturbation, $x_7(kT)$, tends to the non-zero value, $-2s/\alpha$, in the steady state.

16.4. Numerical example

In this section, the behaviour of the manufacturing system discussed in this chapter is illustrated for the particular case when

$$T=16 \text{ hours}, \qquad (16.40\ a)$$

$$H=40 \text{ hours}, \qquad (16.40\ b)$$

$$\alpha=(T/H)=0{\cdot}4, \qquad (16.40\ c)$$

$$\beta_1=1/6{\cdot}63, \qquad (16.40\ d)$$

$$\beta_2=1/4{\cdot}73, \qquad (16.40\ e)$$

$$\beta_3=1/8{\cdot}92, \qquad (16.40\ f)$$

$$\beta_4=1/61{\cdot}9, \qquad (16.40\ g)$$

$$\beta_5=1/3{\cdot}99, \qquad (16.40\ h)$$

$$\beta_6=1/1{\cdot}54, \qquad (16.40\ i)$$

$$\beta_7=1/24{\cdot}21, \qquad (16.40\ j)$$

and

$$\sigma=40. \qquad (16.40\ k)$$

The numerical data given in eqns. (16.40) corresponds to that used by Christensen and Brogan [1] in their study of the system using optimal control theory.

Equations (16.18) and (16.40 k) indicate that the steady-state of the manufacturing system as modelled in § 16.2.1 is given by the equations

$$\left.\begin{aligned} \hat{\boldsymbol{\chi}}' &= [\hat{\chi}_1, 100, 100, 100, 100, 100, \hat{\chi}_7, 100, \hat{\chi}_9], \\ \hat{\boldsymbol{\zeta}}' &= [100, 100, 100, 100, 100, 100, 100]. \end{aligned}\right\} \quad (16.41)$$

The vectors $\hat{\boldsymbol{\chi}}$ and $\hat{\boldsymbol{\zeta}}$ will be regarded as initial conditions in this numerical example : furthermore, the values of $\hat{\chi}_1$, $\hat{\chi}_7$, and $\hat{\chi}_9$ will be arbitrarily chosen in accordance with the equations

$$\left.\begin{aligned} \hat{\chi}_1 &= 150, \\ \hat{\chi}_7 &= 200, \\ \hat{\chi}_9 &= 400. \end{aligned}\right\} \quad (16.42)$$

It is clear that if, for example, the unmodified system is subjected to a step increase in the sales rate, the backlog levels, $\chi_1(kT)$ and $\chi_7(kT)$, and the inventory level, $\chi_9(kT)$, will all be steadily depleted. Deficiencies of behaviour of this type can largely be overcome by introducing local integral feedback in the manner of § 16.2.2 and by implementing control policies of the type synthesized in § 16.3. Thus, for example, eqn. (16.37) assumes the forms

$$z_7(kT) = -1{\cdot}7x_8(kT) - 6{\cdot}05x_9(kT) - 1{\cdot}8x_{10}(kT) \quad (16.43\,a)$$

if $\rho_9 = 0{\cdot}1$ and $\rho_{10} = 0{\cdot}2$,

$$z_7(kT) = -x_8(kT) - 3{\cdot}11875x_9(kT) - 0{\cdot}61875x_{10}(kT) \quad (16.43\,b)$$

if $\rho_9 = 0{\cdot}45$ and $\rho_{10} = 0{\cdot}55$, and

$$z_7(kT) = -0{\cdot}3x_8(kT) - 0{\cdot}8x_9(kT) - 0{\cdot}05x_{10}(kT) \quad (16.43\,c)$$

if $\rho_9 = 0{\cdot}8$ and $\rho_{10} = 0{\cdot}9$.

Figures 16-2, 16-3, and 16-4 show the respective transient responses of the modified manufacturing system when controlled in accordance with the control policy implied by eqns. (16.43 a), (16.43 b), and (16.43 c), and also indicate the manner in which the respective steady states

$$\left.\begin{aligned} \hat{\mathbf{x}}' &= [0, 25, 25, 25, 25, 25, -50, 25, 0, -37{\cdot}5], \\ \hat{\mathbf{x}}' &= [0, 25, 25, 25, 25, 25, -50, 25, 0, -80{\cdot}81], \\ \hat{\mathbf{x}}' &= [0, 25, 25, 25, 25, 25, -50, 25, 0, -650], \end{aligned}\right\} \quad (16.44)$$

are reached when $s = 10$ where, in each case,

$$\hat{\mathbf{z}}' = [25, 25, 25, 25, 25, 25, 25]. \quad (16.45)$$

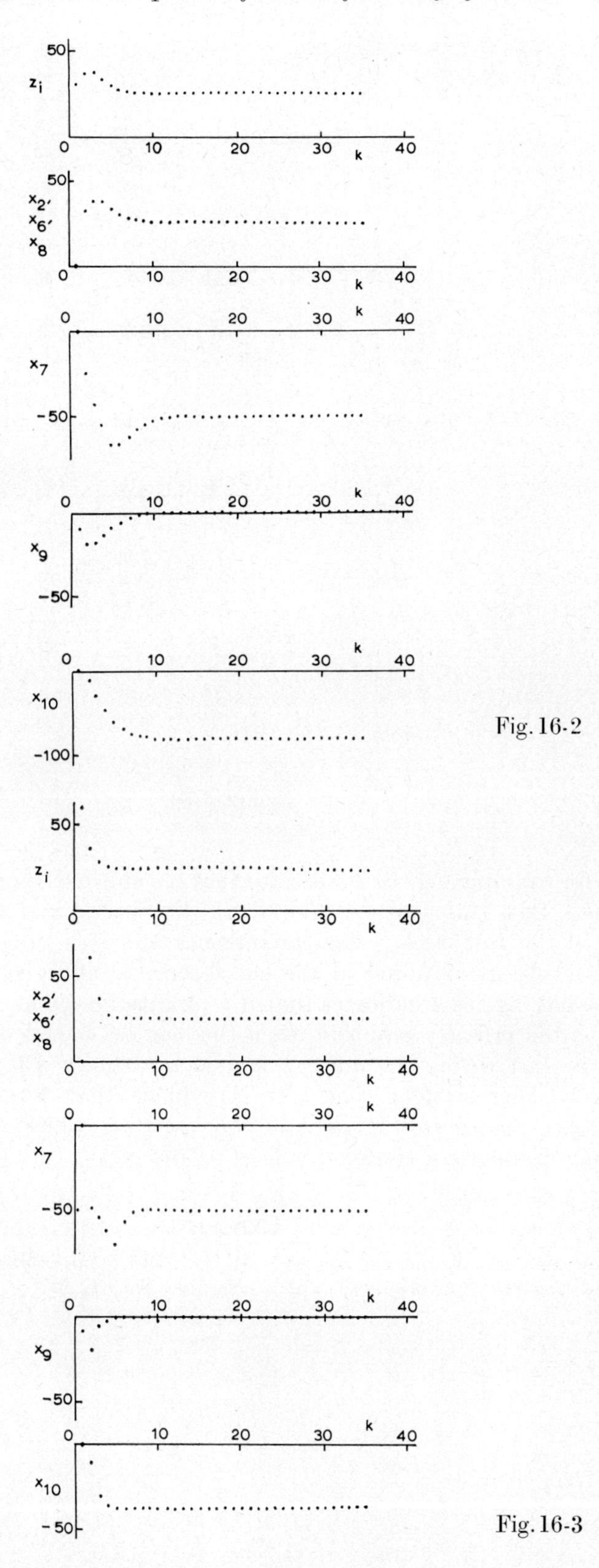

Fig. 16-2

Fig. 16-3

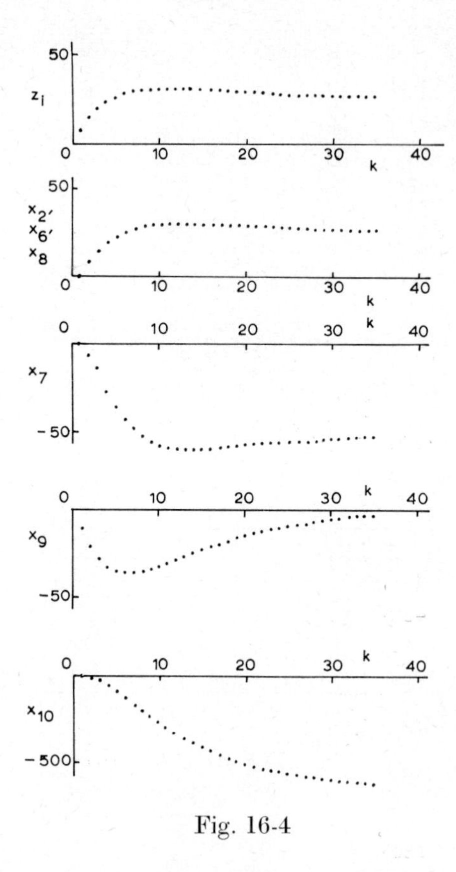

Fig. 16-4

Thus, for example, fig. 16-2 indicates that if a short settling time is the primary requisite, then this can be achieved by choosing ρ_9 and ρ_{10} to lie close to the origin of the unit circle : one consequence of such a choice is that high man-power levels are required in the short term to effect such control. On the other hand, fig. 16-4 indicates that if a progressive build-up in the man-power levels is the primary requisite, then this can be achieved by choosing ρ_9 and ρ_{10} to lie just within the unit circle : such a choice will, of course, lead to a relatively long settling time. It is evident that trade-offs between high short-term man-power levels and long settling times can be achieved by selecting appropriate combinations of ρ_9 and ρ_{10}.

Reference

[1] Christensen, J. L., and Brogan, W. L., 1971, " Modelling and optimal control of a production process ", *Int. J. Systems Sci.*, **1,** 247.

PROBLEMS

Chapter 1

1.1. Determine the free-response characteristics of the systems governed by the following state equations :

(a) $\dot{x}(t) = -3x(t), \quad x(0) = 5$;

(b) $\dot{x}(t) = 2x(t), \quad x(0) = 0{\cdot}4$;

and

(c) $\dot{x}(t) = 0, \quad x(0) = 1$.

1.2. Determine the values of the gain, g, in the control law

$$z(t) = gx(t)$$

such that the systems governed by the following state equations all have a closed-loop eigenvalue equal to -5 :

(a) $\dot{x}(t) = -3x(t) + 2z(t)$,

(b) $\dot{x}(t) = 2x(t) + z(t)$,

and

(c) $\dot{x}(t) = -3z(t)$.

1.3. Determine the free-response characteristics of the closed-loop systems described in Problem 1.2 when the initial conditions are respectively

(a) $x(0) = 5$,

(b) $x(0) = 0{\cdot}4$,

and

(c) $x(0) = 1$.

Compare these response characteristics with the corresponding response characteristics determined in Problem 1.1.

Chapter 2

2.1. Determine the eigenvalues of each of the following plant matrices :

(a) $\mathbf{A} = \begin{bmatrix} 10, & -36 \\ 3, & -11 \end{bmatrix}$,

(b) $\mathbf{A} = \begin{bmatrix} 2, & 1, & -5 \\ 0, & -0{\cdot}5, & 0{\cdot}5 \\ 0, & 0{\cdot}5, & -0{\cdot}5 \end{bmatrix}$,

(c) $\mathbf{A} = \begin{bmatrix} 0, & 1, & 0 \\ 0, & 0, & 1 \\ -6, & -11, & -6 \end{bmatrix}$,

and

(d) $\mathbf{A} = \begin{bmatrix} 0, & -5 \\ 1, & -2 \end{bmatrix}$.

2.2. Determine a modal matrix, $\mathbf{U}$, associated with each of the plant matrices, $\mathbf{A}$, given in Problem 2.1.

2.3. Determine a modal matrix, $\mathbf{V}$, associated with the transpose, $\mathbf{A}'$, of each of the plant matrices, $\mathbf{A}$, given in Problem 2.1.

2.4. Normalize the corresponding pairs of modal matrices, $\mathbf{U}$ and $\mathbf{V}$, determined in Problems 2.2 and 2.3 such that $\bar{\mathbf{V}}'\bar{\mathbf{U}}=\mathbf{I}$ in each case, where $\bar{\mathbf{U}}$ and $\bar{\mathbf{V}}$ are the required normalized modal matrices.

2.5. Verify, using the results of Problem 2.4, that

$$\bar{\mathbf{V}}'\mathbf{A}\bar{\mathbf{U}}=\mathbf{\Lambda}$$

and that

$$\bar{\mathbf{U}}\mathbf{\Lambda}\bar{\mathbf{V}}'=\mathbf{A}$$

in the case of each of the plant matrices given in Problem 2.1.

2.6. Determine the transition matrix and the free-response characteristics of the systems governed by the following state equations :

(a) $$\dot{\mathbf{x}}(t)=\begin{bmatrix}10, & -36\\ 3, & -11\end{bmatrix}\mathbf{x}(t)+\begin{bmatrix}1\\1\end{bmatrix}z(t),\quad \mathbf{x}(0)=\begin{bmatrix}0\\1\end{bmatrix};$$

(b) $$\dot{\mathbf{x}}(t)=\begin{bmatrix}2, & 1, & -5\\ 0, & -0{\cdot}5, & 0{\cdot}5\\ 0, & 0{\cdot}5, & -0{\cdot}5\end{bmatrix}\mathbf{x}(t)+\begin{bmatrix}0\\0\\1\end{bmatrix}z(t),\quad \mathbf{x}(0)=\begin{bmatrix}1\\0\\1\end{bmatrix};$$

(c) $$\dot{\mathbf{x}}(t)=\begin{bmatrix}0, & 1, & 0\\ 0, & 0, & 1\\ -6, & -11, & -6\end{bmatrix}\mathbf{x}(t)+\begin{bmatrix}0\\0\\-1\end{bmatrix}z(t),\quad \mathbf{x}(0)=\begin{bmatrix}5\\0\\0\end{bmatrix};$$

and

(d) $$\dot{\mathbf{x}}(t)=\begin{bmatrix}0, & -5\\ 1, & -2\end{bmatrix}\mathbf{x}(t)+\begin{bmatrix}2\\1\end{bmatrix}z(t),\quad \mathbf{x}(0)=\begin{bmatrix}1\\1\end{bmatrix}.$$

In addition, determine the forced-response characteristics of each of these systems when

(a) $z(t)=k \quad (t\geqslant 0)$

and

(b) $z(t)=kt \quad (t\geqslant 0)$.

2.7. If in the case of the continuous-time systems described in Problem 2.6 the input variable, $z(t)$, is a piecewise-constant function defined by the equation

$$z(t)=z(kT)\quad (kT\leqslant t<(k+1)T\ ;\ k=0,1,2,\ldots),$$

determine the difference state equations of each of the corresponding discrete-time systems in terms of the sampling period, T.

2.8. Determine the free-response characteristics of the systems governed by the following state equations :

$$\text{(a)}\ \mathbf{x}\{(k+1)T\}=\begin{bmatrix}0, & 1\\ 0\cdot 24, & -0\cdot 20\end{bmatrix}\mathbf{x}(kT)+\begin{bmatrix}0\\ 1\end{bmatrix}z(kT),\quad \mathbf{x}(0)=\begin{bmatrix}1\\ 12\end{bmatrix};$$

and

$$\text{(b)}\ \mathbf{x}\{(k+1)T\}=\begin{bmatrix}0, & -0\cdot 75\\ 1, & 2\end{bmatrix}\mathbf{x}(kT)+\begin{bmatrix}1\\ 1\end{bmatrix}z(kT),\quad \mathbf{x}(0)=\begin{bmatrix}10\\ 0\end{bmatrix}.$$

In each case determine $\mathbf{x}(1)$, $\mathbf{x}(2)$, . . ., $\mathbf{x}(5)$.

In addition, determine the forced-response characteristics of each of these systems when subjected to the following two input sequences :

$$\text{(a)}\ z(0)=1,\quad z(1)=1,\quad z(2)=1,\quad z(3)=1,\quad z(4)=1\ ;$$

and

$$\text{(b)}\ z(0)=-2,\quad z(1)=2,\quad z(2)=-2,\quad z(3)=2,\quad z(4)=-2.$$

2.9. Determine the Jordan canonical form of each of the following plant matrices :

$$\text{(a)}\ \mathbf{A}=\begin{bmatrix}-6, & 25\\ -1, & 4\end{bmatrix},$$

$$\text{(b)}\ \mathbf{A}=\begin{bmatrix}-1, & 1, & -3\\ 0, & -1, & 0\\ 0, & 0, & -1\end{bmatrix},$$

$$\text{(c)}\ \mathbf{A}=\begin{bmatrix}-4, & 8\\ -2, & 4\end{bmatrix},$$

and

$$\text{(d)}\ \mathbf{A}=\begin{bmatrix}-\frac{1}{3}, & \frac{1}{2}, & -\frac{1}{6}\\ \frac{7}{3}, & 0, & -\frac{4}{3}\\ \frac{11}{3}, & 2, & -\frac{8}{3}\end{bmatrix}.$$

2.10. Determine a generalized modal matrix, $\mathbf{U}$, associated with each of the plant matrices, $\mathbf{A}$, given in Problem 2.9.

2.11. Determine a generalized modal matrix, $\mathbf{V}$, associated with the transpose, $\mathbf{A}'$, of each of the plant matrices, $\mathbf{A}$, given in Problem 2.9.

2.12. Determine the transition matrix and the free-response characteristics of the systems governed by the following state equations :

$$\text{(a)}\ \dot{\mathbf{x}}(t)=\begin{bmatrix}-6, & 25\\ -1, & 4\end{bmatrix}\mathbf{x}(t)+\begin{bmatrix}0\\ 1\end{bmatrix}z(t),\quad \mathbf{x}(0)=\begin{bmatrix}1\\ 0\end{bmatrix};$$

(b) $\dot{\mathbf{x}}(t)=\begin{bmatrix}-1, & 1, & 3\\ 0, & -1, & 0\\ 0, & 0, & -1\end{bmatrix}\mathbf{x}(t)+\begin{bmatrix}1\\0\\1\end{bmatrix}z(t), \quad \mathbf{x}(0)=\begin{bmatrix}0\\1\\0\end{bmatrix}$;

(c) $\dot{\mathbf{x}}(t)=\begin{bmatrix}-4, & 8\\ -2, & 4\end{bmatrix}\mathbf{x}(t)+\begin{bmatrix}1\\1\end{bmatrix}z(t), \quad \mathbf{x}(0)=\begin{bmatrix}0\\1\end{bmatrix}$;

and

(d) $\dot{\mathbf{x}}(t)=\begin{bmatrix}-\frac{1}{3}, & \frac{1}{2} & -\frac{1}{6}\\ \frac{7}{3}, & 0, & -\frac{4}{3}\\ \frac{11}{3}, & 2, & -\frac{8}{3}\end{bmatrix}\mathbf{x}(t)+\begin{bmatrix}-1\\1\\2\end{bmatrix}z(t), \quad \mathbf{x}(0)=\begin{bmatrix}0\\1\\-1\end{bmatrix}$.

In addition, determine the forced-response characteristics of each of these systems when $z(t)=k$ $(t\geqslant 0)$.

Chapter 3

3.1. In the case of each of the plant matrices given in Problem 2.1, calculate the eigenvalue sensitivity matrices: in addition, calculate $\partial^2\lambda_2/\partial a_{12}{}^2$, $\partial\mathbf{u}_1/\partial a_{12}$, $\partial\mathbf{v}_1'/\partial a_{12}$, $\partial^2\mathbf{u}_1/\partial a_{12}{}^2$, and $\partial^2\mathbf{v}_1'/\partial a_{12}{}^2$.

3.2. If the element, a_{12}, of each of the plant matrices considered in Problem 3.1 is perturbed by an amount $\delta a_{12}=0{\cdot}5$, determine by direct calculation the eigenvalue corresponding to λ_2 and the eigenvectors corresponding to $\mathbf{u}_1$ and $\mathbf{v}_1$. Compare these directly-calculated values with the corresponding first- and second-order estimates.

Chapter 4

4.1. Discuss the mode-controllability and mode-observability characteristics of the systems characterized by the following sets of matrices where, in each case, $\mathbf{J}$ is the Jordan canonical form of the plant matrix, $\mathbf{P}$ is the mode-controllability matrix, and $\mathbf{R}$ is the mode-observability matrix:

(a) $\mathbf{J}=\begin{bmatrix}1, & 0\\ 0, & -2\end{bmatrix}, \quad \mathbf{P}=\begin{bmatrix}3\\4\end{bmatrix}, \quad \mathbf{R}=\begin{bmatrix}1\\0\end{bmatrix}$;

(b) $\mathbf{J}=\begin{bmatrix}1, & 0\\ 0, & -2\end{bmatrix}, \quad \mathbf{P}=\begin{bmatrix}-2, & 0\\ 0, & 1\end{bmatrix}, \quad \mathbf{R}=\begin{bmatrix}2, & 1\\ 1, & -3\end{bmatrix}$;

(c) $\mathbf{J}=\begin{bmatrix}2, & 0, & 0\\ 0, & 0, & 0\\ 0, & 0, & -1\end{bmatrix}, \quad \mathbf{P}=\begin{bmatrix}1, & 0\\ -4, & 2\\ 3, & -5\end{bmatrix}, \quad \mathbf{R}=\begin{bmatrix}-4\\1\\0\end{bmatrix}$;

(d) $\mathbf{J}=\begin{bmatrix}-1, & 1, & 0\\ 0, & -1, & 1\\ 0, & 0, & -1\end{bmatrix}, \quad \mathbf{P}=\begin{bmatrix}0\\0\\2\end{bmatrix}, \quad \mathbf{R}=\begin{bmatrix}1\\0\\0\end{bmatrix}$;

(e) $\mathbf{J}=\begin{bmatrix}0, & 1\\ 0, & 0\end{bmatrix}$, $\mathbf{P}=\begin{bmatrix}1, & 2\\ 0, & 1\end{bmatrix}$, $\mathbf{R}=\begin{bmatrix}4, & 3\\ 1, & 0\end{bmatrix}$;

(f) $\mathbf{J}=\begin{bmatrix}1, & 1, & 0, & 0\\ 0, & 1, & 0, & 0\\ 0, & 0, & 0, & 1\\ 0, & 0, & 0, & 0\end{bmatrix}$, $\mathbf{P}=\begin{bmatrix}2, & 1, & 0\\ -1, & 1, & 0\\ 1, & -3, & 1\\ 0, & 1, & 2\end{bmatrix}$, $\mathbf{R}=\begin{bmatrix}5, & 0\\ 1, & -2\\ 1, & 0\\ 0, & 3\end{bmatrix}$;

(g) $\mathbf{J}=\begin{bmatrix}-1, & 1, & 0\\ 0, & -1, & 0\\ 0, & 0, & -1\end{bmatrix}$, $\mathbf{P}=\begin{bmatrix}2, & 0\\ 1, & -1\\ -2, & 3\end{bmatrix}$, $\mathbf{R}=\begin{bmatrix}2, & 1\\ 0, & 3\\ 4, & 2\end{bmatrix}$;

and

(h) $\mathbf{J}=\begin{bmatrix}-1, & 1, & 0\\ 0, & -1, & 0\\ 0, & 0, & -1\end{bmatrix}$, $\mathbf{P}=\begin{bmatrix}4, & 3\\ 1, & 1\\ -2, & -2\end{bmatrix}$, $\mathbf{R}=\begin{bmatrix}1, & 2\\ 2, & 5\\ -1, & 3\end{bmatrix}$.

4.2. Determine the mode-controllability and mode-observability structure of the systems governed by the following state and output equations:

(a) $\dot{\mathbf{x}}(t)=\begin{bmatrix}10, & -36\\ 3, & -11\end{bmatrix}\mathbf{x}(t)+\begin{bmatrix}1, & 3\\ 1, & 1\end{bmatrix}\mathbf{z}(t)$,

$\mathbf{y}(t)=\begin{bmatrix}0, & 1\\ -1, & 4\end{bmatrix}\mathbf{x}(t)$;

(b) $\dot{\mathbf{x}}(t)=\begin{bmatrix}2, & 1, & -5\\ 0, & -0{\cdot}5, & 0{\cdot}5\\ 0, & 0{\cdot}5, & -0{\cdot}5\end{bmatrix}\mathbf{x}(t)+\begin{bmatrix}0, & 0\\ 0, & 1\\ 1, & 1\end{bmatrix}\mathbf{z}(t)$,

$\mathbf{y}(t)=\begin{bmatrix}1, & 0, & -1\\ 0, & 0, & 1\end{bmatrix}\mathbf{x}(t)$;

(c) $\dot{\mathbf{x}}(t)=\begin{bmatrix}0, & 1, & 0\\ 0, & 0, & 1\\ -6, & -11, & -6\end{bmatrix}\mathbf{x}(t)+\begin{bmatrix}0\\ 0\\ 1\end{bmatrix}z(t)$,

$y(t)=[1, \ 0, \ 0]\mathbf{x}(t)$;

(d) $\dot{\mathbf{x}}(t)=\begin{bmatrix}0, & -5\\ 1, & -2\end{bmatrix}\mathbf{x}(t)+\begin{bmatrix}2\\ 1\end{bmatrix}z(t)$,

$y(t)=[1, \ 1]\mathbf{x}(t)$;

(e) $\dot{\mathbf{x}}(t)=\begin{bmatrix}-6, & 25\\ -1, & 4\end{bmatrix}\mathbf{x}(t)+\begin{bmatrix}0, & 5\\ 1, & 1\end{bmatrix}\mathbf{z}(t)$,

$\mathbf{y}(t)=\begin{bmatrix}1, & -4\\ -1, & 5\end{bmatrix}\mathbf{x}(t)$;

(f) $$\dot{\mathbf{x}}(t) = \begin{bmatrix} -4, & 8 \\ -2, & 4 \end{bmatrix} \mathbf{x}(t) + \begin{bmatrix} 1, & 4 \\ 1, & 2 \end{bmatrix} \mathbf{z}(t),$$

$$y(t) = [-2, \ 3]\mathbf{x}(t);$$

and

(g) $$\dot{\mathbf{x}}(t) = \begin{bmatrix} 1, & 1\cdot5, & -0\cdot5 \\ 0, & 0, & 1 \\ 0, & -2, & 3 \end{bmatrix} \mathbf{x}(t) + \begin{bmatrix} 3, & 2 \\ 3, & 2 \\ 5, & 4 \end{bmatrix} \mathbf{z}(t),$$

$$\mathbf{y}(t) = \begin{bmatrix} 2, & -4, & 2 \\ 0, & 2, & -1 \end{bmatrix} \mathbf{x}(t).$$

Chapter 5

5.1. Synthesize control laws for the second-order system governed by the state equation

$$\dot{\mathbf{x}}(t) = \begin{bmatrix} 1, & -3 \\ -1, & 4 \end{bmatrix} \mathbf{x}(t) + \begin{bmatrix} 0 \\ 1 \end{bmatrix} z(t)$$

such that the eigenvalues of the plant matrix of the resulting closed-loop systems are as follows :

(a) $\rho_1 = -1, \quad \rho_2 = -4$;

and

(b) $\rho_1 = -1 + 2i, \quad \rho_2 = -1 - 2i$.

5.2. Synthesize piecewise-constant control laws for the second-order system governed by the state equation

$$\mathbf{x}\{(k+1)T\} = \begin{bmatrix} 0, & -5 \\ 1, & -2 \end{bmatrix} \mathbf{x}(kT) + \begin{bmatrix} 0 \\ 1 \end{bmatrix} z(kT)$$

such that the eigenvalues of the plant matrix of the resulting closed-loop systems are as follows :

(a) $\rho_1 = 0\cdot5, \quad \rho_2 = -0\cdot5$;

and

(b) $\rho_1 = 0\cdot5 + 0\cdot5i, \quad \rho_2 = 0\cdot5 - 0\cdot5i$.

5.3. Synthesize control laws for the third-order system governed by the state equation

$$\dot{\mathbf{x}}(t) = \begin{bmatrix} -1, & 0, & 0 \\ 0, & -3, & 4 \\ 1, & -3, & 4 \end{bmatrix} \mathbf{x}(t) + \begin{bmatrix} 0 \\ 2 \\ 1 \end{bmatrix} z(t)$$

such that the eigenvalues of the plant matrix of the resulting closed-loop systems are as follows :

(a) $\rho_1 = -1, \quad \rho_2 = -2, \quad \rho_3 = -3$;

and

(b) $\rho_1 = -1, \quad \rho_2 = -1+i, \quad \rho_3 = -1-i.$

5.4. The state equation of a second-order system is

$$\dot{\mathbf{x}}(t) = \begin{bmatrix} -4, & 8 \\ -2, & 4 \end{bmatrix} \mathbf{x}(t) + \begin{bmatrix} 10 \\ 6 \end{bmatrix} z(t).$$

Verify that the Jordan canonical form of the plant matrix of this system is

$$\mathbf{J} = \mathbf{J}_2(0) = \begin{bmatrix} 0, & 1 \\ 0, & 0 \end{bmatrix}$$

and that the system is controllable. Synthesize control laws for the system such that the eigenvalues of the plant matrix of the resulting closed-loop systems are as follows :

(a) $\rho_1 = -1, \quad \rho_2 = -2$;

and

(b) $\rho_1 = -1+2i, \quad \rho_2 = -1-2i.$

5.5. The state equation of a third-order system is

$$\dot{\mathbf{x}}(t) = \begin{bmatrix} 1, & 1{\cdot}5, & -0{\cdot}5 \\ 0, & 0, & 1 \\ 0, & -2, & 3 \end{bmatrix} \mathbf{x}(t) + \begin{bmatrix} 3 \\ 3 \\ 5 \end{bmatrix} z(t).$$

Verify that the Jordan canonical form of the plant matrix of this system is

$$\mathbf{J} = \mathbf{J}_2(1) \oplus \mathbf{J}_1(2) = \begin{bmatrix} 1, & 1, & 0 \\ 0, & 1, & 0 \\ 0, & 0, & 2 \end{bmatrix}$$

and that the system is controllable. Synthesize control laws for the system such that the eigenvalues of the plant matrix of the resulting closed-loop systems are as follows :

(a) $\rho_1 = -1, \quad \rho_2 = -2, \quad \rho_3 = -3$;

and

(b) $\rho_1 = -2, \quad \rho_2 = -1+2i, \quad \rho_3 = -1-2i.$

Chapter 6

6.1. Synthesize control laws for the second-order system governed by the state equation

$$\dot{\mathbf{x}}(t) = \begin{bmatrix} 10, & -36 \\ 3, & -11 \end{bmatrix} \mathbf{x}(t) + \begin{bmatrix} -2, & 5 \\ 3, & -6 \end{bmatrix} \mathbf{z}(t)$$

such that the eigenvalues of the plant matrix of the resulting closed-loop system are $\rho_1 = -2$ and $\rho_2 = -1$ and that, in addition, the sum of the squares of the feedback gains is minimized.

6.2. Synthesize control laws for the second-order system governed by the state equation

$$\dot{\mathbf{x}}(t) = \begin{bmatrix} 11, & -12 \\ 10, & -11 \end{bmatrix} \mathbf{x}(t) + \begin{bmatrix} -1, & -5 \\ 7, & 29 \end{bmatrix} \mathbf{z}(t)$$

such that the eigenvalues of the plant matrix of the resulting closed-loop system are $\rho_1 = -1$ and $\rho_2 = -2$ and that, in addition, the sum of the moduli of the feedback gains is minimized.

6.3. Compare the multi-input control laws determined in Problems 6.1 and 6.2 with the corresponding control laws obtained when

(a) only $z_1(t)$ is used for control

and

(b) only $z_2(t)$ is used for control.

6.4. Synthesize control laws for the second-order system governed by the state equation

$$\dot{\mathbf{x}}(t) = \begin{bmatrix} -9, & 15 \\ -6, & 10 \end{bmatrix} \mathbf{x}(t) + \begin{bmatrix} 2, & -8 \\ -1, & 7 \end{bmatrix} \mathbf{z}(t)$$

such that the eigenvalues of the plant matrix of the resulting closed-loop system are $\rho_1 = -2$ and $\rho_2 = -3$ if the control laws are prescribed to have either of the respective forms

(a) $z_1 = g_{11}x_1, \quad z_2 = g_{21}x_1$;

or

(b) $z_1 = g_{12}x_1, \quad z_2 = g_{22}x_2$.

6.5. Synthesize, using a multi-stage design procedure, control laws for the fourth-order system governed by the state equation

$$\dot{\mathbf{x}}(t) = \begin{bmatrix} \frac{1}{2}, & \frac{1}{2}, & -1, & 1 \\ \frac{1}{2}, & \frac{1}{2}, & -1, & 1 \\ 0, & 0, & -5, & 12 \\ 0, & 0, & -1, & 2 \end{bmatrix} \mathbf{x}(t) + \begin{bmatrix} 2, & 2, & 5 \\ -10, & 0, & 7 \\ 3, & 5, & 15 \\ 1, & 1, & 5 \end{bmatrix} \mathbf{z}(t)$$

such that the eigenvalues of the plant matrix of the resulting closed-loop system are $-\frac{1}{2}$, $-\frac{3}{2}$, $-\frac{5}{2}$, and $-\frac{7}{2}$.

6.6. Design a dyadic controller for the fourth-order system described in Problem 6.5 such that the eigenvalues of the plant matrix of the resulting closed-loop system are $-\frac{1}{2}$, $-\frac{3}{2}$, $-\frac{5}{2}$, and $-\frac{7}{2}$.

Chapter 7

7.1. Design a first-order state observer for the system governed by the state and output equations

$$\dot{\mathbf{x}}(t)=\begin{bmatrix}-1, & 0, & 0\\ 0, & -3, & 4\\ 1, & -3, & 4\end{bmatrix}\mathbf{x}(t)+\begin{bmatrix}1\\ -1\\ 3\end{bmatrix}z(t)$$

and

$$\mathbf{y}(t)=\begin{bmatrix}1, & 0, & 0\\ 0, & 0, & 1\end{bmatrix}\mathbf{x}(t)$$

such that the eigenvalue associated with the observer is -5 and that the remaining eigenvalues of the plant matrix of the closed-loop system incorporating the observer are -2, -3, and -4.

7.2. Design a second-order state observer for the system governed by the state and output equations

$$\dot{\mathbf{x}}(t)=\begin{bmatrix}-1, & 0, & 0\\ 0, & -3, & 4\\ 1, & -3, & 4\end{bmatrix}\mathbf{x}(t)+\begin{bmatrix}1\\ -1\\ 3\end{bmatrix}z(t)$$

and

$$y(t)=[1, \ 0, \ 0]\mathbf{x}(t)$$

such that the eigenvalues associated with the observer are -5 and -10, and that the remaining eigenvalues of the plant matrix of the closed-loop system incorporating the observer are -2, $-2+2i$, and $-2-2i$.

7.3. If in the case of the system described in Problem 7.1, the input variable, $z(t)$, and the output vector, $\mathbf{y}(t)$, are piecewise-constant functions defined by the equations

$$\left.\begin{aligned} z(t)&=z(kT)\\ &\text{and}\\ \mathbf{y}(t)&=\mathbf{y}(kT)\end{aligned}\right\}\ (kT\leqslant t<(k+1)T\ ;\ k=0, 1, 2, \ldots),$$

design a first-order discrete-time state observer, in the case when the sampling period, T, is 0·2, such that the eigenvalue associated with the observer is 0·1, and that the remaining eigenvalues of the plant matrix of the closed-loop system incorporating the observer are either

(a) 0·5, $-0{\cdot}5$, and 0·25

or

(b) 0·5, $0{\cdot}5+0{\cdot}5i$, and $0{\cdot}5-0{\cdot}5i$.

Chapter 8

8.1. Synthesize dyadic control laws incorporating appropriate integral feedback for the system governed by the state equation

$$\dot{\mathbf{x}}(t) = \begin{bmatrix} 10, & -36 \\ 3, & -11 \end{bmatrix} \mathbf{x}(t) + \begin{bmatrix} 9, & 6 \\ 7, & 5 \end{bmatrix} \mathbf{z}(t) + \mathbf{d}(t)$$

such that the eigenvalues of the plant matrix of the resulting closed-loop system are $\rho_1 = -0{\cdot}2$, $\rho_2 = -1$, and $\rho_3 = -3$ and such that $\mathbf{x}(t) \to \mathbf{0}$ as $t \to \infty$ when

(a) $\mathbf{d} = \begin{bmatrix} 1 \\ 0 \end{bmatrix}$

and

(b) $\mathbf{d} = \begin{bmatrix} 0 \\ -2 \end{bmatrix}$.

8.2. Synthesize, using a multi-stage design procedure, control laws incorporating integral feedback for the system governed by the state equation

$$\dot{\mathbf{x}}(t) = \begin{bmatrix} 11, & -12 \\ 10, & -11 \end{bmatrix} \mathbf{x}(t) + \begin{bmatrix} 4, & 3 \\ 3, & 2 \end{bmatrix} \mathbf{z}(t) + \begin{bmatrix} 2 \\ -3 \end{bmatrix}$$

such that the eigenvalues of the plant matrix of the resulting closed-loop system are $\rho_1 = -0{\cdot}2$, $\rho_2 = -0{\cdot}4$, $\rho_3 = -2$, and $\rho_4 = -3$ and such that $\mathbf{x}(t) \to \mathbf{0}$ as $t \to \infty$.

Chapter 9

9.1. Synthesize a control law for the second-order system governed by the state equation

$$\dot{\mathbf{x}}(t) = \begin{bmatrix} 0, & 1 \\ -2, & 3 \end{bmatrix} \mathbf{x}(t) + \begin{bmatrix} 0 \\ 1 \end{bmatrix} z(t)$$

such that the eigenvalues of the resulting closed-loop system are $\rho_1 = -1$ and $\rho_2 = -3$. Investigate the sensitivity characteristics of the closed-loop system.

9.2. Synthesize a piecewise-constant control law for the third-order system governed by the state equation

$$\mathbf{x}\{(k+1)T\} = \begin{bmatrix} 2, & 1, & -5 \\ 0, & -0{\cdot}5, & 0{\cdot}5 \\ 0, & 0{\cdot}5, & -0{\cdot}5 \end{bmatrix} \mathbf{x}(kT) + \begin{bmatrix} 5 \\ 0 \\ 2 \end{bmatrix} z(kT)$$

such that the eigenvalues of the resulting closed-loop system are $\rho_1 = 0{\cdot}2$, $\rho_2 = 0{\cdot}8$, and $\rho_3 = -0{\cdot}5$. Investigate the sensitivity characteristics of the closed-loop system.

INDEX